The Biochemistry of Glycoproteins and Proteoglycans

The Biochemistry of Glycoproteins and Proteoglycans

Edited by
William J. Lennarz
The Johns Hopkins University School of Medicine
Baltimore, Maryland

PLENUM PRESS • NEW YORK AND LONDON

Library of Congress Cataloging in Publication Data

Main entry under title:

The Biochemistry of glycoproteins and proteoglycans.

Includes index.
1. Glycoproteins. 2. Proteoglycans. 3. Biological chemistry. I. Lennarz, William J.
QP552.G59B56   574.1'924   79-9176
ISBN 0-306-40243-2

QP
552
.G59
B56

© 1980 Plenum Press, New York
A Division of Plenum Publishing Corporation
227 West 17th Street, New York, N.Y. 10011

Printed in the United States of America

Contributors

Gilbert Ashwell • National Institute of Arthritis, Metabolism and Digestive Diseases, National Institutes of Health, Bethesda, Maryland

Paul H. Atkinson • Departments of Pathology and Developmental Biology and Cancer, Albert Einstein College of Medicine, Bronx, New York

John Hakimi • Departments of Pathology and Developmental Biology and Cancer, Albert Einstein College of Medicine, Bronx, New York

Rosalind Kornfeld • Department of Medicine, Washington University School of Medicine, St. Louis, Missouri

Stuart Kornfeld • Department of Medicine, Washington University School of Medicine, St. Louis, Missouri

William J. Lennarz • Department of Physiological Chemistry, The Johns Hopkins University School of Medicine, Baltimore, Maryland

Elizabeth F. Neufeld • National Institute of Arthritis, Metabolism and Digestive Diseases, National Institutes of Health, Bethesda, Maryland

Lennart Rodén • University of Alabama in Birmingham, Birmingham, Alabama

Saul Roseman • Department of Biology and the McCollum Pratt Institute, The Johns Hopkins University, Baltimore, Maryland

Harry Schachter • Department of Biochemistry, Hospital for Sick Children, Toronto, Canada

Pamela Stanley • Department of Cell Biology, Albert Einstein College of Medicine, Bronx, New York

Douglas K. Struck • Department of Physiological Chemistry, The Johns Hopkins University School of Medicine, Baltimore, Maryland

Preface

Although glycoproteins and proteoglycans have been a subject of research for many years, it is only during the last five or so years that they have aroused the interest of a very broad cross section of investigators in the biological sciences. The reason for this expanded interest in these molecules is simple: not only are glycoproteins and proteoglycans ubiquitous, but many are molecules with well-defined and important biological functions. The list of molecules that fall into this category grows daily; interferon, immunoglobulins, certain hormones, many cell surface receptors, and viral coat proteins are but a few examples. Thus, investigators with interests as diverse as viral replication, cell–cell interactions, polyisoprenoid synthesis, secretory processes, hormone responses, embryonic development, and immunology have become concerned with glycoproteins and proteoglycans.

The objective of this book is to summarize the current state of knowledge on the biochemistry of these molecules. Coverage is by no means encyclopedic; rather the thrust is to emphasize the recent advances. The first chapter deals primarily with structural work on the oligosaccharide chains of glycoproteins, but it will be apparent in it and in the succeeding two chapters on biosynthesis that not only do structural studies aid biosynthetic investigations, but that studies on biosynthesis often play a major role in elucidation of structure. The fourth chapter emphasizes the utility of the genetic approach in elucidation of the structure and function of cell surface glycoproteins, while the fifth deals with the important and complex question of the possible role of these molecules in the regulation of growth of normal and transformed cells. In the sixth chapter another potential functional role of cell surface molecules, namely, in the binding and uptake of glycosylated ligands, is discussed.

Finally, in the last chapter, a comprehensive description of the current state of knowledge of the biosynthesis and the catabolism of the proteoglycans is presented.

William J. Lennarz

Baltimore

Contents

Chapter 4

Surface Carbohydrate Alterations of Mutant Mammalian Cells Selected for Resistance to Plant Lectins

Pamela Stanley

Chapter 5

Alterations in Glycoproteins of the Cell Surface

Paul H. Atkinson and John Hakimi

Chapter 6

Carbohydrate Recognition Systems for Receptor-Mediated Pinocytosis

Elizabeth F. Neufeld and Gilbert Ashwell

Chapter 7

✓ *Structure and Metabolism of Connective Tissue Proteoglycans*

Lennart Rodén

Structure of Glycoproteins and Their Oligosaccharide Units

Rosalind Kornfeld and Stuart Kornfeld

1. INTRODUCTION

The presence of oligosaccharide chains covalently attached to the peptide backbone is the feature that distinguishes glycoproteins from other proteins and accounts for some of their characteristic physical and chemical properties. Glycoproteins occur in fungi, green plants, viruses, bacteria, and in higher animal cells where they serve a variety of functions. Connective tissue glycoproteins, such as the collagens and proteoglycans of various animal species, are structural elements as are the cell wall glycoproteins of yeasts and green plants. The submaxillary mucins and the glycoproteins in the mucous secretions of the gastrointestinal tract, which consist of numerous oligosaccharide chains attached at closely spaced intervals to a peptide backbone, serve as lubricants and protective agents. The body fluids of vertebrates are rich in glycoproteins secreted from various glands and organs. Constituents of blood plasma which are glycoproteins include the transport proteins transferrin, ceruloplasmin, and transcobalamin I as well as the immunoglobulins, all the known clotting factors, and many of the components of complement. Follicle-stimulating hormone, luteinizing hormone, and thyroid-stimulating hormone (secreted by the pituitary) and chorionic gonadotropin are all glycoproteins

Rosalind Kornfeld and Stuart Kornfeld • Department of Medicine, Washington University School of Medicine, St. Louis, Missouri 63110.

as are the enzymes ribonuclease and deoxyribonuclease (secreted by the pancreas) and α-amylase (secreted by the salivary glands). Fungi secrete a number of glycoprotein enzymes, for example, Taka-amylase and invertase. Another group of glycoproteins are those which occur as integral components of cell membranes in a variety of species. Enveloped viruses contain surface glycoproteins that are involved in the attachment of the virus to its host, and in eukaryotic cells the histocompatibility antigens are membrane glycoproteins. There is a growing body of evidence to suggest that cell surface glycoproteins are involved in a number of physiologically important functions such as cell–cell interaction, adhesion of cells to substratum, and migration of cells to particular organs, for example, the "homing" of lymphocytes to the spleen and the metastasis of tumor cells to preferred sites.

Since the carbohydrate moieties of glycoproteins are presumed to account for some of the biological functions displayed by glycoproteins, a great deal of effort has been exerted in recent years to determine the structure of the oligosaccharide chains of both soluble and membrane glycoproteins. Concurrently a number of investigators have recently elucidated the pathway of biosynthesis of the N-glycosidically linked oligosaccharide chains via lipid–oligosaccharide intermediates. These two lines of investigation are complementary in that now a sufficient number of complete oligosaccharide structures are known for specific glycoproteins to indicate what sorts of lipid-linked oligosaccharide intermediates might be their precursors and, conversely, the nature of the lipid-linked oligosaccharide intermediates so far discovered suggests what types of structure might reasonably be expected to occur in completed glycoproteins. In this discussion of glycoprotein oligosaccharide structure, two different views must be taken. From one point of view the structures of the oligosaccharide chains of many glycoproteins are very similar and fall into a few broad categories, for example, the O-glycosidically linked oligosaccharides of various mucinous glycoproteins and the N-glycosidically linked oligosaccharides of various serum glycoproteins. The other point of view focuses on the fine structural differences that distinguish one member of such a group from another member. For example, the difference of a single sugar residue determines whether a blood group substance glycoprotein has A-type specificity or B-type specificity, and the presence of an *N*-glycolyl group on neuraminic acid instead of an *N*-acetyl group is responsible for the antibody response seen in serum sickness (Higashi *et al.*, 1977).

This chapter is organized according to the broad categories, and in each section specific examples will be given to point out the fine structural differences that occur. The aim here is to give an overview rather than a comprehensive review of all the structures now known, and the reader is referred to recent reviews (Marshall, 1972; Spiro, 1973; Hughes, 1973; Montreuil, 1975; Kornfeld and Kornfeld, 1976) for additional new material and to the monograph edited by Gottschalk (1972) for an in-depth discussion of the historical background and earlier structural work on glycoproteins.

2. ISOLATION AND STRUCTURAL ANALYSIS OF GLYCOPEPTIDES

2.1. Methods for Isolation of Glycopeptides

Customarily one first digests a glycoprotein with a proteolytic en-zyme such as pronase to degrade the protein portion and produce gly-copeptides having only a few amino acid residues. Alternatively it may be useful to prepare subtilisin or tryptic peptides or cyanogen bromide fragments from a glycoprotein as a first step to enable one to map the position of oligosaccharides on the peptide chain and then to degrade those fragments further with pronase. Since many glycoproteins have more than one oligosaccharide chain per molecule, the glycopeptides must be separated and purified. A variety of techniques such as gel filtration, ion-exchange chromatography, and paper electrophoresis have been used. More recently the use of lectins covalently attached to an insoluble support (e.g., Sepharose or agarose beads) has proved to be a valuable tool in separating glycopeptides on the basis of their carbohy-drate structure. Ogata et al. (1975) tested a group of glycopeptides of known structure for their ability to bind to columns of concanavalin A–Sepharose and found that only those glycopeptides containing two or more mannose residues capable of interacting with concanavalin A, that is, are terminal or substituted only at C-2, had affinity for the column and were subsequently eluted with 0.1 M α-methyl mannoside. Similarly, Krusius et al. (1976) have used concanavalin A–Sepharose columns to separate glycopeptides containing N-linked oligosaccharides with a single C-2 substituted mannose (e.g., fetuin glycopeptide), which do not bind, from those containing two C-2 substituted mannose residues (e.g., trans-ferrin glycopeptide), which bind and are eluted with 15 mM α-methyl glucoside, and from those containing many terminal and/or C-2 substi-tuted mannose residues (e.g., ovalbumin glycopeptide), which bind and are eluted with 200 mM α-methyl glucoside. Other lectins with different sugar specificities may also be linked to various insoluble supports, and these affinity adsorbents, many of which are now commercially available, should provide an even broader spectrum of fractionation possibilities for purifying individual glycopeptides with carbohydrate chains of differ-ing structure. At some point during the fractionation procedure it is useful to introduce a radioactive label into the glycopeptide; this greatly facilitates the detection of the glycopeptide at all stages of its analysis. One method involves acetylating the free amino groups in the peptide with [^{14}C]acetic anhydride (Koide and Muramatsu, 1974) and another involves limited oxidation of a terminal sugar residue followed by reduction with NaB[^{3}H$_4$]. Ashwell and his co-workers have used the latter technique to label sialic acid residues following mild periodate oxidation (Van Lenten and Ashwell, 1972) and to label galactose residues after oxidation with galactose oxidase (Morell and Ashwell, 1972).

2.2. Methods Used in Determination of Oligosaccharide Structure

Once the purified glycopeptide is in hand, the sugar and amino acid composition should be determined. Gas–liquid chromatography offers a very sensitive technique for sugar analysis and a number of different derivatives have been used, for example, the trimethyl silyl derivatives and alditol acetates (Laine *et al.*, 1972). Alternatively one may use the automated sugar analyzers that employ ion-exchange chromatography with borate buffers (Lee, 1972). This preliminary analysis will provide clues as to which sugar and amino acid are involved in the linkage of the oligosaccharide to the peptide. Oligosaccharides linked N-glycosidically from *N*-acetylglucosamine to asparagine may be distinguished from chains linked O-glycosidically to the hydroxyl groups of serine or threonine by treating the glycopeptide with mild alkali in the presence of sodium borohydride. This procedure selectively cleaves the O-linked chains from the peptide via a β-elimination reaction with concomitant reduction of the linkage sugar to a sugar alcohol; the released oligosaccharide alcohol may then be separated and analyzed. The use of borotritide will result in introduction of ^{3}H into the linkage sugar alcohol, thus facilitating its detection in subsequent analysis.

Removal of N-linked oligosaccharide chains from the peptide requires much stronger conditions and can result in degradation. Lee and Scocca (1972) have described the use of hot 1 N sodium hydroxide in the presence of sodium borohydride, and heating in anhydrous hydrazine containing 1% hydrazine sulfate has also been used (Fukuda *et al.*, 1976). A much gentler approach that also has the virtue of greater specificity is to use one of the recently described endo-β-*N*-acetylglucosaminidases. These enzymes cleave the linkage between the *N*-acetylglucosamine residues of the chitobiose unit present in many N-linked oligosaccharides, releasing a carbohydrate chain with *N*-acetylglucosamine at the free reducing end and leaving behind a residue of *N*-acetylglucosamine attached to asparagine in the peptide chain. A variety of these endoglycosidases have now been reported with different specificities for the type of oligosaccharide chains that may be attached to the chitobiose (see Chapter 5). The endo-β-*N*-acetylglucosaminidase H, purified by Tarentino and Maley (1974) from culture filtrates of *Streptomyces griseus* (later identified as *S. plicatus* by Tarentino and Maley, 1976), has specificity for neutral oligosaccharides, particularly of the high-mannose type (Tarentino *et al.*, 1974). Tai *et al.* (1977*b*) have examined a series of glycopeptides of known structure as endo-β-N-acetylglucosaminidase substrates and shown that the C_{II} enzyme from *Clostridium perfringens* has a specificity very similar to that of H enzyme but different from that of endo-β-N-acetylglucosaminidase D isolated from culture filtrates of *Diplococcus pneumoniae* by Koide and Muramatsu (1974). The latter enzyme, unlike C_{II} or H enzymes is able to act on side-chain-free complex-type oligosaccharides having a $(Man)_3(GlcNAc)_2 \longrightarrow$ Asn sequence or on complex oligosaccharides which contain a single side chain and one terminal mannose (Ito *et al.*, 1975). Oligosaccharide chains released by these endoglycosidases may be radiolabeled by reducing the free N-acetylglucosamine residue with sodium borotritide.

Once the oligosaccharide linkage to peptide has been established the next step is to determine the sequence of sugars in the chain, their anomeric configuration (α or β), and their position of linkage to the underlying sugar residue. Sequential degradation of the sugar chains from their nonreducing ends with various exoglycosidases (e.g., neuraminidase, β-galactosidase, β-N-acetylglucosaminidase, α-mannosidase, etc.) will reveal not only the sequence of sugars but also their anomeric configuration based on the specificity of the glycosidase for α or β linkage. Another approach is to fragment the oligosaccharide into smaller saccharides using mild acid hydrolysis, acetolysis, or hydrazinolysis followed by nitrous acid deamination. Since different glycosidic bonds vary in their susceptibility to cleavage by these reagents, depending on which sugar and hydroxyl group are involved, different types of fragments are produced with each technique. If the composition and sequence of a sufficient number of fragments is then determined, it is often possible to reconstruct the sequence of the original oligosaccharide.

Since every hexose has four hydroxyl groups and each hexosamine has three hydroxyl groups available for substitution, there are many linkage positions possible for a given sequence and many opportunities for branching to occur in oligosaccharide chains. Therefore the intact glycopeptide, released oligosaccharide, or short saccharide fragments must be subjected either to periodate oxidation or methylation to discover which hydroxyl groups are substituted. Periodate can oxidize vicinal unsubstituted hydroxyl groups to produce characteristic degradation products from the sugars. Periodate oxidation followed by borohydride reduction and mild acid hydrolysis, to cleave the acetal linkage of the oxidized and reduced sugar residues, is known as Smith degradation (Spiro, 1966) and can produce fragmentation of oligosaccharides in which an internal sugar residue is susceptible to oxidation. In oligosaccharides which do not have oxidizable internal sugars, sequential rounds of Smith degradation can also provide information on sugar sequence by destroying the terminal unsubstituted sugars exposed in each sucessive round of degradation. Methylation analysis, however, is the most specific method for deducing sugar linkages because every unsubstituted hydroxyl can be methylated and the resultant methylated sugars identified. The methylation technique described by Hakomori (1964), using methyl iodide and methylsulfinyl carbanion in dimethyl sulfoxide, results in complete methylation of even small amounts of oligosaccharide in a single step. This technique, combined with the gas–liquid chromatographic/mass spectrometric system devised by Björndal *et al.* (1967*a,b*; 1970) for separating and identifying small amounts of methylated sugars as their alditol acetates, has resulted in widespread application of the method to glycoprotein structure work. An exciting new approach to the analysis of oligosaccharide structure is high-resolution ^{1}H NMR spectroscopy. This procedure, used in conjunction with methylation analysis, can provide extensive structural data in a very short time (Fournet *et al.*, 1978).

3. GLYCOPEPTIDES CONTAINING OLIGOSACCHARIDES LINKED O-GLYCOSIDICALLY TO THE PEPTIDE

3.1. Oligosaccharides Linked through N-Acetylgalactosamine to Serine and Threonine

In many glycoproteins, particularly the mucins, the oligosaccharide chains are attached to the peptide by an O-glycosidic linkage from N-acetylgalactosamine to serine and/or threonine. These oligosaccharides are most conveniently studied after cleavage from the peptide by a β-elimination reaction carried out in alkaline borohydride. The linkage sugar is converted to a sugar alcohol and identified after hydrolysis of the oligosaccharide. The structures of a group of glycopeptides in which N-acetylgalactosamine is the linkage sugar are shown in Table 1. In structure A, found in various submaxillary mucins, the N-acetylgalactosamine is substituted at C-6 by either N-acetyl neuraminic acid or N-glycolyl neuraminic acid. Structure B, in which the N-acetylgalactosamine is substituted at C-3 by a β-galactosyl residue, occurs in a number of different glycoproteins and also serves as the core region of a variety of larger oligosaccharide chains. The "antifreeze glycoprotein" of Antarctic fish (DeVries *et al.*, 1971; Shier *et al.*, 1975) contains 31 disaccharide units attached to threonine in the repeating peptide sequence of (Thr-Ala-Ala). In a human IgA1 immunoglobulin (Baenziger and Kornfeld, 1974*b*) the disaccharide is linked to four serine residues in the

Table 1. Glycopeptides Linked through N-Acetylgalactosamine to the Hydroxyl Group of Serine and Threonine[a]

Structure	Glycoprotein
A $\quad$ NANA $\xrightarrow{\alpha2,6}$ GalNAc $\longrightarrow$ Ser(Thr) (NGNA)	Submaxillary mucins
B $\quad$ Gal $\xrightarrow{\beta1,3}$ GalNAc $\longrightarrow$ Ser(Thr)	"Antifreeze" glycoprotein of antarctic fish: Human IgA1; β subunit HCG; cartilage keratin sulfate; epiglycanin of TA$_3$-HA cells; lymphocyte, RBC, and milk fat globule membranes
C $\quad$ Gal $\xrightarrow{\beta1,3}$ GalNAc $\longrightarrow$ Ser(Thr) $\quad\quad\;\uparrow \alpha2,3$ $\quad\quad$ NANA	Bovine kininogen; epiglycanin of TA$_3$-HA cells; B$_{16}$ melanoma cells
D $\quad$ Gal $\xrightarrow{\beta1,3}$ GalNAc $\longrightarrow$ Ser(Thr) $\quad\quad\;\uparrow \alpha2,3 \quad \uparrow \alpha2,6$ $\quad\quad$ NANA $\quad\;$ NANA	Fetuin: human RBC membrane sialoglycoprotein; bovine kininogen; rat brain
E $\;$ Gal $\xrightarrow{\beta1,3(4)}$ GlcNAc $\xrightarrow{1,2(4,6)}$ Gal $\xrightarrow{\beta1,3(4)}$ GalNAc $\longrightarrow$ Ser(Thr) $\;\;\uparrow \alpha2,3$ NANA	Epiglycanin
F $\quad$ Gal $\xrightarrow{1,3}$ GlcNAc $\xrightarrow{1,3}$ Gal $\xrightarrow{1,3}$ GalNAc $\longrightarrow$ Ser(Thr) $\quad\quad\quad\quad\quad\quad\quad\quad\quad\quad\quad\quad\;\uparrow 1,6$ $\quad\quad\quad\quad\quad\quad\quad\quad\quad\quad\quad$ GlcNAc $\quad\quad\quad\quad\quad\quad\quad\quad\quad\quad\quad\quad\;\uparrow 1,4$ $\quad\quad\quad\quad\quad\quad\quad\quad\quad\quad\quad\quad$ Gal	Human gastric mucin; core region of human and hog blood group substances

[a] References given in the text.

proline-rich hinge-region glycopeptide that occurs in α_1-type heavy chains but is deleted in the α_2-type (Frangione and Wolfenstein-Todel, 1972). Similarly the β subunit of human chorionic gonadotropin is glycosylated on serine residues 118, 121, and 123 in a proline-rich region of the peptide chain (Bahl, 1969; Bahl *et al.*, 1972). Glycopeptide B also occurs in keratan sulfate from bovine cartilage (Bray *et al.*, 1967; Kieras, 1974) and in the membranes of lymphocytes (Newman *et al.*, 1976), erythrocytes, and milk-fat globules (Glockner *et al.*, 1976). The mucinous glycoprotein epiglycanin isolated by Codington *et al.* (1975) from the surface of TA_3-HA cancer cells contains six different oligosaccharide chains of increasing length including structures B, C, and E. Structure C, in which the galactose residue of the disaccharide of structure B is sialylated at C-3, has also been shown to occur in fragment 1 released from bovine plasma high-molecular-weight kininogen by the action of plasma kallikrein (Endo *et al.*, 1977) and in pronase digests of B16 melanoma cells (Bhavanandan *et al.*, 1977). Structure D, containing a second *N*-acetyl neuraminic acid residue attached to C-6 of the linkage *N*-acetylgalactosamine, was first described by Thomas and Winzler (1969), who isolated it from the sialoglycoprotein of human erythrocyte membranes. Subsequently the same glycopeptide was isolated from fetuin by Spiro and Bhoyroo (1974), from rat brain by Finne (1975), and from bovine kininogen by Endo *et al.* (1977). Structure E, present in epiglycanin, consists of an elongation of the core disaccharide by a NANA $\longrightarrow$ Gal $\longrightarrow$ GlcNAc $\longrightarrow$ sequence. Structure F, isolated by Oates *et al.* (1974) from human gastric mucin consists of the core disaccharide substituted on C-3 of galactose by Gal $\xrightarrow{1,3}$ GlcNAc and on C-6 of the *N*-acetylgalactosamine by Gal $\xrightarrow{1,4}$ GlcNAc. This same structure has been found in the inner region of various human and hog blood group substances, whose complex structures have been worked out by Kabat and his co-workers and by Watkins and Morgan and their co-workers (Watkins, 1972) over the last twenty years. Figure 1 is the composite "megalosaccharide" structure of blood group substance put forward by Feizi *et al.* (1971) which encompasses both their findings and those of others and indicates which terminal sugar residues (or combination of residues) confer blood group specificity for A, B, H, Le[a], and Le[b].

Table 2 shows the structures of a series of glycopeptides found in various submaxillary mucins. Structure 1, containing a sialic acid attached to C-6 of the linkage *N*-acetylgalactosamine occurs in all submaxillary mucins studied. All of the other structures contain the Gal $\xrightarrow{\beta 1,3}$ GalNAc $\longrightarrow$ core disaccharide which in 2, 3, and 4 is substituted on C-6 of *N*-acetylgalactosamine by a sialic acid residue. The addition of a fucosyl residue to the terminal galactose residue, as in structure 3 found in canine and A[-] and A[+] porcine submaxillary mucins, confers H-type

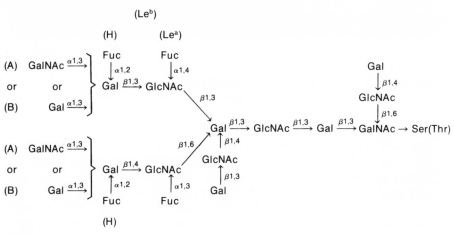

Figure 1. Composite "megalosaccharide" structure proposed for blood group substance by Feizi *et al.* (1971).

Table 2. Glycopeptides Found in Various Submaxillary Mucins[a]

	Structure	Sources of submaxillary mucins
1	GalNAc ⟶ Ser(Thr) ↑ $\alpha2,6$ SA	A⁻ and A⁺ porcine, bovine, ovine, and canine
2	Gal $\xrightarrow{\beta1,3}$ GalNAc ⟶ Ser(Thr) ↑ $\alpha2,6$ SA	Canine
3	Gal $\xrightarrow{\beta1,3}$ GalNAc ⟶ Ser(Thr) ↑ $\alpha1,2$ ↑ $\alpha2,6$ Fuc SA	A⁻ and A⁺ porcine, canine
4	GalNAc $\xrightarrow{\alpha1,3}$ Gal $\xrightarrow{\beta1,3}$ GalNAc ⟶ Ser(Thr) ↑ $\alpha1,2$ ↑ $\alpha2,6$ Fuc SA	A⁺ porcine
5	SA $\xrightarrow{2,6}$ Gal $\xrightarrow{1,3}$ GalNAc ⟶ Ser(Thr) ↑ $\alpha1,2$ [Fuc]±	A⁻H⁻ porcine
6	GlcNAc $\xrightarrow{\beta1,6}$ Gal $\xrightarrow{\beta1,3}$ GalNAc ⟶ Ser(Thr) SO₄ ↑ $\alpha1,2$ [Fuc]±	Canine
7	Gal $\xrightarrow{\beta}$ GlcNAc $\xrightarrow{\beta1,6}$ Gal $\xrightarrow{\beta1,3}$ GalNAc ⟶ Ser(Thr) ↑ $\alpha1,2$ SO₄ ↑ $\alpha1,2$ Fuc [Fuc]	Canine

[a] Data for the A⁻ and A⁺ porcine mucin are from Carlson (1968) and for the A⁻H⁻ porcine mucin from Baig and Aminoff (1972). Data for the canine mucin are from Lombart and Winzler (1974), and for bovine and ovine mucins from Bertolini and Pigman (1970). SA stands for sialic acid, which is *N*-glycolylneuraminic acid in the porcine mucins and *N*-acetylneuraminic acid in the other mucins.

blood group reactivity. The further addition of a terminal GalNAc $\xrightarrow{\alpha 1,3}$ as in structure 4, found by Carlson (1968) only in A$^+$ porcine submaxillary mucins, confers A-type blood group reactivity. From hogs that displayed neither A- nor H-type blood group reactivity, Baig and Aminoff (1972) isolated a submaxillary mucin containing structure 5, which has the basic core structure with sialic acid attached to galactose rather than N-acetylgalactosamine. This sialic acid residue blocks the expression of H activity by the fucosyl residue when it is present at C-2 of galactose. Lombart and Winzler (1974) reported that canine submaxillary mucins have, in addition to the sialic acid-containing glycopeptides, the nonsialylated, sulfated glycopeptides shown in structures 6 and 7. These glycopeptides have variable amounts of fucose and are characterized by the presence of an N-acetylglucosamine residue, sulfated at either C-3 or C-4, attached to galactose of the core disaccharide. The glycopeptides shown in Tables 1 and 2 have structural features that are identical, such as the core disaccharide, the linkage of sialic acid always being 2,6 to N-acetylgalactosamine, and the linkage of fucose always being $\alpha 1,2$ to galactose. They also display structural differences, especially in the periphery, in the form of additional sugars and different linkages. For example, in Table 1 the sialic acid in structures C, D, and E is linked $\alpha 2,3$ to galactose, but sialic acid is linked $\alpha 2,6$ to galactose in the A$^-$H$^-$ porcine submaxillary mucin shown in structure 5 of Table 2.

3.2. Oligosaccharides Linked to Serine and Threonine through Sugars Other Than N-Acetylgalactosamine

Oligosaccharides linked O-glycosidically from mannose to a serine or threonine residue in the peptide occur most commonly in glycoproteins from yeasts and fungi. The yeast mannan from wild-type *Saccharomyces cerevisiae*, when treated with mild alkali by Nakajima and Ballou (1974a), underwent a β-elimination reaction which liberated oligosaccharide chains containing one to four mannose residues. The mannotetraose Man $\xrightarrow{\alpha 1,3}$ Man $\xrightarrow{\alpha 1,2}$ Man $\xrightarrow{\alpha 1,2}$ Man had the same structure as the mannotetraose liberated from the polysaccharide component of yeast mannan by acetolysis and, furthermore, a mutant strain of yeast lacking $\alpha 1,3$-mannosyltransferase activity produced no mannotetraose upon β-elimination, indicating that the same enzyme must act to elongate oligosaccharide chains attached to serine and threonine as well as the side chains in the polysaccharide. The mannotriose sequence Man $\xrightarrow{\alpha 1,2}$ Man $\xrightarrow{\alpha 1,2}$ Man that occurs in mannan was also found by Raizada *et al.* (1975) to occur in the envelope glycoprotein of *Cryptococcus laurentii* along with the following pentasaccharide sequence:

$$\text{Man} \xrightarrow{\alpha 1,2} \text{Man} \xrightarrow{\alpha 1,6} \text{Man} \xrightarrow{\alpha 1,3} \text{Man} \longrightarrow \text{Ser(Thr)}$$
$$\uparrow$$
$$\text{Xyl}^{\beta 1,2}$$

containing xylose. Sentandreu and Northcote (1969) also found mannose oligosaccharides linked to serine or threonine in the cell wall glycopeptides from baker's yeast. Glucose and galactose may also occur in mannose-linked oligosaccharides. For example, the mycodextranase secreted by *Penicillium melinii* was shown (Rosenthal and Nordin, 1975) to contain some 25 carbohydrate side chains attached to serine or threonine with the structures Glc $\xrightarrow{\alpha 1,2}$ Man and Man $\xrightarrow{\alpha 1,2}$ Glc $\xrightarrow{\alpha 1,2}$ Man. Similarly the glucamylase from *Aspergillus niger* (Pazur *et al.*, 1971) has alkali-labile, mannose-linked side chains containing galactose, glucose, and mannose.

The cuticle collagen of the clamworm *Nereis* was shown by Spiro and Bhoyroo (1971) to give rise to a glucuronic acid-containing glycopeptide with the structure GlcUa $\xrightarrow{\alpha 1,6}$ Man $\longrightarrow$ Thr, as well as galactose-containing glycopeptides similar to those found by Muir and Lee (1969, 1970) in the cuticle collagen of the earthworm *Lumbricus terrestris* and shown to have the structures Gal $\xrightarrow{\alpha 1,2}$ Gal $\longrightarrow$ Ser(Thr) and Gal $\xrightarrow{\alpha 1,2}$ Gal $\xrightarrow{\alpha 1,2}$ Gal $\longrightarrow$ Ser(Thr). Hallgren *et al.* (1975) reported the occurrence of the glycopeptide Glc $\xrightarrow{\beta 1,3}$ Fuc $\longrightarrow$ Thr in normal human urine, and recently the same glycopeptide, as well as the simpler Fuc $\longrightarrow$ Thr(Ser), was isolated from rat tissue (Larriba *et al.*, 1977). In the linkage region of the proteoglycans, discussed in Chapter 7, xylose is the connecting sugar in O-glycosidic linkage to serine or threonine. So the attachment of oligosaccharide to serine and threonine can involve a pentose, a methyl pentose, or a hexose as well as the amino sugar *N*-acetylgalactosamine.

3.3. Oligosaccharides Linked to the Hydroxyl Group of Hydroxylysine and Hydroxyproline

Collagens and basement membranes contain yet another type of glycoprotein linkage structure consisting of galactose and glucosyl-galactose disaccharides linked O-glycosidically to the hydroxyl group of hydroxylsine (Hyl) in the peptide. This linkage region was first detected by Butler and Cunningham (1966) in guinea pig skin tropocollagen, and its structure was shown by Spiro (1967) to be Glc $\xrightarrow{\beta 1,2}$ Gal $\longrightarrow$ Hyl in the glycopeptide isolated from glomerular basement membrane. Since then the same disaccharide unit has been found in a number of collagens, and Morgan *et al.* (1970) have shown that glycosylated hydroxylysine residues in vertebrate collagens always occur within the characteristic amino acid sequence Gly-X-Hyl-Y-Arg. Subsequently Isemura *et al.* (1973) found the same sequence in glycopeptides from collagen of the invertebrate cuttlefish. Mahieu *et al.* (1973) isolated the disaccharide-containing glycopeptide from human glomerular basement membrane and showed that it carries the major antigenic determinants of humoral and cellular anti-basement-membrane autoimmunization in studies on seven patients with anti-glomerular-basement-membrane antibody-mediated glomerulonephritis.

Plant cell walls contain hydroxyproline-rich glycoproteins, called extensins, which probably function as structural proteins analogous to collagens of animal cells. Glycopeptides isolated from various plant cell walls have been shown to contain arabinose linked covalently to the hydroxyl group of hydroxyproline (Lamport, 1969; Heath and Northcote, 1971). More recently Lamport *et al.* (1973) reported that a glycopeptide from extensin also contained galactosylserine and proposed the following structure for it:

$$
\text{Ser—Hyp—Hyp—Hyp—Hyp—Ser—Hyp—Lys}
$$
$$
\uparrow \qquad \uparrow \qquad \uparrow \qquad \uparrow \qquad \uparrow \qquad \uparrow \qquad \uparrow
$$
$$
\text{Gal} \quad (\text{Ara})_4 \quad (\text{Ara})_4 \quad (\text{Ara})_4 \quad (\text{Ara})_4 \quad (\text{Ara})_4 \quad \text{Gal}
$$

Allen *et al.* (1976) have isolated a glycopeptide with a similar structure from potato lectin (Allen and Neuberger, 1973).

4. GLYCOPEPTIDES CONTAINING THE N-ACETYLGLUCOSAMINYL–ASPARAGINE LINKAGE

Another group of glycoproteins contain oligosaccharide chains linked N-glycosidically from *N*-acetylglucosamine to the amide nitrogen of asparagine in the peptide. These oligosaccharide chains are of two general types: the simple, which contain only mannose and *N*-acetylglucosamine residues, and the complex, which in addition to those sugars also contain sialic acid, galactose, and fucose. A single glycoprotein may contain both simple and complex oligosaccharide chains, as do thyroglobulin (Arima and Spiro, 1972) and IgM (Shimizu *et al.*, 1971; Hickman *et al.*, 1972) or it may contain both complex N-linked chains and O-linked chains as do fetuin (Spiro, 1973), human IgA (Baenziger and Kornfeld, 1974*a,b*), and the human erythrocyte membrane sialoglycoprotein (Thomas and Winzler, 1969, 1971; Kornfeld and Kornfeld, 1970).

4.1. Structure of the Core Region

The N-acetylglucosaminyl–asparagine linkage, GlcNAc $\xrightarrow{\beta}$ Asn, was isolated from hen ovalbumin and shown to be identical to chemically synthesized 2-acetamido-1-N-(4′-L-aspartyl-)2-deoxy-β-D-glucopyranosylamine by Marks *et al.* (1963) and Marshall and Neuberger (1964). It was then discovered by Tarentino *et al.* (1970) that bovine RNase B, which has a simple-type oligosaccharide chain $(Man)_6(GlcNAc)_2$ attached to Asn 34, contains the di-N-acetylchitobiose sequence GlcNAc $\xrightarrow{\beta1,4}$ GlcNAc $\xrightarrow{\beta}$ Asn in its linkage region. The structure of the sequence was identical to that of the compound synthesized chemically by Spinola and Jeanloz (1970). Subsequently both ovalbumin and RNase B glycopeptides were shown by Sukeno *et al.* (1971, 1972) to have a mannose residue linked $\beta1,4$ to the di-N-acetylchitobiose so that the core region had the structure Man $\xrightarrow{\beta1,4}$ GlcNAc $\xrightarrow{\beta1,4}$ GlcNAc $\longrightarrow$ Asn. This finding arose from the observation that treatment of those glycopeptides with α-mannosidase removed all but one mannose residue which could then be removed with hen oviduct β-mannosidase. Similarly Sugahara *et al.* (1972) used snail β-mannosidase to release the core mannose of ovalbumin glycopeptide, and Li and Lee (1972) used pineapple β-mannosidase to release the core mannose of Taka-amylase (*Aspergillus* α-amylase) and ovalbumin glycopeptides. Now that investigators are careful to use highly purified α-mannosidase preparations, β-linked core mannose residues have been found to occur in a number of glycoproteins. Furthermore the various endo-β-N-acetylglucosaminidases described in Section 2.2, which can split the di-N-acetylchitobiose bond, have been very useful in indicating the presence of that sequence in the core region. Those glycoproteins which have been shown to contain the N-acetylglucosaminyl–asparagine linkage as well as those now known to contain the core structure of mannosyl-di-N-acetylchitobiose were tabulated by Kornfeld and Kornfeld (1976) in a recent review.

4.2. Structure of "Simple"-Type Oligosaccharides

The "simple"-type oligosaccharide chains which contain only mannose and N-acetylglucosamine residues are more appropriately called "high-mannose" chains for they are simple neither in structure nor in analysis of structure. Ovalbumin, a classic example of a glycoprotein containing a high-mannose oligosaccharide, has been studied for many years and was shown by Huang *et al.* (1970) to give rise to a series of glycopeptides which varied in compostion from $(Man)_5(GlcNAc)_2$-Asn to $(Man)_6GlcNAc_4$-Asn. In the larger glycopeptides the extra N-acetylglucosamine residues were shown to be located at the nonreducing ends of the mannose chains. In recent work from Kobata's laboratory (Tai *et al.*, 1975*b*, 1977*a,b*) the complete structure of a similar series of ovalbumin glycopeptides has been reported. These are shown in Figure 2 in structures A, B, C, and D, which have compositions ranging from $(Man)_4(GlcNAc)_2$-Asn to $(Man)_7(GlcNAc)_2$-Asn, and in structures G and H, which have the composition $(Man)_5(GlcNAc)_4$-Asn. Also shown, in structure 2E, is the $(Man)_8(GlcNAc)_2$-Asn glycopeptide isolated from membranes of Chinese hamster ovary cells by Li and Kornfeld (1979). Figure 2F shows the structure of the high-mannose glycopeptide (or A glycopeptide) from bovine thyroglobulin which has been worked out by Ito *et al.* (1977) and has the composition $(Man)_9(GlcNAc)_2$-Asn. The remarkable feature of the series of glycopeptides A through F is that not only do they all contain the same β-mannosyl-di-*N*-acetylchitobiose core structure but they all have the same branching pattern of the outer chain α-mannosyl residues. The structure of 2C, the $(Man)_6(GlcNAc)_2$-Asn glycopeptide, which has also been found in *Asperigillus oryzae* α-amylase (Yamaguchi *et al.*, 1971), in Chinese hamster ovary cell membranes (Li and Kornfeld, 1979), and in human IgM (Chapman and Kornfeld, 1979), corresponds exactly to the structure of the inner $(Man)_6(GlcNAc_2$-Asn portion of glycopeptides D, E, and F. Notice also that outer chain branching is confined to the mannose residue linked to C-6 of the core β-mannosyl residue. The $(Man)_{12}(GlcNAc)_2$-Asn glycopeptide isolated by Nakajima and Ballou (1974*b*) from the mannan–protein linkage region of the yeast *Saccharomyces cerevisiae* has a structure virtually identical to this series except for some heterogeneity in terminal mannose linkages. The structures 2G and 2H, isolated from the larger ovalbumin glycopeptide fractions, are classified by Tai *et al.* (1977*a*) as hybrid structures that have features of both the high-mannose and complex-type oligosaccharide chains. The core region and α-mannose residues in 2G and 2H are like structure 2B but the extra terminal *N*-acetylglucosamine residues have presumably been attached by specific β-*N*-acetylglucosaminyltransferases similar to those which elongate the outer chains of complex oligosaccharides. More recently, Yamashita *et al.* (1978) have described two other glycopeptides from ovalbumin which are similar to 2G and 2H but also contain a terminal galactose residue attached to an outer chain *N*-acetylglucosamine residue. The constancy of the arrangement of the mannosyl residues in all the glycopeptides of divergent origin shown in Figure 2 can probably be ascribed to their sharing a common biosynthetic pathway via oligosaccharide–lipid intermediates (see Section 4.4). On the other hand, some high-mannose oligosaccharide chains have different structures, as exemplified by the glycopeptides shown in Figure 3. Glycopeptide 3A, isolated from lima bean lectin by Misaki and Goldstein (1977) has a structure similar to that of glycopeptide 2A except for the

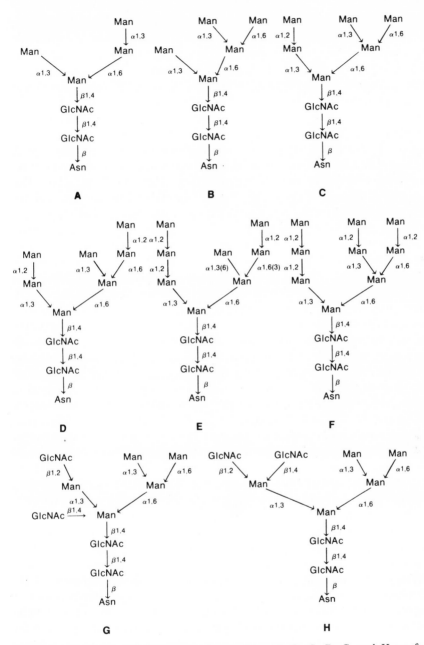

Figure 2. Structures of high-mannose glycopeptides. A, B, C, D, G, and H are from ovalbumin (Tai *et al.*, 1975*b*; 1977*a,b*); B, C, E, and F are from Chinese hamster ovary cell membranes (Li and Kornfeld, 1979); and F is the A glycopeptide from bovine thyroglobulin (Ito *et al.*, 1977).

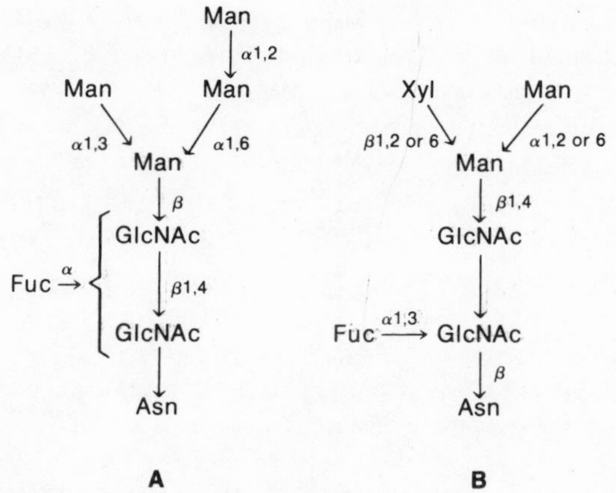

Figure 3. Structures of two unusual high-mannose glycopeptides. A is from lima bean lectin (Misaki and Goldstein, 1977) and B is from pineapple stem bromelain (Fukuda *et al.*, 1976).

presence of an α-fucose residue linked to one of the *N*-acetylglucosamine residues in the core and the linkage of the terminal mannose to C-2 of the mannose which is linked to C-6 of the core mannose. Structure 3B has been proposed for the glycopeptide from pineapple stem bromelain by Fukuda *et al.* (1976) based on their observations as well as those of earlier investigators (Yasuda *et al.*, 1970; Lee and Scocca, 1972). It is unusual in containing a xylose residue and the branch-point mannose is substituted at C-2 and C-6 rather than C-3 and C-6 as in the other glycopeptides. The structural data do not place the β-xylosyl and α-fucosyl residues unequivocally, and they may be linked either as shown or with their positions reversed.

4.3. Structure of "Complex"-Type Oligosaccharides

The "complex"-type asparagine-linked oligosaccharides contain a variable number of outer chains linked to a core of mannose and *N*-acetylglucosamine. The core region most often found contains two α-mannose residues attached to a β-mannosyl-di-*N*-acetylchitobiose unit, the same structure that occurs in the inner region of typical high-mannose oligosaccharides. The sequence of the outer chains is most often sialic acid ⟶ Gal ⟶ GlcNAc ⟶, as shown in structure A in Figure 4, which is a composite encompassing structural features seen in a number of glycopeptides isolated from different glycoproteins, all of which have the same basic structure. The galactose to *N*-acetylglucosamine linkage is usually β1,4, but an exception is bovine prothrombin, which contains both Gal $\xrightarrow{\beta 1,4}$ and Gal $\xrightarrow{\beta 1,3}$ GlcNAc sequences (Mizuochi et al., 1979). Both the galactose and N-acetylglucosamine residues in the latter sequence may be substituted with sialic acid residues. The ± in front of the α-fucosyl residue and terminal β-N-acetylglucosaminyl residue indicates that these residues may or may not be present—depending on the source of the glycopeptide. Similarly there may be microheterogeneity in the terminal sialic acid residues so that a full complement of two residues is not always present. Furthermore, the linkage to galactose differs in different glycopeptides; in some the sialic acid is attached to C-3 of galactose and in others to C-6. Otherwise the sequence and linkages are identical. To be specific, the human IgG glycopeptide Hum B-3 studied by Baenziger (1975) has all the features shown in 4A, and the sialic acids are linked to C-6 of galactose, whereas the human IgA1 glycopeptide IIA (Baenziger and Kornfeld, 1974a) has a fucose, and only a single sialic acid residue linked to C-6 of the galactose in the outer chain attached to C-3 of the branch mannose, but does contain the terminal N-acetylglucosamine residue attached to the branch mannose. The human IgE glycopeptides B-2 and B-3 (Baenziger et al., 1974) have no terminal N-acetylglucosamine on the branch mannose and contain two sialic acid residues linked to C-6 of galactose, whereas the B-1 glycopeptide has

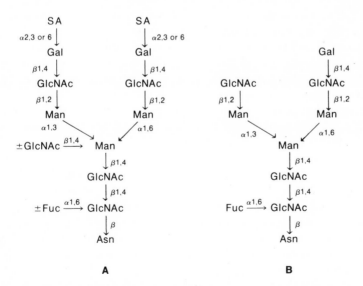

A   **B**

Figure 4. Structure of glycopeptides with complex N-linked oligosaccharides. A is a composite structure characteristic of many glycopeptides as described in the text, and B is from human IgG M^{c}M (Kornfeld et al., 1971).

only one sialic acid residue on the outer chain that is attached to C-3 of the branch mannose. The structure of human transferrin glycopeptide (Spik *et al.*, 1975) has neither fucose nor the terminal N-acetylglucosamine residue, and the two sialic acid residues are linked to C-6 of galactose. Further evidence of the common occurrence of this oligosaccharide structure in glycoproteins is the finding that the tissues and urine of patients with various lysosomal glycosidase deficiencies contain oligosaccharides of this structure in varying stages of degradation, depending on the enzyme deficiency. For example Michalski *et al.* (1977) reported a patient with neuraminidase deficiency whose urine contained oligosaccharides with the structure of 4A except that the Fucose $\longrightarrow$ GlcNAc $\longrightarrow$ Asn moiety was missing as though cleaved by an endoglycosidase, and the terminal N-acetylglucosamine on the branch mannose was missing. The most plentiful oligosaccharide found had sialic acids linked to C-6 of galactose, another had sialic acids linked to C-3, and two others were of mixed type, one having the C-6 linkage on the chain attached to C-3 of the branch mannose and the other having it on the chain attached to C-6 of the branch mannose. Similarly the water-soluble storage products accumulated in the livers of patients with G_{M1} gangliosidosis (β-galactosidase deficiency) and G_{M2} gangliosidosis (β-N-acetylglucosaminidase deficiency) contain oligosaccharides with the same structure as those found in the neuraminidase-deficient patient's urine, except in the G_{M1} case no sialic acid residues were present and in the G_{M2} case neither sialic acid nor galactose residues were present (Wolfe *et al.*, 1974; Ng Ying Kin and Wolfe, 1974).

Structure 4B is the glycopeptide isolated from a human IgG myeloma protein (M^{c}M) by Kornfeld *et al.* (1971) and later shown (Baenziger, unpublished) to have the galactose-terminated outer chain attached to C-6 of the branch mannose in accord with the data of Tai *et al.* (1975a) on the structure of bovine IgG glycopeptide. They showed that in the bovine IgG glycopeptide containing less than two full residues of galactose, a full residue of galactose was present on the outer chain linked to C-6 of the branch mannose and a partial residue was on the chain attached to C-3 of the branch mannose, i.e., like structure 4B with a ±Gal on the terminal GlcNAc. These findings are important because they indicate that so-called microheterogeneity is not random. It is noteworthy that partial sialylation, on the other hand, is confined to the outer chain attached to C-6 of the branch mannose in the IgA1 glycopeptide IIA (Baenziger and Kornfeld, 1974a) and in a urinary oligosaccharide from a neuraminidase-deficient patient (Michalski *et al.*, 1977). The most likely explanation for the nonrandom attachment of various terminal sugars in complex-type oligosaccharides is that the glycosyltransferases responsible for their addition have specificity requirements that go beyond the acceptor sugar residue to encompass other structural features on the oligosaccharides. Thus, structure 4B may be a suitable acceptor for a galactosyltransferase that adds galactose to the terminal N-acetylglucos-amine residue but may not act as an acceptor for sialyltransferases until that second galactosyl residue has been added. These considerations emphasize the importance of the accurate determination of oligosaccharide fine structure. The structures found not only give us clues about their biosynthetic assembly but also provide the substrates needed to define the specificity of various glycosyltransferases.

Other glycoproteins contain more highly branched complex oligosaccharide chains and several examples are shown in Figure 5. These structures all contain the mannosyl-di-N-acetylchitobiose core region and the core β-mannose is substituted at C-3 and C-6 by two α-mannosyl residues just like the structures in Figure 4. However, the structures in Figure 5 also have a third outer chain linked to one of the α-mannose residues and 5A and 5D have a fourth outer chain linked to one of the N-acetylglucosamine residues. Figure 5A is the structure proposed by Schwarzmann et al. (1978) for a glycopeptide from desialyzed α-1 acid glycoprotein. The additional Gal $\longrightarrow$ GlcNAc $\longrightarrow$ chains are attached to C-4 of mannose and to C-3 of an N-acetylglucosamine (the five •GlcNAc residues are all possible candidates). Fournet et al. (1978) have reported that α_1-acid glycoprotein contains bi-, tri-, and tetraantennary structures. In the latter instance the mannose residue linked α1,6 to the β-linked mannose is substituted at positions 2 and 6 rather than at 2 and 4. These workers also found that in some species an outer N-acetylglu-cosamine is substituted at the 3 position by an α-linked residue. The glycopeptide from vesicular stomatitis virus glycoprotein shown in Figure 5B (Reading et al., 1978) has a third outer chain linked to C-4 of mannose, but the porcine thyroglobulin B glycopeptide in structure 5C (Kondo et al., 1977; Toyoshima et al., 1972, 1973) and the calf thymocyte membrane glycopeptide B-3 in structure 5D (Kornfeld, 1978) have the branch-point α-mannose residue substituted at C-3 and C-6. However, the branch-point α-mannose residue is linked to C-6 of the core mannose in structure 5C and to C-3 of the core mannose in structure 5D. The thymocyte membrane glycopeptide (5D) also has an additional SA $\longrightarrow$ Gal $\longrightarrow$ GlcNAc $\longrightarrow$ chain attached to C-3 of an N-acetylglucosamine (the •GlcNAc residues are possible candidates) and differs from all other structures by having the sequence Gal $\xrightarrow{\beta1,3}$ Gal at the nonreducing ends of two chains. Recently, Baenziger and Fiete (1979) have deduced the complete structure of the complex oligosaccharide of fetuin, which is like structure 5B except that it has no fucose and the sialic acid of the middle branch is linked α2,6 to galactose.

The structures shown in Figure 6 differ from the more typical

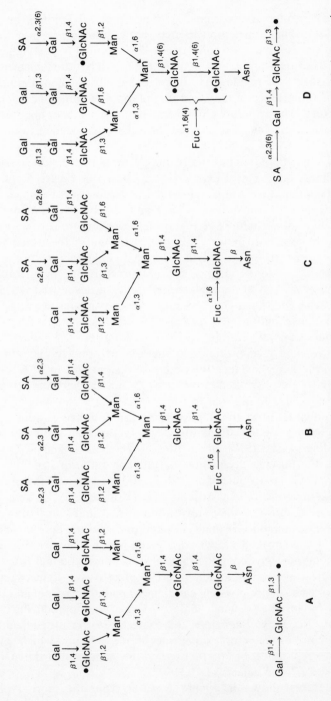

Figure 5. Structures of glycopeptides with highly branched complex N-linked oligosaccharides. A is from desialized α_1-acid glycoprotein (Schwarzmann *et al.*, 1978); B is from vesicular stomatitis virus glycoprotein (Reading *et al.*, 1978); C is the porcine thyroglobulin B glycopeptide (Kondo *et al.*, 1977; Toyoshima *et al.*, 1972, 1973); and D is glycopeptide B-3 from calf thymocyte membranes (Kornfeld, 1978).

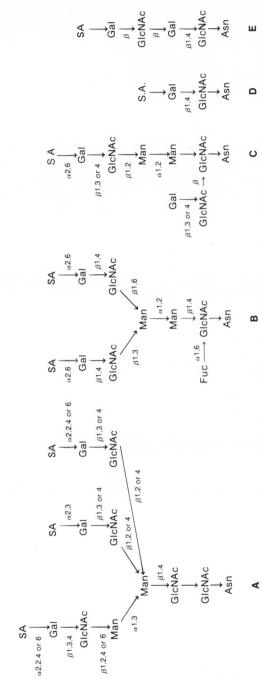

Figure 6. Structures of some glycopeptides with atypical complex N-linked oligosaccharides. A is glycopeptide I from a rabbit liver membrane glycoprotein (Kawasaki and Ashwell, 1976); B is from a human alveolar glycoprotein (Bhattacharyya and Lynn, 1977); C is from human erythrocyte membrane sialoglycoprotein (Kornfeld and Kornfeld, 1970, 1971); and D and E are from human urine (Sugahara *et al.*, 1976).

complex-type oligosaccharide structures seen in Figures 4 and 5 in that they contain less mannose. The structure of glycopeptide I isolated by Kawasaki and Ashwell (1976) from rabbit liver membrane, shown in 6A, contains only two mannose residues, one of which is substituted on C-2, C-3, and C-4 with outer chains. Structure 6B, the glycopeptide isolated from a human alveolar glycoprotein by Bhattacharyya and Lynn (1977), has only two mannose residues and only one N-acetylglucosamine residue in the core region. Similarly structure 6C, a glycopeptide isolated from human erythrocyte membrane sialoglycoprotein (Kornfeld and Kornfeld, 1970, 1971) contains only two mannose residues and, although it has two N-acetylglucosamine residues in the core, they are not linked in the typical mannosyl-di-N-acetylchitobiose sequence. Structures 6D and E were isolated from the urine of a patient with aspartylglycosylaminuria (Sugahara et al., 1976), and they have very different structures from other glycopeptides containing the aspartylglucosamine linkage in that they contain no mannose and consist of relatively short unbranched oligosaccharide chains. Even if they are degradation products of some larger, branched structure, the original material must have had an unusual structure. Recently, Hara et al. (1978) have reported that the complex branched-chain oligosaccharides of the α-subunit of several pituitary hormones contain the unusual sequence Man $\xrightarrow{\beta}$ GalNAc $\longrightarrow$ GlcNAc $\longrightarrow$ rather than chitobiose.

Complex-type oligosaccharides containing outer branches with Fuc $\xrightarrow{\alpha 1,2,3 \text{ or } 6}$ Gal, Gal $\xrightarrow{\beta 1,4}$ (Fuc $\xrightarrow{\alpha 1,3}$) GlcNAc, and Gal $\xrightarrow{\beta 1,4}$ (Fuc $\xrightarrow{\alpha 1,6}$) GlcNAc groupings have been detected in several glycoproteins (Tsay et al., 1975; Nishigaki et al., 1978; Strecker et al., 1977; Fournet et al., 1978; Purkayastha et al., 1979).

4.4. Structure of the Oligosaccharide–Lipid Intermediate

Recent experiments from several laboratories (Robbins et al., 1977; Tabas et al., 1978; Hunt et al., 1978) have demonstrated that both simple- and complex-type oligosaccharides are derived from a single high-molecular-weight lipid-linked oligosaccharide (see Chapters 2 and 3). This finding provides an explanation for the observation that the core regions of these two types of oligosaccharide units are identical in so many instances. The structure of the precursor molecule is shown in Figure 7. It differs from the $(Man)_9(GlcNAc)_2$-Asn unit of thyroglobulin and Chinese hamster ovary cell membranes (Figure 2) only in that it contains three glucose residues in the outer chain attached to C-3 of the core β-mannose residue. In vivo pulse-chase experiments with [³H]mannose have shown that this oligosaccharide is transferred *en bloc* from the lipid carrier to the newly synthesized protein. Within a few minutes of the transfer, processing of the oligosaccharide begins with the removal of the glucose residues. By 15–30 min the major intermediate is $(Man)_9(GlcNAc)_2$-Asn (Kornfeld *et al.*, 1978; Hubbard and Robbins, 1979). The subsequent removal of additional mannose residues is presumed to give rise to the various high-mannose oligosaccharides. It is also possible that additional mannose residues could be added to the processing intermediates, leading to further diversity among the high-mannose oligosaccharides, but such additions have not yet been demonstrated. When the final product is a complex-type oligosaccharide, additional mannose residues are clipped, and the sugar residues which form the outer branches are added. A proposal for the sequence of processing is shown in Figure 8. In this scheme, it is postulated that the acceptor for the first outer chain *N*-acetylglucosamine residue is $(Man)_5(GlcNAc)_2$-Asn rather than $(Man)_3(GlcNAc)_2$-Asn. The basis for this postulate is the finding that $(Man)_5(GlcNAc)_2$-Asn accumulates in variant lines of CHO cells which are deficient in the particular UDP-GlcNAc:glycoprotein *N*-acetylglucosaminyltransferase activity which transfers *N*-acetylglucosamine residues to C-2 of the mannose residue which is linked $\alpha1,3$ to the β-mannose residue (Tabas *et al.*, 1978; Robertson *et al.*, 1978). This indicates that the removal of the last two mannose residues from the precursor oligosaccharide is coordinated with the addition of the first outer *N*-acetylglucosamine residue. The simplest explanation is that the addition of the first outer *N*-acetylglucosamine residue is the signal for a specific α-mannosidase to remove the final two α-mannose residues. Direct evidence for this reaction has recently been obtained by Tabas and Kornfeld (1978). This sequence of processing provided an explanation for the unusual oligosaccharide units present on bovine rhodopsin (Liang *et al.*, 1979; Fukuda *et al.*, 1979). In Chapter 3 oligosaccharide processing is

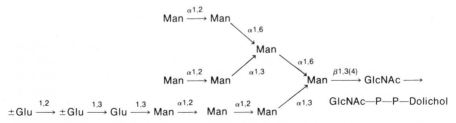

Figure 7. Proposed structure of the lipid-linked oligosaccharide precursor in glycoprotein synthesis (Li *et al.*, 1978).

discussed in detail with emphasis on the specific glycosyltransferases involved in the reactions.

Since there is a common protein-bound oligosaccharide precursor which can be processed to form either a high-mannose or a complex-type chain, one might predict that a particular asparagine residue in a given protein could sometimes carry a high-mannose chain and sometimes a complex chain. Although this type of heterogeneity is rarely seen, it does occur in bovine pancreatic RNase. This protein can have either a high-mannose oligosaccharide (as in RNase B) or a complex-type oligosac-charide (as in RNase C and D) at Asn-34 (Plummer, 1968). Presumably some of the RNase molecules are partially processed while others are completely processed. The hydrid oligosaccharide chains from ovalbumin (Figure 2G and 2H) may represent molecules in which chain elongation began but processing of the final mannose residues did not occur. The factors which govern the degree of processing have yet to be elucidated.

The explanation for the development of processing on the polypep-tide may be related to the generation of diversity of complex oligosac-charide structures. Thus many primitive organisms such as yeast and fungi contain high-mannose oligosaccharides, while complex-type oligo-

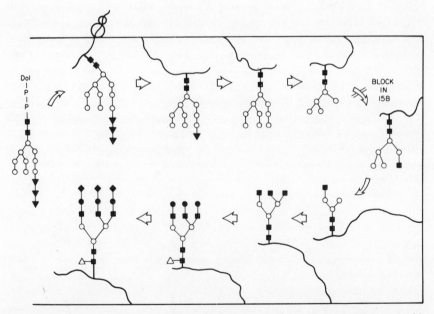

Figure 8. Proposed sequence for the processing of peptide-bound N-linked oligosaccharide chains: ■, N-acetylglucosamine residues; ○, mannose residues; ▼, glucose residues; ●, galactose residues; ◆, sialic acid residues; and △, fucose residues (Kornfeld et al., 1978).

saccharides appear to have developed at a later time in evolution. This observation suggests that high-mannose oligosaccharides are the ancestors of complex-type oligosaccharides and that processing represents a mechanism for synthesizing complex-type oligosaccharides without the necessity for developing a second lipid-linked oligosaccharide pathway.

The reader should consult Chapters 2 and 3 for a more complete discussion of the biosynthesis of the oligosaccharide units of glycoproteins.

5. ROLE OF OLIGOSACCHARIDES IN GLYCOPROTEIN SYNTHESIS

Having examined the structural details of the oligosaccharide chains of glycoproteins, it seems appropriate to step back and view them in their relationship to the whole glycoprotein structure. The techniques of protein sequencing have advanced rapidly, and now the complete amino acid sequence of a number of glycoproteins has been determined and the precise location of the carbohydrate chains has been assigned. From this information, as well as the sequences known for isolated glycopeptides, Marshall (1972) has pointed out that N-linked oligosaccharide chains are always attached to an asparagine residue which occurs in a sequence Asn-X-Ser(Thr) but that every such sequence in a protein is not glycosylated. No regular pattern has emerged for sequences surrounding serine and threonine residues that are glycosylated, but they often occur in proline-rich regions of the polypeptide chain. One of the first proteins to be sequenced was bovine ribonuclease which in the B form contains a high-mannose oligosaccharide on Asn-34 (Hirs et al., 1960; Plummer and Hirs, 1964). Furthermore the four S–S cross bridges have been placed, providing a picture of the folding of the polypeptide, and finally X-ray crystallography at 2 Å resolution (Richards and Wyckoff, 1971) has provided the data to construct a three-dimensional model of this protein. The conformation of the RNase molecule revealed by these studies shows that Asn-34 occurs on the surface freely exposed to the surrounding medium, and Puett (1973) has concluded from circular dichroism measurements that the oligosaccharide in RNase B has little effect on the overall conformation of the protein. Porcine RNase also contains 124 amino acid residues of very similar sequence and the same S–S cross-linking as bovine RNase (Jackson and Hirs, 1970a) but much more carbohydrate, including a high-mannose oligosaccharide at Asn-34 and two complex oligosaccharides at Asn-21 and Asn-76 which account for about 35% of the molecular weight (Jackson and Hirs, 1970b). After removing most of the carbohydrate chains by treatment with exoglycosidases, Wang and Hirs (1977) compared the modified porcine RNase to the native glycosylated protein for a number of properties and concluded that the larger oligosaccharide chains had no influence on the reoxidation and refolding rate or on the overall conformational stability of the molecule. However, circular dichroism measurements and spectral analysis revealed that the larger oligosaccharides present in the native RNase influenced the environment around several tyrosine side chains to effect a stabilization of the surface structure.

The notion that the carbohydrate side chains play a role in the conformation of glycoproteins is also strengthened by the recent report of Silverton et al. (1977) on the three-dimensional structure of an intact human immunoglobulin deduced from X-ray crystallographic data. Using the low-resolution 6 Å electron-density map obtained on intact Dob, a human IgG cryoglobulin, combined with known domain coordinates obtained at higher resolution on various IgG fragments and aided by a computer search, they obtained the complete quaternary structure of the Dob IgG molecule including placement of the complex oligosaccharide chains attached at Asn-297 of each heavy chain. Their space-filling model is reproduced in Figure 9 where the hexose units of the carbohydrate are shown as large black spheres, and the branched oligosaccharide chain is seen to wrap partly around the C_H2 domain and prevent its making contact with the C_H1 domain. The fact that the carbohydrate occupies a fixed position in the molecule and plays a major role in the interaction of the Fc and Fab regions suggests that its absence would alter the conformation of IgG and perhaps its functional properties as well. Support for this suggestion comes from a report by Koide et al. (1977), in which extensive glycosidase digestion was used to remove most of the carbohydrate from a rabbit IgG antibody against sheep erythrocytes. They found that the antigen combining site of the Fab region was unaffected in the sugar-depleted IgG. However, the Fc region was affected, since it could no longer interact with Fc receptors either on monocytes or on cytotoxic lymphoid cells and had lost its capacity to mediate complement fixation.

A striking example of the requirement for glycosylation for normal glycoprotein function is the finding that the formation of several enveloped viruses is inhibited by agents capable of blocking glycosylation. In several of these systems evidence has been obtained that the oligosaccharide units of the viral glycoproteins have a major effect on the properties of these proteins. Schwarz et al. (1976) used tunicamycin to block the glycosylation of influenza virus hemagglutinin and found that the carbohydrate-free hemagglutinin precursor was degraded intracellularly by proteases. This suggests that the oligosaccharide units of the hemag-

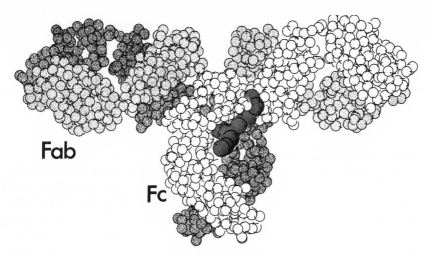

Figure 9. Space-filling view of the Dob IgG molecule taken from Silverton *et al.* (1977). One complete heavy chain is in white and the other is dark gray; the two light chains are lightly shaded. The large black spheres represent the individual hexose units of the complex carbohydrate. Drawing courtesy of Dr. D. R. Davies.

glutinin protect it against nonspecific cleavage and possibly have a positive role in identifying the normal sites of cleavage. Leavitt *et al.* (1977) demonstrated that although the nonglycosylated glycoproteins of Sindbis virus and vesicular stomatitis virus are stable in the infected cell, they are unable to migrate to the cell surface. The nonglycosylated glycoproteins of both viruses were found to be insoluble in nonionic detergents owing to aggregation, whereas their glycosylated counterparts were soluble (Gibson *et al.*, 1979). These data indicate that in some cases the oligosaccharide units of the viral glycoproteins can have a profound effect on the properties of these molecules. Other studies on the effects of inhibition of glycosylation on glycoprotein synthesis, secretion, and function are discussed in Chapter 2.

6. REFERENCES

Allen, A. K., and Neuberger, A., 1973, Purification and properties of the lectin from potato tubers, a hydroxyproline-containing glycoprotein, *Biochem. J.* **135**:307.

Allen, A. K., Desai, N. N., and Neuberger, A., 1976, The structure of a glycopeptide isolated from potato lectin, Abstracts of the 10th Int. Congress of Biochem., Hamburg, p. 503.

Arima, T., and Spiro, R. G., 1972, Studies of the carbohydrate units of thyroglobulin, *J. Biol. Chem.* **247**:1836.

Baenziger, J. U., 1975, Determination of the complete sequence of the oligosaccharide units present on IgG, IgE, and IgA₁ myeloma proteins, Ph.D. thesis, Washington University, St. Louis, Mo., 136 pp.

Baenziger, J. U., and Fiete, D., 1979, Structure of the complex oligosaccharides of fetuin, *J. Biol. Chem.* **254:**789.

Baenziger, J., and Kornfeld, S., 1974*a*, Structure of the carbohydrate units of IgA₁ immunoglobulin. I. Composition, glycopeptide isolation, and structure of the asparagine-linked oligosaccharide units, *J. Biol. Chem.* **249:**7260.

Baenziger, J., and Kornfeld, S., 1974*b*, Structure of the carbohydrate units of IgA₁ immunoglobulin. II. Structure of the O-glycosidically linked oligosaccharide units, *J. Biol. Chem.* **249:**7270.

Baenziger, J., Kornfeld, S., and Kochwa, S., 1974, Structure of the carbohydrate units of IgE immunoglobulin. II. Sequence of the sialic acid-containing glycopeptides, *J. Biol. Chem.* **249:**1897.

Bahl, O. P., 1969, Human chorionic gonadotropin. II Nature of the carbohydrate units, *J. Biol. Chem.* **244:**575.

Bahl, O. P., Carlson, R. B., Bellisario, R., and Swaminathan, N., 1972, Human chorionic gonadotropin: Amino acid sequence of the α and β subunits, *Biochem. Biophys. Res. Commun.* **48:**416.

Baig, M. M., and Aminoff, D., 1972, Glycoproteins and blood group activity. I. Oligosaccharides of serologically inactive hog submaxillary glycoproteins, *J. Biol. Chem.* **247:**6111.

Bertolini, M., and Pigman, W., 1970, The existence of oligosaccharides in bovine and ovine submaxillary mucins, *Carbohydr. Res.* **14:**53.

Bhattacharyya, S. N., and Lynn, W. S., 1977, Structural studies on the oligosaccharides of a glycoprotein isolated from alveoli of patients with alveolar proteinosis, *J. Biol. Chem.* **252:**1172.

Bhavanandan, V. P., Umemoto, J., Banks, J. R., and Davidson, E. A., 1977, Isolation and partial characterization of sialoglycopeptides produced by a murine melanoma, *Biochemistry* **16:**4426.

Björndal, H., Lindberg, B., and Svensson, S., 1967*a*, Gas–liquid chromatography of partially methylated alditols as their acetates, *Acta Chem. Scand.* **21:**1801.

Björndal, H., Lindberg, B., and Svensson, S., 1967*b*, Mass spectrometry of partially methylated alditol acetates, *Carbohydr. Res.* **5:**433.

Björndal, H., Hellerqvist, C. G., Lindberg, B., and Svensson, S., 1970, Gas–liquid chromatography and mass spectrometry in methylation analysis of polysaccharides, *Angew. Chem. Int. Ed. Engl.* **9:**610.

Bray, B. A., Lieberman, R., and Meyer, K., 1967, Structure of human skeletal keratosulfate: The linkage region, *J. Biol. Chem.* **242:**3373.

Butler, W. T., and Cunningham, L. W., 1966, Evidence for the linkage of a disaccharide to hydroxylsine in tropocollagen, *J. Biol. Chem.* **241:**3882.

Carlson, D. M., 1968, Structures and immunochemical properties of oligosaccharides isolated from pig submaxillary mucins, *J. Biol. Chem.* **243:**616.

Chapman, A., and Kornfeld, R., 1979, Structure of the high mannose oligosaccharides of a human IgM myeloma protein, *J. Biol. Chem.* **254:**816.

Codington, J. F., Linsley, K. B., Jeanloz, R. W., Irimura, T., and Osawa, T., 1975, Immunochemical and chemical investigations of the structure of glycoprotein fragments obtained from epiglycanin, a glycoprotein at the surface of the TA-3Ha cancer cell, *Carbohydr. Res.* **40:**171.

DeVries, A. L., Vandenheede, J., and Feeney, R. E., 1971, Primary structure of freezing-point-depressing glycoproteins, *J. Biol. Chem.* **246:**305.

Endo, Y., Yamashita, K., Han, Y. N., Iwanaga, S., and Kobata, A., 1977, The carbohydrate structure of a glycopeptide released by action of plasma kallikrein on bovine plasma high-molecular-weight kininogen, *J. Biochem. Tokyo* **82**:545.

Feizi, T., Kabat, E. A., Vicari, G., Anderson, B., and Marsh, W. L., 1971, Immunochemical studies on blood groups, *J. Immunol.* **106**:1578.

Finne, J., 1975, Structure of the O-glycosidically linked carbohydrate units of rat brain glycoproteins, *Biochim. Biophys. Acta* **412**:317.

Fournet, B., Montreuil, J., Strecker, G., Dorland, L., Haverkamp, J., Vliegenthart, J. F. G., Binette, J. P., and Schmid, K., 1978, Determination of the primary structures of 16 asialo-carbohydrate units derived from human plasma α_1-acid glycoprotein by 360-MHz ^{1}H NMR spectroscopy and permethylation analysis, *Biochemistry* **17**:5206.

Frangione, B., and Wolfenstein-Todel, C., 1972, Partial duplication in the "hinge" region of IgA$_1$ myeloma proteins, *Proc. Natl. Acad. Sci. U.S.A.* **69**:3673.

Fukuda, M., Kondo, T., and Osawa, T., 1976, Studies on the hydrazinolysis of glycoproteins, core structures of oligosaccharides obtained from porcine thyroglobulin and pineapple stem bromelain, *J. Biochem. Tokyo* **80**:1223.

Fukuda, M., Papermaster, D., and Hargrave, P. A., 1979, Rhodopsin carbohydrate: Structure of small oligosaccharides attached at two sites near the amino-terminus, *J. Biol. Chem.* **254**:8201.

Gibson, R., Schlesinger, S., and Kornfeld, S., 1979, The nonglycosylated glycoprotein of VSV is temperature-sensitive and undergoes intracellular aggregation at elevated temperatures, *J. Biol. Chem.* **254**:3600.

Gottschalk, A. (ed.), 1972, *Glycoproteins,* 2nd ed., Elsevier, Amsterdam.

Glockner, W. M., Newman, R. A., Dahr, W., and Uhlenbruck, G., 1976, Alkali-labile oligosaccharides from glycoproteins of different erythrocyte and milk fat globule membranes, *Biochim. Biophys. Acta* **443**:402.

Hakomori, S., 1964, A rapid permethylation of glycolipid, and polysaccharide catalyzed by methylsulfinyl carbanion in dimethyl sulfoxide, *J. Biochem. Tokyo* **55**:205.

Hallgren, P., Lundblad, A., and Svensson, S., 1975, A new type of carbohydrate protein linkage in a glycopeptide from normal human urine, *J. Biol. Chem.* **250**:5312.

Hara, K., Ruthnam, P., and Saxena, B., 1978, Structure of the carbohydrate moieties of α subunits of human follitropin, lutropin and thyrotropin, *J. Biol. Chem.* **253**:1582.

Heath, M. F., and Northcote, D. H., 1971, Glycoprotein of the wall of sycamore tissue-culture cells, *Biochem. J.* **125**:952.

Hickman, S., Kornfeld, R., Osterland, C. K., and Kornfeld, S., 1972, The structure of the glycopeptides of a human γM-immunoglobulin, *J. Biol. Chem.* **247**:2156.

Higashi, H., Naiki, M., Matuo, S., and Okouchi, K., 1977, Antigen of "serum sickness" type of antibodies in human sera: Identification as gangliosides with *N*-glycolyl-neuraminic acid, *Biochem. Biophys. Res. Commun.* **79**:388.

Hirs, C. H. W., Moore, S., and Stein, W. H., 1960, The sequence of the amino acid residues in performic acid-oxidized ribonuclease, *J. Biol. Chem.* **235**:633.

Huang, C. C., Mayer, H. E., Jr., and Montgomery, R., 1970, Microheterogeneity and paucidispersity of glycoproteins. Part I. The carbohydrates of chicken ovalbumin, *Carbohydr. Res.* **13**:127.

Hubbard, S. C., and Robbins, P. W., 1979, Synthesis and processing of protein-linked oligosaccharides *in vivo*, *J. Biol. Chem.* **254**:4568.

Hughes, R. C., 1973, Glycoproteins as components of cellular membranes, *Prog. Biophys. Mol. Biol.* **26**:189.

Hunt, L. A., Etchison, J. R., and Summers, D. F., 1978, Oligosaccharide chains are trimmed during synthesis of the envelope glycoprotein of vesicular stomatitis virus, *Proc. Natl. Acad. Sci. U.S.A.* **75**:754.

Isemura, M., Ikenaka, T., and Matsushima, Y., 1973, Comparative study of carbohydrate–

protein complexes. I. The structures of glycopeptides derived from cuttlefish skin collagen, *J. Biochem. Tokyo* **74:**11.

Ito, S., Muramatsu, T., and Kobata, A., 1975, Release of galactosyl oligosaccharides by endo-β-N-acetylglucosaminidase D, *Biochem. Biophys. Res. Commun.* **63:**938.

Ito, S., Yamashita, K., Spiro, R. G., and Kobata, A., 1977, Structure of a carbohydrate moiety of a unit A glycopeptide of calf thyroglobulin, *J. Biochem. Tokyo* **81:**1621.

Jackson, R. L., and Hirs, C. H. W., 1970a, The primary structure of porcine pancreatic ribonuclease. II. The amino acid sequence of the reduced S-aminoethylated protein, *J. Biol. Chem.* **245:**637.

Jackson, R. L., and Hirs, C. H. W., 1970b, The primary structure of porcine pancreatic ribonuclease. I. The distribution and sites of carbohydrate attachment, *J. Biol. Chem.* **245:**624.

Kawasaki, T., and Ashwell, G., 1976, Carbohydrate structure of glycopeptides isolated from an hepatic membrane-binding protein specific for asialoglycoproteins, *J. Biol. Chem.* **251:**5292.

Kieras, F. J., 1974, The linkage region of cartilage keratan sulfate, *J. Biol. Chem.* **249:**7506.

Koide, N., and Muramatsu, T., 1974, Endo-β-N-acetylglucosaminidase acting on carbohydrate moieties of glycoproteins: Purification and properties of the enzyme from *Diplococcus pneumoniae*, *J. Biol. Chem.* **249:**4897.

Koide, N., Nose, M., and Muramatsu, T., 1977, Recognition of IgG by F_c receptor and complement: Effects of glycosidase digestion, *Biochem. Biophys. Res. Commun.* **75:**838.

Kondo, T., Fukuda, M., and Osawa, T., 1977, The structure of unit B-type glycopeptides from porcine thyroglobulin, *Carbohydr. Res.* **58:**405.

Kornfeld, R., 1978, Structure of the oligosaccharides of three glycopeptides from calf thymocyte plasma membranes, *Biochemistry* **17:**1415.

Kornfeld, R., and Kornfeld, S., 1970, The structure of phytohemagglutinin receptor site from human erythrocytes, *J. Biol. Chem.* **245:**2536.

Kornfeld, R., and Kornfeld, S., 1976, Comparative aspects of glycoprotein structure, *Annu. Rev. Biochem.* **45:**217.

Kornfeld, R., Keller, J., Baenziger, J., and Kornfeld, S., 1971, The structure of the glycopeptide of human γG myeloma proteins, *J. Biol. Chem.* **246:**3259.

Kornfeld, S., and Kornfeld, R., 1971, The structure of phytohemagglutinin receptor sites, in: *Glycoproteins of Blood Cells and Plasma* (G. A. Jamieson and T. J. Greenwalt, eds.), pp. 50–68, Lippincott, Philadelphia.

Kornfeld, S., Li, E., and Tabas, I., 1978, Characterization of the processing intermediates in the synthesis of the complex oligosaccharide unit of the vesicular stomatitis virus G protein, *J. Biol. Chem.* **253:**7771.

Krusius, T., Finne, J., and Rauvala, H., 1976, The structural basis of the different affinities of two types of acidic N-glycosidic glycopeptides for concanavalin A–Sepharose, *FEBS Lett.* **71:**117.

Laine, R. A., Esselman, W. J., and Sweeley, C. C., 1972, Gas–liquid chromatography of carbohydrates, *Methods Enzymol.* **28B:**159.

Lamport, D. T. A., 1969, The isolation and partial characterization of hydroxyproline-rich glycopeptides obtained by enzymic degradation of primary cell walls, *Biochemistry* **8:**1155.

Lamport, D. T. A., Katona, L., and Roerig, S., 1973, Galactosylserine in extensin, *Biochem. J.* **135:**125.

Larriba, G., Klinger, M., Sramek, S., and Steiner, S., 1977, Novel fucose-containing components from rat tissues, *Biochem. Biophys. Res. Commun.* **77:**79.

Leavitt, R., Schlesinger, S., and Kornfeld, S., 1977, Impaired intracellular migration and altered solubility of nonglycosylated glycoproteins of vesicular stomatitis virus and Sindbis virus, *J. Biol. Chem.* **252**:9018.

Lee, Y. C., 1972, Analysis of sugars by automated liquid chromatography, *Methods Enzymol.* **28B**:63.

Lee, Y. C., and Scocca, J. R., 1972, A common structural unit in asparagine–oligosaccharides of several glycoproteins from different sources, *J. Biol. Chem.* **247**:5753.

Li, E., and Kornfeld, S., 1979, Structural studies of the major high mannose oligosaccharide units from Chinese hamster ovary cell glycoproteins, *J. Biol. Chem.* **254**:1600.

Li, E., Tabas, I., and Kornfeld, S., 1978, Structure of the lipid-linked oligosaccharide precursor of the complex-type oligosaccharides of the vesicular stomatitis virus G protein, *J. Biol. Chem.* **253**:7762.

Li, Y. T., and Lee, Y. C., 1972, Pineapple α- and β-D-mannopyranosidases and their action on core glycopeptides, *J. Biol. Chem.* **247**:3677.

Liang, C.-J., Yamashita, K., Muellenberg, C. G., Shichi, H., and Kobata, A., 1979, Structure of the carbohydrate moieties of bovine rhodopsin, *J. Biol. Chem.* **254**:6414.

Lombart, C. G., and Winzler, R. J., 1974, Isolation and characterization of oligosaccharides from canine submaxillary mucin, *Eur. J. Biochem.* **49**:77.

Mahieu, P. M., Lambert, P. H., and Maghuin-Rogister, G. R., 1973, Primary structure of a small glycopeptide isolated from human glomerular basement membrane and carrying a major antigenic site, *Eur. J. Biochem.* **40**:599.

Marks, G. S., Marshall, R. D., and Neuberger, A., 1963, Carbohydrates in protein. 6. Studies on the carbohydrate–peptide bond in hen's egg albumin, *Biochem. J.* **87**:274.

Marshall, R. D., 1972, Glycoproteins, *Annu. Rev. Biochem.* **41**:673.

Marshall, R. D., and Neuberger, A., 1964, Carbohydrates in protein. VIII. The isolation of 2-acetamide-1-(1-β-aspartamido)-1,2-dideoxy-β-D-glucose from hen's egg albumin, *Biochemistry* **3**:1596.

Michalski, J. C., Strecker, G., Fournet, B., Cantz, M., and Spranger, J., 1977, Structures of sialyl-oligosaccharides excreted in the urine of a patient with mucolipidosis I, *FEBS Lett.* **79**:101.

Misaki, A., and Goldstein, I. J., 1977, Glycosyl moiety of the lima bean lectin, *J. Biol. Chem.* **252**:6995.

Mizuochi, T., Yamashita, K., Fujikawa, K., Kisiel, W., and Kobata, A., 1979, The carbohydrate of bovine prothrombin, *J. Biol. Chem.* **254**:6419.

Montreuil, J., 1975, Recent data on the structure of the carbohydrate moiety of glycoproteins—Metabolic and biologic implications, *Pure Appl. Chem.* **42**:431.

Morell, A. G., and Ashwell, G., 1972, Tritium labeling of glycoproteins that contain terminal galactose residues, *Methods Enzymol.* **28B**:205.

Morgan, P. H., Jacobs, H. G., Segrest, J. P., and Cunningham, L. W., 1970, A comparative study of glycoptroteins derived from selected vertebrate collagens, *J. Biol. Chem.* **245**:5042.

Muir, L., and Lee, Y. C., 1969, Structures of the D-galactose oligosaccharides from earthworm cuticle collagen, *J. Biol. Chem.* **244**:2343.

Muir, L, and Lee, Y. C., 1970, Glycopeptides from earthworm cuticle collagen, *J. Biol. Chem.* **245**:502.

Nakajima, T., and Ballou, C.E., 1974*a*, Characterization of the carbohydrate fragments obtained from *Saccharomyces cerevisiae* mannan by alkaline degradation, *J. Biol. Chem.* **249**:7679.

Nakajima, T., and Ballou, C. E., 1974*b*, Structure of the linkage region between the polysaccharide and protein parts of *Saccharomyces cerevisiae* mannan, *J. Biol. Chem.* **249**:7685.

Newman, R. A., Glockner, W. M., and Uhlenbruck, G., 1976, Immunochemical detection of the Thomsen–Friedenreich antigen (T-antigen) on the pig lymphocyte plasma membrane, *Eur. J. Biochem.* **64:**373.

Ng Ying Kin, N. M. K., and Wolfe, L. S., 1974, Oligosaccharides accumulating in the liver from a patient with GM_2-gangliosidosis variant O (Sandhoff–Jatzkewitz disease), *Biochem. Biophys. Res. Commun.* **59:**837.

Nishigaki, M., Yamashita, K., Matsuda, I., Arashima, S., and Kobata, A., 1978, Urinary oligosaccharides of fucosidosis, *J. Biochem. Tokyo* **84:**823.

Oates, M. D., Rosbottom, A. C., and Schrager, J., 1974, Further investigations into the structure of human gastric mucin: The structural configuration of the oligosaccharide chains, *Carbohydr. Res.* **34:**115.

Ogata, S., Muramatsu, T., and Kobata, A., 1975, Fractionation of glycopeptides by affinity column chromatography on concanavalin A–Sepharose, *J. Biochem. Tokyo* **78:**687.

Pazur, J. H., Knull, H. R., and Cepure, A., 1971, Glycoenzymes: Structure and properties of the two forms of glucoamylase from *Aspergillus niger, Carbohydr. Res.* **20:**83.

Plummer, T. H., Jr., 1968, Glycoproteins of bovine pancreatic juice, *J. Biol. Chem.* **243:**5961.

Plummer, T. H., Jr., and Hirs, C. H. W., 1964, On the structure of bovine pancreatic ribonuclease B: Isolation of a glycopeptide, *J. Biol. Chem.* **239:**2530.

Puett, D., 1973, Conformational studies on a glycosylated bovine pancreatic ribonuclease, *J. Biol. Chem.* **248:**3566.

Purkayastha, S., Rao, C. V. N., and Lamm, M. E., 1979, Structure of the carbohydrate chain of free secretory component from human milk, *J. Biol. Chem.* **254:**6583.

Raizada, M. K., Schutzbach, J. S., and Ankel, H., 1975, *Cryptococcus laurentii* cell envelope glycoprotein, *J. Biol. Chem.* **250:**3310.

Reading, C. L., Penhoet, E. E., and Ballou, C. E., 1978, Carbohydrate structure of vesicular stomatitis virus glycoprotein, *J. Biol. Chem.* **253:**5600.

Richards, F. M., and Wyckoff, H. W., 1971, Bovine pancreatic ribonuclease, in: *The Enzymes* (P. D. Boyer, ed.), 3rd ed., Vol. 4, pp. 647–806, Academic Press, New York.

Robbins, P. W., Hubbard, S. C., Turco, S. J., and Wirth, D. F., 1977, Proposal for a common oligosaccharide intermediate in the synthesis of membrane glycoproteins, *Cell* **11:**893.

Robertson, M. A., Etchison, J. R., Robertson, J. S., Summers, D. F., and Stanley, P., 1978, Specific changes in the oligosaccharide moieties of VSV grown in different lectin-resistant CHO cells, *Cell* **13:**515.

Rosenthal, A. L., and Nordin, J. H., 1975, Enzymes that hydrolyze fungal cell wall polysaccharides, *J. Biol. Chem.* **250:**5295.

Schwarz, R. T., Rohrschneider, J. M., and Schmidt, M. F. G., 1976, Suppression of glycoprotein formation of semliki forest, influenza, and avian sarcoma virus by tunicamycin, *J. Virol.* **19:**782.

Schwarzmann, G., Hatcher, V. B., and Jeanloz, R. W., 1978, Purification and structural elucidation of several carbohydrate side chains from α_1-acid glycoprotein, *J. Biol. Chem.* **253:**6983.

Sentandreu, R., and Northcote, D. H., 1969, The characterization of oligosaccharides attached to threonine and serine in a mannan glycopeptide obtained from the cell wall of yeast, *Carbohydr. Res.* **10:**584.

Shier, W. T., Lin, Y., and De Vries, A. L., 1975, Structure of the carbohydrate of antifreeze glycoproteins from an Antarctic fish, *FEBS Lett.* **54:**135.

Shimizu, A., Putnam, F. W., Paul, C., Clamp, J. R., and Johnson, I., 1971, Structure and role of the five glycopeptides of human IgM immunoglobulins, *Nature (London), New Biol.* **231:**73.

Silverton, E. W., Manuel, A. N., and Davies, D. R., 1977, Three-dimensional structure of an intact human immunoglobulin, *Proc. Natl. Acad. Sci. U.S.A.* **74:**5740.

Spik, G., Bayard, B., Fournet, B., Strecker, G., Bouquelet, S., and Montreuil, J., 1975, Complete structure of the two carbohydrate units of human serotransferrin, *FEBS Lett.* **50:**296.

Spinola, M., and Jeanloz, R. W., 1970, The synthesis of a di-*N*-acetylchitobiose asparagine derivative, 2-acetamido-4-*O*-(2-acetamido-2-deoxy-β-D-glucopyranosyl)-1-*N*-(4-L-aspartyl-2-deoxy-β-D-glucopyranosylamine, *J. Biol. Chem.* **245:**4158.

Spiro, R. G., 1966, Characterization of carbohydrate units of glycoproteins, *Methods Enzymol.* **8:**26.

Spiro, R. G., 1967, The structure of the disaccharide unit of the renal glomerular basement membrane, *J. Biol. Chem.* **242:**4813.

Spiro, R. G., 1973, Glycoproteins, *Adv. Protein Chem.* **27:**349.

Spiro, R. G., and Bhoyroo, V. D., 1971, The carbohydrate of invertebrate collagens: A glucuronic acid–mannose disaccharide unit, *Fed. Proc.* **30:**1223.

Spiro, R. G., and Bhoyroo, V. D., 1974, Structure of the O-glycosidically linked carbohydrate units of fetuin, *J. Biol Chem.* **249:**5704.

Strecker, G., Fournet, B., Spik, G., Montreuil, J., Durand, P., and Tondeur, M., 1977, Structure of 9 oligosaccharides and glycopeptides rich in fucose excreted in the urine of two patients with fucosidosis, *C. R. Acad. Sci. (Paris)* **284D:**85.

Sugahara, K., Okumura, T., and Yamashina, I., 1972, Purification of β-mannosidase from a snail, *Achatina fulica,* and its action on glycopeptides, *Biochim. Biophys. Acta* **268:**488.

Sugahara, K., Funakoshi, S., Funakoshi, I., Aula, P., and Yamashina, I., 1976, Characterization of one neutral and two acidic glycoasparagines isolated from the urine of patients with asparatyl-glycosylaminuria, *J. Biochem. Tokyo* **80:**195.

Sukeno, T., Tarentino, A. L., Plummer, T. H., Jr., and Maley, F., 1971, On the nature of α-mannosidase-resistant linkages in glycoproteins, *Biochem. Biophys. Res. Commun.* **45:**219.

Sukeno, T., Tarentino, A. L., Plummer, T. H., Jr., and Maley, F., 1972, Purification and properties of α-D and β-D-mannosidases from hen oviduct, *Biochemistry* **11:**1493.

Tabas, I., and Kornfeld, S., 1978, Identification of an α-D-mannosidase activity involved in a late stage of processing of complex-type oligosaccharides, *J. Biol. Chem.* **253:**7779.

Tabas, I., Schlesinger, S., and Kornfeld, S., 1978, The processing of high mannose oligosaccharides to form complex type oligosaccharides in the newly synthesized polypeptides of the vesicular stomatitis virus G protein and the IgG heavy chain, *J. Biol. Chem.* **253:**716.

Tai, T., Ito, S., Yamashita, K., Muramatsu, T., and Kobata, A., 1975*a*, Asparagine-linked oligosaccharide chains of IgG: A revised structure, *Biochem. Biophys. Res. Commun.* **65:**968.

Tai, T., Yamashita, K., Ogata-Arakawa, M., Koide, N., Muramatsu, T., Iwashita, S., Inoue, Y., and Kobata, A., 1975*b*, Structural studies of two ovalbumin glycopeptides in relation to the endo-β-*N*-acetylglucosaminidase specificity, *J. Biol. Chem.* **250:**8569.

Tai, T., Yamashita, K., Ito, S., and Kobata, A., 1977*a*, Structures of the carbohydrate moiety of ovalbumin glycopeptide III and the difference in specificity of endo-β-*N*-acetylglucosaminidases C$_{II}$ and H, *J. Biol. Chem.* **252:**6687.

Tai, T., Yamashita, K., and Kobata, A., 1977*b*, The substrate specificities of endo-β-*N*-acetylglucosaminidases C$_{II}$ and H, *Biochem. Biophys. Res. Commun.* **78:**434.

Tarentino, A. L., and Maley, F., 1974, Purification and properties of an endo-β-*N*-acetylglucosaminidase from *Streptomyces griseus, J. Biol. Chem.* **249:**811.

Tarentino, A. L., and Maley, F., 1976, Purification and properties of an endo-β-N-acetyl-glucosaminidase from hen oviduct, *J. Biol. Chem.* **251**:6537.

Tarentino, A., Plummer, T. H., Jr., and Maley, F., 1970, Studies on the oligosaccharide sequence of ribonuclease B, *J. Biol. Chem.* **245**:4150.

Tarentino, A. L., Plummer, T. H., and Maley, F., 1974, The release of intact oligosaccharides from specific glycoproteins by endo-β-N-acetylglucosaminidase H, *J. Biol. Chem.* **249**:818.

Thomas, D. B., and Winzler, R. J., 1969, Structural studies on human erythrocyte glycoproteins: Alkali-labile oligosaccharides, *J. Biol. Chem.* **244**:5943.

Thomas, D. B., and Winzler, R. J., 1971, Structure of glycoproteins of human erythrocytes: Alkali-stable oligosaccharides, *Biochem. J.* **124**:55.

Toyoshima, S., Fukuda, M., and Osawa, T., 1972, Chemical nature of the receptor site for various phytomitogens, *Biochemistry* **11**:4000.

Toyoshima, S., Fukuda, M., and Osawa, T., 1973, The presence of β-mannosidic linkage in acidic glycopeptide from porcine thyroglobulin, *Biochem. Biophys. Res. Commun.* **51**:945.

Tsay, G. C., Dawson, G., and Sung, S.-S. J., 1975, Structure of the accumulating oligosaccharide in fucosidosis, *J. Biol. Chem.* **251**:5852.

Van Lenten, L., and Ashwell, G., 1972, Tritium-labeling of glycoproteins that contain sialic acid, *Methods Enzymol.* **28B**:209.

Wang, C. F. F., and Hirs, C. H. W., 1977, Influence of the heterosaccharides in porcine pancreatic ribonuclease on the conformation and stability of the protein, *J. Biol. Chem.* **252**:8358.

Watkins, W. M., 1972, Blood-group specific substances, in: *Glycoproteins* (A. Gottschalk, ed.), 2nd ed., Part B, p. 830, Elsevier, Amsterdam.

Wolfe, L. S., Senior, R. G., and Ng Ying Kin, N. M. K., 1974, The structures of oligosaccharides accumulating in the liver of G_{M1}-gangliosidosis, type I, *J. Biol. Chem.* **249**:1828.

Yamaguchi, H., Ikenaka, T., and Matsushima, Y., 1971, The complete sequence of a glycopeptide obtained from Taka-amylase A, *J. Biochem. Tokyo* **70**:587.

Yamashita, K., Tachibana, Y., and Kobata, A., 1978, The structures of the galactose-containing sugar chains of ovalbumin, *J. Biol. Chem.* **253**:3862.

Yasuda, Y., Takahashi, N., and Murachi, T., 1970, The composition and structure of carbohydrate moiety of stem bromelain, *Biochemistry* **9**:25.

The Function of Saccharide–Lipids in Synthesis of Glycoproteins

Douglas K. Struck and William J. Lennarz

1. INTRODUCTION

As discussed in Chapter 1, one of the major classes of glycoproteins is that in which the carbohydrate chain is linked to the polypeptide chain by means of an N-glycosidic bond. As shown in Figure 1, this N-glycosidic bond is between an *N*-acetylglucosaminyl residue and the amido nitrogen of an asparagine residue in the polypeptide chain.

This linkage is not restricted to a specific functional type of glycoprotein; it is present in membrane proteins, enzymes, secretory proteins without enzymatic functions, and immunoglobulins. Although there appear to be a few exceptions (see Chapter 1), in virtually all cases the carbohydrate chain of these glycoproteins has a common pentasaccharide core (Figure 2). The complete carbohydrate chains, containing additional residues attached to the α-mannosyl units, have been classified into two groups, i.e., those containing no additional Man* residues but instead GlcNAc, Gal, and SA (complex type, Figure 2), and those containing only additional Man residues (polymannose type, Figure 2). However, it is becoming clear that this is probably an oversimplification. For example, yeast mannan, which is really a glycoprotein, contains many dozens of Man residues attached to the core structure (Nakajima and Ballou,

Douglas K. Struck and William J. Lennarz • Department of Physiological Chemistry, The Johns Hopkins University School of Medicine, Baltimore, Maryland 21205.

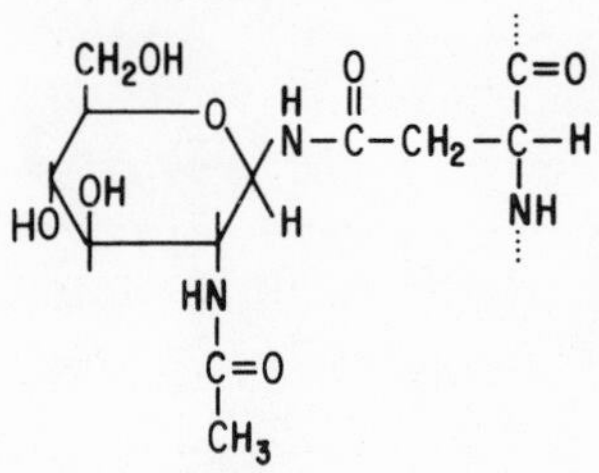

Figure 1. Structure of the linkage between carbohydrate and protein in N-glycosidically linked glycoproteins.

1974), whereas RNase B contains only one or two additional Man residues (Baynes and Wold, 1976).

In any case, the existence of a common core structure in many of the N-glycosidically linked glycoproteins suggests a common mechanism of synthesis of at least the internal region of the saccharide chain. Until recent years, most studies directed towards elucidation of the sequence of synthesis of the carbohydrate chain have been carried out *in vivo* using tissue slices pulse-labeled with radioactive sugars (Spiro and Spiro, 1966; Herscovics, 1969, 1970; Melchers, 1970, 1973). Needless to say, such *in vivo* pulse-labeling experiments have limitations. First, there can be considerable interconversion of the labeled sugars administered to the animal or the tissue slices. Second, the final results only suggest the order of

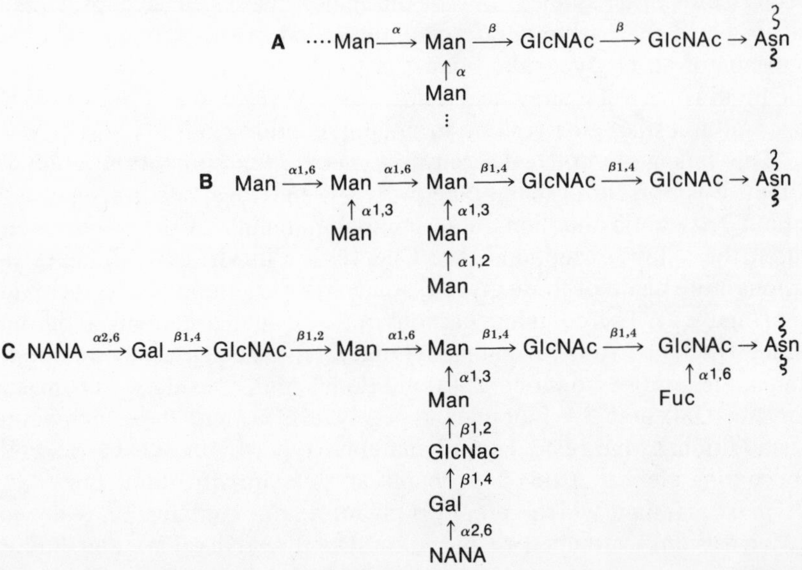

Figure 2. Basic structural features of N-glycosidically linked glycoproteins. The common core structure (A) is found in both the complex type (B) and polymannose type (C) chain. A more detailed discussion of these structural types is found in Chapter 1.

addition of the sugars, but not the mechanism whereby the assembly occurs. Nevertheless, the conclusion of these *in vivo* labeling studies was generally consistent with the known structural features of the glycoproteins. That is, labeling of the internal sugars, e.g., mannose, occurred early relative to the addition of the more distal sugars, such as SA and Gal. Moreover, by virtue of the fact that the distal sugars (GlcNAc, Gal, and SA) could be removed enzymatically from certain glycoproteins, it was possible to carry out *in vitro* studies on the assembly of the terminal portion of the oligosaccharide chains. These studies showed that the distal sugars (GlcNAc, Gal, and SA) of the complex-type carbohydrate chains were added to the core, one sugar at a time, from their respective nucleotide derivatives (see Chapter 3). No evidence for assembly of the core oligosaccharide by a similar mechanism was obtained.

Thus, although it was clear that glycosyl transferases were involved in the terminal phases of protein glycosylation, what remained was elucidation of the mechanism of assembly of the inner, core oligosaccharide. Studies since 1970 have demonstrated that the core oligosaccharide is preassembled on the lipid carrier, dolichol phosphate; only after assembly on the lipid carrier is complete is the oligosaccharide transferred to the protein acceptor. The reader may consult earlier reviews of these studies (Lennarz, 1975; Waechter and Lennarz, 1976; Lucas and Waechter, 1976; Hemming, 1977).

2. DOLICHOL

The family of polyisoprenols known as the dolichols, which were discovered almost 20 years ago (Pennock *et al.*, 1960), are found in the tissues of a variety of eukaryotes. The chemistry of these compounds, which occur as the free alcohols, the phosphomonoesters, and as esters with long-chain fatty acids have been reviewed in detail elsewhere (Hemming, 1974). The structure of dolichol phosphate is shown in Figure 3. Although a sensitive assay for the alcohol is available (Keller and Adair, 1977), at present there is no direct, simple quantitative assay for the dolichol phosphate. Such an assay would be invaluable in studying the distribution of dolichol phosphate in various membranes and its possible role in regulation of glycoprotein synthesis (see Section 8). Regarding the

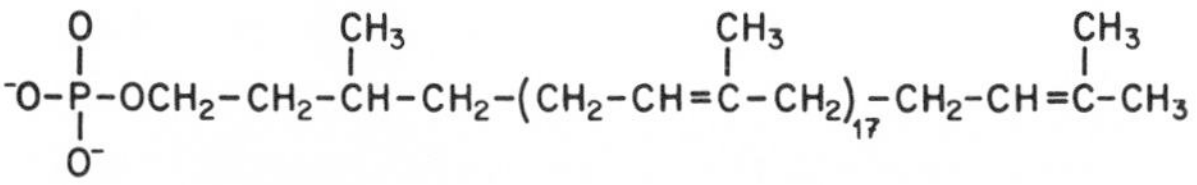

$$\text{^{-}O}-\underset{\underset{\text{O}^-}{|}}{\overset{\overset{\text{O}}{\|}}{\text{P}}}-\text{OCH}_2-\text{CH}_2-\underset{\overset{|}{\text{CH}_3}}{\text{CH}}-\text{CH}_2-\left(\text{CH}_2-\text{CH}=\underset{\overset{|}{\text{CH}_3}}{\text{C}}-\text{CH}_2\right)_{17}-\text{CH}_2-\text{CH}=\underset{\overset{|}{\text{CH}_3}}{\text{C}}-\text{CH}_3$$

Figure 3. Structure of dolichol phosphate.

disposition of dolichol phosphate in biological membranes, it is of interest to consider the extreme chain length of this compound; excluding the polymer of isoprene, rubber, dolichol is probably the longest natural hydrocarbon made up of a single repeating unit. A molecular model of dolichol in fully extended form is shown in Figure 4. The length of this form approaches 100 Å, although a helixlike form that is only somewhat over 70 Å in length has been suggested (Keenan *et al.*, 1977). For comparison, the chain length (25 Å) of oleic acid, a common constituent of the hydrophobic chain of phospholipids, is shown. It is obvious the extended form of dolichol is longer than the thickness of the hydrophobic core of most biological membranes (40–60 Å). It seems likely, therefore, that this carrier of activated sugars does not exist in membranes in its fully extended form. Clearly it will be of great importance to understand the structural orientation of dolichol phosphate in membranes so that its role as a carrier can be defined in detail (see Section 7).

3. MONOGLYCOSYL DERIVATIVES OF DOLICHOL PHOSPHATE

3.1. Mannosylphosphoryldolichol and N-Acetylglucosaminylpyrophosphoryldolichol

In the mid-1960s it became clear that a variety of complex glycans (e.g., peptidoglycan, lipopolysaccharide, mannan, complex polysaccharides) that exist either in or outside the cytoplasmic membrane of bacteria are synthesized by a mechanism involving preassembly of activated, hydrophobic derivatives of various saccharides (see review by Scher and Lennarz, 1972). The hydrophobic portion of these activated intermediates was found to be the polyisoprenol, undecaprenol, which contains 11 isoprene units. Based on these findings, it was postulated that analogous compounds might exist in eukaryotic cells and that they might be involved in glycoprotein or proteoglycan synthesis. In 1970, Behrens and Leloir (1970) provided the first firm evidence for this hypothesis. They reported

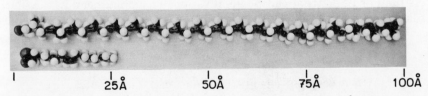

| | 25Å | 50Å | 75Å | 100Å |

Figure 4. Molecular model of dolichol. In fully extended form it is 100 Å in length. Oleic acid (25 Å) is shown for comparison.

that liver homogenates synthesized a compound analogous to the mannosylphosphorylundecaprenol previously studied in bacteria (Scher *et al.*, 1968). The hydrophobic moiety of the compound synthesized in liver preparations was shown to be a dolichol, containing approximately 19 isoprene units. The major structural differences between dolichol (see Figure 3) and undecaprenol are that the former contains a much longer hydrophobic chain and that the α-isoprenoid unit of this chain is saturated. Subsequent detailed structural studies established that mannosylphosphoryldolichol (Figure 5) contains one mannose unit per phosphate residue and that the phosphate group is linked at C-1 (Baynes *et al.*, 1973; Evans and Hemming, 1973); the anomeric configuration of the glycosidic linkage is β (Herscovics *et al.*, 1975). Both the eukaryotic and prokaryotic intermediate appear to be synthesized by the same type of reaction, involving transfer of a mannosyl unit from GDP-mannose to the polyprenol phosphate, as shown below:

$$\text{GDP-Mannose} + \text{Dolichol-P} \rightleftharpoons \text{Mannosyl-P-Dolichol} + \text{GDP}$$

The ΔG for the reaction has not been established, but it is clear that it must not be very far from zero since the reaction can readily be reversed by addition of GDP (Richards and Hemming, 1972; Waechter *et al.*, 1973).

The specificity of the mannosyltransferase of retina for the polyprenol phosphate has been studied, and it is clear that the enzyme shows a marked preference for dolichol phosphate (Kean, 1977). Shorter polyprenols in which the α-isoprene unit is unsaturated (e.g., ficaprenol or undecaprenol phosphate) are poor substrates. A variety of other polyprenol-related compounds also show very low acceptor activity. Studies on liver enzymes that catalyze the synthesis of mannosylphosphoryldolichol as well as the formation of an analogous compound, glucosylphosphoryldolichol (see below), have shown that the major requirement is that the first isoprene unit be saturated: There appears to be much less specificity for the length of the polyprenol chain (Mankowski *et al.*, 1975, 1977). In yeast the specificity for the polyisoprenol unit is even less pronounced; changes in the chain length, as well as in saturation of the α-isoprene unit, had little effect on mannosylphosphorylpolyprenol synthesis (Pless and Palamarczyk, 1978). However, a marked preference for

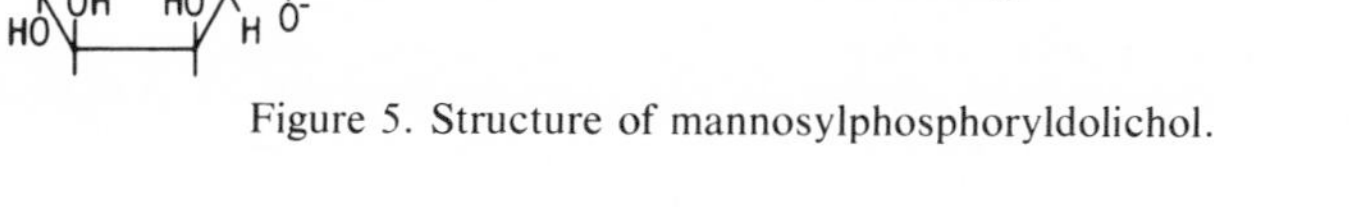

Figure 5. Structure of mannosylphosphoryldolichol.

mannosyl lipids containing an α-saturated isoprene unit was observed in experiments measuring their utilization as mannosyl donors to protein. A similar lack of chain-length specificity apparently exists in synthesis of N-acetylglucosaminylpyrophosphorylpolyprenol; the same family of dolichol homologues that are found in yeast participate in the synthesis of this compound *in vitro* (Reuvers *et al.*, 1978a).

Two other monoglycosyl derivatives of dolichol in which the glycosyl group appears to be linked to dolichol by a phosphodiester bond are glucosylphosphoryldolichol and xylosylphosphoryldolichol (see Waechter and Lennarz, 1976). It has been shown that both can serve as glycosyl donors in assembly of an oligosaccharide–lipid (see Section 4). However, it is now clear that the key "primer" glycosyl lipid that serves as the acceptor for other glycosyl units in assembly of oligosaccharide–lipid is one containing a single GlcNAc residue linked to dolichol via a pyrophosphate bridge. The enzymatic synthesis of this compound has been demonstrated in membranes from a wide variety of tissues (see review by Waechter and Lennarz, 1976; Waechter and Harford, 1977; Heifetz and Elbein, 1977a). The anomeric configuration of the O-glycosidic linkage of the GlcNAc unit in this molecule is α, and it is clear that GlcNAc-1-PO$_4$ (rather than GlcNAc) is transferred from UDP-GlcNAc to dolichol phosphate:

$$\text{UDP-GlcNAc} + \text{Dol-P} \longrightarrow \text{GlcNAc-}\alpha\text{-P-P-Dol} + (\text{UMP})$$

3.2. Other Glycosyl Polyprenol Derivatives

Finally, the existence of another class of lipid-linked sugars, in which the hydrophobic moiety is retinol, rather than dolichol, should be mentioned. A mannosyl derivative of retinol phosphate (Rosso *et al.*, 1975), subsequently shown to be β-mannosylphosphorylretinol (Rosso *et al.*, 1976), has been isolated, and several lines of evidence for its participation in glycosylation of proteins has been reported (DeLuca *et al.*, 1973, 1975; Rosso *et al.*, 1977). Although there is little question about the ability of liver enzymes to catalyze synthesis of mannosylphosphorylretinol, it is unclear whether or not the retinol phosphate merely serves as a poor analogue for dolichol phosphate. The results of studies with liver (DeLuca *et al.*, 1973) and retina (Kean, 1977) enzyme preparations indicate that dolichol phosphate is a far better acceptor of a mannosyl unit than retinol phosphate. *In vivo* administration of retinyl palmitate has been shown to stimulate incorporation of mannose into mannosylphosphoryldolichol as well as into mannosylphosphorylretinol (Hassell *et al.*, 1978). Another glycosyl derivative of retinol phosphate, galactosylphosphorylretinol, has been shown to be synthesized by enzyme preparations of mouse mastocytoma tissue (Peterson *et al.*, 1976). In addition stimulation of galactose incorporation from UDP-Gal into proteins by exogenous dolichol phosphate has been reported in homogenates of NIL 2E cells (McEvoy *et al.*, 1977). Hopefully, further studies will provide insight into the question of the specificity of these reactions for polyprenol phosphates and the possible role of galactosylphosphoryl lipids in the glycosylation of proteins.

4. ASSEMBLY OF OLIGOSACCHARIDE-LIPIDS

Leloir and co-workers reported the first evidence for the synthesis of an oligosaccharide containing Glc, Man, and GlcNAc that was linked by a pyrophosphate bridge to dolichol (Behrens *et al.*, 1973, and references cited therein; Tabora and Behrens, 1977). The Glc-containing oligosaccharide-lipid was postulated to have the following structure, where x may be as great as 14:

$$(Glc)_2\text{-}(Man)_x\text{-}(GlcNAc)_2\text{-P-P-Dol}$$

Although more recently Glc-containing oligosaccharide-lipids have been studied by other laboratories (see below), much of the initial work on assembly of oligosaccharide-lipids in a variety of systems dealt with a simple oligosaccharide-lipid with the general structure:

$$(\alpha\text{-Man})_{4-6}\text{-}\beta\text{-Man-}\beta\text{-GlcNAc-GlcNAc-P-P-Dol}$$

To date no rigorous chemical analysis has been reported that directly proves the existence of pyrophosphate linkage between the oligosaccharide and the lipid, but this seems highly likely since it has been shown that mild acid hydrolysis of the oligosaccharide-lipid yields dolichol pyrophosphate as well as an oligosaccharide (Herscovics *et al.*, 1977a).

A scheme for assembly of this oligosaccharide based on studies in a variety of membrane preparations is outlined in Figure 6. As discussed above, in the initial step GlcNAc-1-PO_4 is transferred from UDP-GlcNAc to Dol-P. The product, GlcNAc-P-P-Dol, has been isolated from several systems (see Waechter and Lennarz, 1976; Ghalambor *et al.*, 1974; Chen and Lennarz, 1977) and in at least one case shown to be identical to chemically synthesized α-GlcNAc-P-P-Dol (Warren and Jeanloz, 1974). The second step, elongation of isolated GlcNAc-P-P-Dol to the disaccharide-lipid, β-GlcNAc-GlcNAc-P-P-Dol, by addition of a GlcNAc unit from UDP-GlcNAc has also been demonstrated in several systems (Herscovics *et al.*, 1978; Chen and Lennarz, 1977; Reuvers *et al.*, 1977), and the disaccharide-lipid has been shown to be identical to the chemically

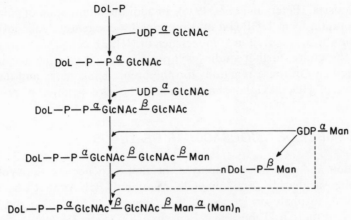

Figure 6. Pathway for assembly of oligosaccharide–lipids.

synthesized compound (Herscovics *et al.*, 1978). Enzymatic synthesis of the trisaccharide–lipid containing one unit of mannosyl β-linked to the distal GlcNAc unit of GlcNAc-GlcNAc-P-P-Dol (Levy *et al.*, 1974; Wedgwood *et al.*, 1974; Nakayama *et al.*, 1976; Heifetz and Elbein, 1977*b*) appears to involve direct transfer of a mannosyl unit from GDP-mannose. Man-P-Dol is ineffective as a mannosyl donor at this step (Levy *et al.*, 1974; Heifetz and Elbein, 1977*b*). However, it is clear that Man-P-Dol (or GDP-Man, which can be converted to Man-P-Dol) can serve as a mannosyl donor in the next step, namely, elongation of β-Man-β-GlcNAc-GlcNAc-P-P-Dol to oligosaccharide-P-P-Dol (Levy *et al.*, 1974; Chen and Lennarz, 1976). The sequence of reactions involved in this elongation process has not been elucidated. Although all of the mannosyl units introduced appear to be linked α, recent evidence suggests that GDP-Man (see dashed lines, Figure 6), as well as Man-P-Dol, may serve as a mannosyl donor for some of the units (Chambers *et al.*, 1977; Forsee *et al.*, 1977). Clearly, in order to understand the details of this phase of the biosynthetic process, enzymatic and chemical studies on the partially elongated oligosaccharide–lipid must be undertaken. In fact, to fully document the many enzymatic reactions involved in the overall synthesis of the oligosaccharide–lipid from GDP-Man and UDP-GlcNAc, it will be necessary to isolate solubilized, purified forms of each of the enzymes, free of any endogenous saccharide–lipid intermediates. Efforts to accomplish this are underway, and thus far soluble forms of the transferases catalyzing synthesis of GlcNAc-P-P-Dol, GlcNAc-GlcNAc-P-P-Dol, β-Man-β-GlcNAc-GlcNAc-P-P-Dol, and Man-P-Dol have been reported (Heifetz and Elbein, 1976*a*,*b*).

In both the mouse myeloma and hen oviduct enzyme systems the oligosaccharide synthesized *in vitro* from UDP-GlcNAc and GDP-Man contains seven to nine monosaccharide units. The anomeric configuration of the Man and GlcNAc units in the oligosaccharide-lipid synthesized by hen oviduct membranes was determined to be that shown in Figure 6 by use of purified exo- and endoglycosidases (Chen, *et al.*, 1975). It is clear that the anomeric configuration of the internal linkages is identical to that found in many of the Asn-linked secretory glycoproteins (cf. Figure 2). It should be emphasized, however, that the range in the chain length of the oligosaccharide-lipid synthesized *in vitro* varies widely, depending on the enzyme system. Thus, for example, in aorta (Chambers *et al.*, 1977) and mammary (White, 1978) membrane preparations oligosaccharides ranging in saccharide content from five to ten units are formed. At the present time the question of how the chain length of the oligosaccharide chain of oligosaccharide-lipid is regulated is only beginning to be understood. It is important to elucidate this process because, as noted in Section 1, different glycoproteins contain different numbers of α-mannose units, ranging from two in proteins with a complex-type side chain to at least five in those with a polymannose-type side chain.

In this connection, the existence of Glc-containing oligosaccharide-lipids has assumed particular importance. Several laboratories have shown that, in the presence of UDP-Glc, the oligosaccharide-lipid synthesized *in vitro* is increased in size by one to three additional hexose units and contains Glc (Herscovics *et al.*, 1977b; Robbins *et al.*, 1977a; Scher *et al.*, 1977). Several groups have shown that Glc-P-Dol, synthesized from UDP-Glc and Dol-P, can serve as the glycosyl donor in this elongation process (Behrens *et al.*, 1971; Herscovics *et al.*, 1977b; Parodi, 1976; Scher *et al.*, 1977; Waechter and Scher, 1978). In addition it has been found that even larger oligosaccharide-lipids containing Glc can be isolated directly from a variety of tissues (Spiro *et al.*, 1976a). Schematic representations encompassing the known structural features of the oligosaccharide-lipid isolated from calf thyroid (Spiro *et al.*, 1976b) and that obtained from VSV infected cells (Li *et al.*, 1978) are shown in Figure 7.

It is not yet possible to define the final steps in assembly of the Glc-containing oligosaccharide-lipid as they relate to the pathway shown in Figure 6. Because the demonstration of the synthesis of Glc-containing oligosaccharide-lipids *in vitro*, as well as their occurrence *in vivo*, indicates that they, rather than Glc-free oligosaccharide-lipids may be the natural substrate involved in glycosylation of proteins (see Section 5), it will be important to have more detailed information on the biosynthesis of the Glc-containing oligosaccharide-lipid.

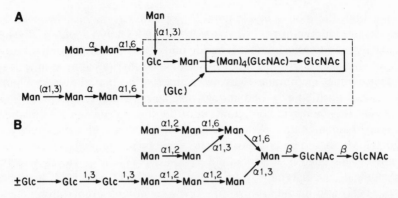

Figure 7. Structural features of the oligosaccharide isolated from the oligosaccharide–lipid from calf thyroid (A) and from VSV-infected cells (B).

5. TRANSFER OF OLIGOSACCHARIDE CHAINS TO ENDOGENOUS MEMBRANE PROTEINS

Following the early observation that the radioactivity in a large Glc-containing oligosaccharide–lipid was incorporated into endogenous proteins, more detailed studies on the transfer using $(\alpha\text{-Man})_{4-6}\text{-}\beta\text{-Man-}\beta\text{-}$GlcNAc-GlcNAc-P-P-Dol were undertaken (Waechter and Lennarz, 1976). Experiments using [14C]-Man and [3H]-GlcNAc-labeled oligosaccharide–lipid prepared from oviduct membrane showed that both labels were transferred to the protein, thus suggesting that the oligosaccharide chain was transferred *en bloc*. This finding was corroborated in both the myeloma and oviduct membrane systems by structural analysis of the labeled oligosaccharide chain attached to the protein. The chain was shown to be resistant to release under conditions that result in β-elimination of saccharides linked O-glycosidically to seryl or threonyl residues. Since the chain was released only after strong reductive alkaline hydrolysis, it was concluded that the linkage must be via an N-glycosidic bond, presumably to Asn. Comparison of the properties of the reduced oligosaccharide chain released from the protein by this procedure with the reduced oligosaccharide prepared from oligosaccharide-P-P-dolichol indicated they were indistinguishable. Thus the transfer reaction can be depicted as shown in Figure 8. It should be noted, however, that formation of the putative second product of the reaction, dolichol pyrophosphate, has not been directly demonstrated. Moreover, at least in the case of membrane protein acceptors, it has not been unequivocally established that the linkage is to an Asn residue.

Little is known about the endogenous proteins glycosylated *in vitro* aside from their apparent molecular weight as determined by SDS-PAGE.

Molecular weight values ranging from 25,000 in hen oviduct membranes (Pless and Lennarz, 1975) to 145,000 in calf brain membranes (Waechter and Harford, 1977) have been reported, depending on the enzyme system. Moreover, in most membrane systems a multiplicity of polypeptides are labeled.

Recently the specificity of the transferase (or transferases) that catalyze transfer of the saccharide from saccharide–lipid to endogenous proteins has been investigated. Studies using exogenous β-GlcNAc-GlcNAc-P-P-Dol or β-Man-β-GlcNAc-GlcNAc-P-P-Dol as donors in the oviduct membrane system showed that both of these compounds serve as direct donors of their saccharide chain (Chen and Lennarz, 1976, 1977). Similarly, yeast membranes have been shown to catalyze transfer of β-GlcNAc-GlcNAc-P-P-Dol (Reuvers *et al.*, 1977) as well as Man- and GlcNAc-containing trisaccharide–lipid (presumably β-Man-β-GlcNAc-GlcNAc-lipid) (Nakayama *et al.*, 1976) directly to protein. Thus far, no evidence for direct transfer of a single GlcNAc unit from GlcNAc-P-P-Dol to membrane proteins has been obtained. If, in fact, only one transferase is involved in transfer it may recognize only the first two monosaccharide units (i.e., the chitobiosyl unit) on the oligosaccharide-lipid. Were this the case it would provide an explanation for why, in addition to $(\alpha$-Man$)_{4-6}$-β-Man-β-GlcNAc-GlcNAc-P-P-Dol, β-Man-GlcNAc-GlcNAc-P-P-Dol and GlcNAc-GlcNAc-P-P-Dol can also serve as substrates. However, it should be emphasized that in order to make a truly meaningful comparison of the relative activity of these saccharide–lipids as substrates several experimental problems remain to be resolved. First, to determine the absolute molar amounts of each saccharide transferred to protein, the specific radioactivity of the saccharide–lipid must be known. Second, the pool size of any endogenous saccharide-lipid of the

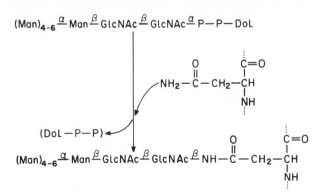

Figure 8. Transfer of the oligosaccharide chain from its lipid carrier to the asparagine side chain of a protein.

same structure, which may dilute the specific activity of labeled, added saccharide–lipid, must be determined.

Nevertheless, the observation that shorter saccharide–lipids can be transferred to protein prior to elongation raises the question of the possible existence of an alternative pathway, involving addition of mannosyl units after transfer of di- or trisaccharide units to protein. Several lines of evidence argue against this alternative. In the case of studies in the oviduct membrane system, attempts to complete the elongation of all the membrane proteins prelabeled with either the disaccharide or the trisaccharide unit by the addition of GDP-Man were unsuccessful (Chen and Lennarz, 1976, 1977). Although a fraction of the saccharide chains of the proteins were elongated to an oligosaccharide, this could be readily explained by elongation of the di- or trisaccharide–lipid to oligosaccharide–lipid prior to transfer of the saccharide unit to protein. This idea is further strengthened by studies that showed that no oligosaccharide-containing protein was synthesized when a preformed, purified disaccharide–protein was incubated with GDP-Man and oviduct membranes (Struck and Lennarz, unpublished studies). Similar observations have been made in the yeast system; no further elongation of β-Man-GlcNAc-GlcNAc units attached to protein could be observed upon addition of GDP-Man (Lehle and Tanner, 1978).

As noted earlier, another class of saccharide–lipids, which is larger in size than the oligosaccharide discussed previously and contains Glc in addition to Man and GlcNAc, has been reported. An estimate of the relative rate of enzymatic incorporation of this oligosaccharide, compared to the Glc-free oligosaccharide into membrane protein has been made (Turco *et al.*, 1977). Some of the problems discussed above in making this type of comparison meaningful have been overcome, and the conclusion drawn by the authors is that the larger Glc-containing oligosaccharide–lipid is eight times more active than the Glc-free oligosaccharide–lipid as a donor of its oligosaccharide chain to endogenous membrane protein acceptors. Transfer of a large Glc-containing oligosaccharide from lipid to endogenous protein acceptors has also been demonstrated in thyroid (Staneloni and Leloir, 1978; Spiro *et al.*, 1978*a*) and oviduct membranes (Chen, 1978; Chen and Lennarz, 1978*a*). In addition transfer to a structurally characterized exogenous substrate, α-lactalbumin (see Section 6), has been demonstrated (Chen, 1978; Chen and Lennarz, 1978*a*). Studies on the glycoproteins synthesized in VSV and Sindbis infected cells, led to the proposal that a single, Glc-containing oligosaccharide–lipid is the true donor of saccharide chains *in vivo* (Robbins *et al.*, 1977*b*). This possibility is consistent with other studies on the assembly of viral-coat glycoproteins (Tabas *et al.*, 1978; Hunt *et al.*, 1978) that show that the initial saccharide chain linked to protein is a large oligosaccharide and that it is further processed by cleavage of hexose units to produce a short mannose-containing chain that ultimately is converted to the complex-type chain (see Chapters 3 and 4). The processing of large oligosaccharide chains once they are attached to the protein apparently is not limited to viral glycoproteins, since it has been observed in myeloma cells that secrete IgG (Tabas *et al.*, 1978), in thyroid gland (Staneloni and Leloir, 1978), and in chick fibroblasts (Hubbard *et al.*, 1978). Although the detailed steps in this processing sequence are not yet known, enzymatic excision of Man (Tabas and Kornfeld, 1978) and Glc (Spiro *et al.*, 1978b; Hubbard *et al.*, 1978; Chen, 1978; Chen and Lennarz, 1978b) residues from the large Glc-containing oligosaccharide linked to protein has been detected in membrane preparations from several sources. Further discussion of the processing of oligosaccharide chains is discussed in Chapter 3.

6. TRANSFER OF OLIGOSACCHARIDE CHAINS FROM OLIGOSACCHARIDE-LIPIDS TO EXOGENOUS, SOLUBLE PROTEINS

In view of the virtual identity of the oligosaccharide chains of the oligosaccharide-lipid and the oligosaccharide chains of secretory proteins, it was surprising that studies involving use of membranes from either myeloma, which produces large amounts of kappa-type (K-type) light-chain immunoglobulins or from oviduct, which produces large amounts of ovalbumin, failed to provide any evidence for the glycosylation of those proteins. Instead, as noted above, only a variety of uncharacterized membrane proteins were labeled. These results suggest that the unglycosylated form of these secretory proteins is not present in these membrane preparations. Consistent with this possibility, an unglycosylated form of K-type light chain secreted by myeloma cells treated with 2-deoxyglucose (Eagon and Heath, 1977) can be glycosylated *in vitro* by a myeloma membrane preparation (Eagon *et al.*, 1975). This study, along with *in vivo* experiments with oviduct slices treated with tunicamycin to block glycosylation of ovalbumin (see Section 9), provide strong evidence that secretory proteins, as well as membrane proteins, are glycosylated via the lipid-linked pathway.

6.1. Existence of a Tripeptide Acceptor Sequence

It is clear that to study in detail the mechanism of glycosylation and the structural requirements of the protein substrate, one must be able to utilize exogenous protein substrates of known amino acid sequence. A number of years ago it was noted that in N-glycosidically linked glycoproteins the Asn bearing the carbohydrate chain invariably was part of the tripeptide sequence -Asn-X-Ser- or -Asn-X-Thr-, where X represents one of the 20 amino acids (Eylar, 1965; Marshall, 1974). However, it should be emphasized that not all proteins containing the sequence -Asn-X-Ser/Thr- are glycosylated. Some have never been found in a glycosylated form. Others, for example, bovine pancreatic RNase, are found in both an unglycosylated (RNase A) and a glycosylated form (RNase B).

One possible explanation for why certain proteins containing -Asn-X-Ser/Thr- are not glycosylated is that specific amino acids must be present in the X position for the protein to be a substrate. However, based on an earlier survey (Marshall, 1974) it appeared that, with the possible exception of Pro or Asp as X, glycoproteins with X being one of the remaining 18 amino acids were known to exist. Recently, a computer-facilitated examination of the sequence of 767 proteins revealed the presence of 342 -Asn-X-Ser/Thr- tripeptide sequences (Pless, Saffen, and Lennarz, unpublished studies). Because the occurrence of Asn-linked glycoproteins in prokaryotes is extremely rare, further analysis considered only proteins from eukaryotes. The results of examination of these proteins, in which the total number of tripeptide sequences is 159, are summarized in Table 1.

Of these 159 tripeptide sequences, 49 are found to exist in a glycosylated form, whereas 110 are not. The difficulty in making any firm correlations between which of the possible 20 amino acids is permitted in the X position, when one has only 49 examples of glycosylated proteins and 20 possible X substituents, is obvious. The only clear-cut statement based on a reasonable sample size can be made where X is Asp. No glycosylated proteins containing this residue in the X position of the tripeptide are known, whereas ten lacking carbohydrates are known. In other cases the sample size is probably too small to be meaningful, i.e., there are no instances of glycoproteins containing Trp as X, but only three instances of it in unglycosylated proteins containing the tripeptide sequence. Further refinement of the sample size, for example, by excluding all homologous proteins (e.g., RNases from different eukaryotes) that contain the tripeptide in the same position in the overall amino acid sequence, does not provide any clearer picture. Thus, in agreement with the early analysis, these data indicate that virtually any amino acid (except perhaps Asp) can occur in the X position of a N-glycosidically linked glycoprotein. Thus, although the sequence -Asn-X-Ser/Thr- is necessary for glycosylation, apparently X does not determine whether or not it is glycosylated.

In an attempt to elucidate the factors that might determine which -Asn-X-Ser/Thr- site can be glycosylated, the procedure of Chou and

Table 1. Analysis of Sequence Data of 159
Tripeptide Sequences Found in Glycoproteins
and Proteins from Eukaryotic Sources

Amino acid in X position	CHO | -Asn-X-Ser/Thr-	-Asn-X-Ser/Thr-
Ala	7	5
Val	3	9
Leu	4	25
Ile	7	4
Pro	1	7
Phe	1	6
Trp	1	0
Met	3	3
Gly	1	7
Ser	4	8
Thr	6	1
Cys	0	5
Tyr	0	3
Asn	1	4
Gln	1	4
Asp	0	10
Glu	0	1
Lys	4	5
Arg	0	1
His	5	2
Total	49	110

Fasman (1974a,b) has been used to estimate the secondary structure of the peptide segment around the carbohydrate peptide linkage of 28 different N-glycosidically linked glycoproteins (Aubert *et al.*, 1976). This analysis suggested that in somewhat more than half of the glycoproteins the glycosylated asparagine residues were localized at a β-turn in the polypeptide chain. In another study, using a similar approach, 30 of 31 glycosylated residues were found to be in sequences that favor β-turn or loop structures. In the Chou and Fasman treatment Asn, Ser, and Thr are considered to be residues that do not favor α-helix formation. It is not unexpected, therefore, that glycosylated tripeptides containing these amino acids might be found in more disordered regions in the final protein. Finally, it is clear from an extensive structural survey of a wide variety of glycosylated and unglycosylated RNases (Lenstra *et al.*, 1977) that attempts to deduce the factors regulating glycosylation at the molecular level by analyzing the final secreted products are not likely to be fruitful. A wide variety of factors, including the relative rates of translation, glycosylation, posttranslational folding, packaging, and secretion may obscure the issue. Perhaps the most striking example of how the analysis of glycoproteins formed *in vivo* provides little insight in the molecular factors controlling glycosylation is found in the case of bovine pancreatic RNase. Although this enzyme is secreted in both an unglycosylated (RNase A) and glycosylated form (RNase B), both forms have the identical amino acid sequence.

6.2. In Vitro Glycosylation of Denatured Soluble Proteins

Because of the limitations of *in vivo* studies discussed above, an attempt was made to study the structural features necessary for exogenous proteins of known amino acid sequence to serve as substrates for the oligosaccharide transferase in oviduct membranes (Pless and Lennarz, 1977). Initial studies showed that ovalbumin (which contains at least one free -Asn-X-Ser/Thr- in addition to the one that already is glycosylated), RNase A, and α-lactalbumin, all of which are potential substrates, were inactive in their native state. In contrast, after denaturation, reduction, and derivatization of the sulfhydryl groups, all three were found to serve as acceptors of the oligosaccharide chain of the oligosaccharide-lipids. Control proteins, lacking an -Asn-X-Ser/Thr-, as well as a protein already glycosylated at -Asn-X-Ser/Thr-, i.e., RNase B, were inactive as substrates. Of particular interest was the finding that several proteins that do contain -Asn-X-Ser/Thr- acceptor sites were inactive as substrates even after denaturation, reduction, and derivatization.

It should be noted that enzymes in liver (Khalkhali and Marshall, 1975; Khalkhali *et al.*, 1977), serum (Khalkhali and Marshall, 1976), and yeast (Khalkhali *et al.*, 1976) have been shown to catalyze conversion of native RNase A to a glycosylated form of RNase containing one GlcNAc unit. Although the mechanism of this monoglycosylation has not been investigated, the finding that the reaction studied in yeast membranes is inhibited 70% by the presence of tunicamycin (see below) suggests that a lipid intermediate may be involved (Khalkhali *et al.*, 1976). The reason why native RNase A serves as a substrate in these enzyme preparations, but not in oviduct, remains to be elucidated.

The initial study on the lipid-mediated glycosylation of exogenous proteins by the oviduct membrane system has been extended in two ways. First, a larger number of potential acceptor proteins, derivatized in a variety of ways, have been tested as substrates (Kronquist and Lennarz, 1978). Second, polypeptide fragments of several of the potential acceptor proteins have been prepared and tested as substrates. The results of the studies on 13 reduced and derivatized proteins containing one or more -Asn-X-Ser/Thr- sites are summarized in Table 2. In all cases the native, underivatized protein was inactive as a substrate. However,

Table 2. Summary of Denatured, Derivatized Proteins Used as Substrates
for Oligosaccharide Transferase

Protein	S derivative	Potential site(s) of carbohydrate attachment
Active		
Ovalbumin	$-SO_3^-$	-Asn$_{311}$-Leu-Ser-
α-Lactalbumin	$-SO_3^-$	-Asn$_{45}$-Gln-Ser-; -Asn$_{74}$-Ile-Ser-
	$-CH_2COO^-$	
	$-CH_2CONH_2$	
	$-CH_2CH_2NH_3^+$	
RNase A	$-SO_3^-$	-Asn$_{34}$-Leu-Thr-
	$-CH_2COO^-$	
	$-CH_2CONH_2$	
DNase	$-CH_2COO^-$	-Asn$_{103}$-Asp-Ser-
	$-CH_2CH_2NH_3^+$	
Prolactin	$-CH_2COO^-$	-Asn$_{31}$-Leu-Ser-
	$-CH_2CH_2NH_3^+$	
Triosephosphate isomerase	$-CH_2COO^-$	-Asn$_{195}$-Val-Ser-
Inactive		
DNase	$-SO^-$	-Asn$_{103}$-Asp-Ser-
Catalase	$-CH_2COO^-$	-Asn$_{242}$-Leu; -Ser-; -Asn$_{437}$-Val-
	$-CH_2CH_2NH_3^+$	Thr-; -Asn$_{479}$-Phe-Ser-
Concanavalin A	$-CH_2COO^-$	-Asn$_{118}$-Ser-Thr-; -Asn$_{162}$-Gly-
	$-CH_2CH_2NH_3^+$	Ser-
Elastase	$-CH_3COO^-$	-Asn$_{66}$-Gly-Thr-; -Asn$_{123}$-Asu-
	$-CH_2CH_2NH_3^+$	Ser-; -Asn$_{215}$-Val-Thr-
Glyceraldehyde-3P-	$-CH_2COO^-$	-Asn$_{146}$-Ala-Ser-
dehydrogenase	$-CH_2CH_2NH_3^+$	
Trypsinogen	$-CH_2COO^-$	-Asn$_{151}$-Ser-Ser-
	$-CH_2CH_2NH_3^+$	
Carboxypeptidase A	$-SO_3^-$	-Asn$_{90}$-Pro-Ser-
Alcohol dehydrogenase	$-SO_3^-$	-Asn$_{300}$-Leu-Ser-

after denaturation six of the 13 proteins were glycosylated *in vitro*. Moreover, the nature of the substituent attached to the sulfhydryl groups does
not appear to be the controlling factor. Thus, although in the case of α-
lactalbumin, for example, there are differences in their relative acceptor
activity, all four derivatized forms of α-lactalbumin (SO_3^-, CH_2COO^-,
CH_2CONH_2, $CH_2CH_2NH_3^+$) are active. The one exception to this appears
to be DNase: S-Carboxymethylated and S-aminoethylated DNase are
active, whereas sulfitolyzed DNase is inactive.

The remaining seven proteins, which are inactive regardless of the
nature of the alkyl group, show no positive or negative correlation in
terms of either molecular weight or amino acid sequence around the
-Asn-X-Ser/Thr- that would allow one to distinguish them as a group from

the proteins that are active as acceptors of the carbohydrate chain. However, a major clue to why these proteins are not glycosylated may have been provided by the finding that cleavage of the polypeptide chain of concanavalin A or catalase with CNBr provides a mixture of fragments that are active as acceptors (Kronquist and Lennarz, 1978). These results suggest that these proteins may be inactive not because they lack specific structural requirements, but rather because remaining domains of order in the denatured protein prevent interaction of the transferase with the -Asn-X-Ser/Thr- acceptor site. Perhaps cleavage with CNBr produces fragments lacking these remaining domains of order.

More detailed studies, with separated, characterized fragments of α-lactalbumin provide strong support for the idea that the -Asn-X-Ser/Thr- acceptor tripeptide sequence is not only necessary but that it is sufficient for a peptide to serve as a substrate (Struck *et al.*, 1978*a*). A series of fragments of S-aminoethylated α-lactalbumin ranging in size from a CNBr fragment containing 123 residues down to a hexapeptide have been prepared. A study of these as carbohydrate acceptors has revealed that all except the hexapeptide, which contains Asn at the amino terminal end, serve as acceptors. Thus, it is clear that most, if not all of the polypeptide chain of α-lactalbumin is superfluous in relation to its ability to serve as a substrate for the oligosaccharide transferase. Similar studies on even smaller peptides from RNase A, and on synthetic peptides, indicate that tripeptides of the type Asn-X-Ser/Thr serve as acceptors providing that the amino terminus of the Asn residue is blocked with an acetyl group and the carboxy terminus is blocked by formation of an amide (Hart *et al.*, 1979).

In summary, these results suggest that for some proteins disruption of the tertiary structure is sufficient to allow enzymatic attachment of carbohydrates. Other denatured proteins containing the -Asn-X-Ser/Thr- site may possess additional restrictions on glycosylation imposed by their secondary structure. At least in the cases studied thus far these restrictions are removed when the polypeptide chain is fragmented. These findings indicate that the critical factor necessary for the *in vitro* formation of the N-glycosidic linkage is adequate exposure of the acceptor tripeptide sequence of the membrane-bound transferases of the lipid-linked pathway. If this condition also determines glycosylation *in vivo*, one could hypothesize that glycoproteins arise during evolution by point mutations that produce an acceptor tripeptide sequence in a region of a protein which becomes accessible, at some point, to the final transferase of the lipid-linked pathway. This accessibility and thus, glycosylation, would be controlled by higher orders of structure (i.e., secondary or tertiary) imposed by the primary sequence of the protein. Thus, aside from the presence of a tripeptide sequence of the type -Asn-X-Ser/Thr-, the primary sequence of the protein *per se* would not directly regulate glycosylation.

Finally, work in several laboratories may eventually provide support, at the level of the intact cell, for the hypothesis that accessibility of a given tripeptide sequence is the major factor in determining its glycosylation *in vivo*. It has recently been shown that when polyadenylated mRNA from the MOPC-46B plasmacytoma, which secretes a glycosylated K-chain, is injected into *Xenopus laevis* oocytes, synthesis of a glycosylated K-chain apparently identical to the MOPC-46B product can be detected (Jilka *et al.*, 1977). Conversely, oocytes injected with mRNA from MOPC-321, which produces an unglycosylated K-chain, failed to incorporate isotopically labeled sugars into immunoprecipitable K-chain. These results indicate that the signals for glycosylation have remained stable during vertebrate evolution and that the use of mutant mRNAs in the oocyte injection system may provide further information on the requirements for *in vivo* glycosylation. There are also indications that deletion mutants derived from mouse myeloma cell lines may also lead to the determination of which features of the primary sequence of immunoglobulin heavy chains are directly involved in carbohydrate attachment (Birshtein *et al.*, 1974; Weitzman and Scharf, 1976). These preliminary investigations provide hope that studies on the regulation of glycosylation may soon be facilitated by the type of detailed genetic analysis that has proven so useful in elucidating the factors which control transcription and translation.

7. A MODEL FOR GLYCOSYLATION OF MEMBRANE AND SECRETORY GLYCOPROTEINS

Following the pioneering studies of Palade on the function of the rough endoplasmic reticulum in synthesis of secretory proteins, Redman and Sabatini (1966) provided the first evidence suggesting that secretory proteins pass through the membrane of the endoplasmic reticulum as they are synthesized and are deposited in the cisternal space upon polypeptide chain completion. Blobel and co-workers (Blobel and Dobberstein, 1975a,b; Devillers-Thiery *et al.*, 1975) have shown that insertion of several secretory proteins into the membrane involves the formation of an inital hydrophobic sequence of amino acid residues termed a "signal peptide." In addition to these studies dealing with secretory proteins, recently it has been shown that viral membrane glycoproteins also are synthesized on the rough endoplasmic reticulum (Katz *et al.*, 1977; Toneguzzo and Ghosh, 1978). Moreover the results of Katz *et al.* (1977)

suggest that the glycoprotein is inserted through the membrane so that its carbohydrate chain is located in the cisternal space. Similar findings have been reported for the synthesis of the secretory glycoprotein, rat α-lactalbumin (Lingappa *et al.*, 1978).

Regarding glycosylation of newly translated proteins, several studies (Kiely *et al.*, 1976; Bergman and Kuehl, 1977) have provided strong evidence that this process is temporally closely connected to translation, since polypeptides that are newly completed, but still attached to tRNA, are already glycosylated. On this basis one would predict that the enzymes involved in saccharide–lipid synthesis and transfer of the oligosaccharide chain from dolichol pyrophosphate to the protein would be found in the rough endoplasmic reticulum, where translation takes place. Although one group (Vargas and Carminatti, 1977) has reported that both rough endoplasmic reticulum and smooth endoplasmic reticulum contain the oligosaccharide transferase, another (Czichi and Lennarz, 1977) has shown that, using sugar nucleotide as the initial glycose donor, the overall lipid-linked system is highly enriched in the rough endoplasmic reticulum. This conclusion is supported by recent observations that indicate mannose is added to the extension peptide of procollagen within the lumen of the rough endoplasmic reticulum (Anttinen *et al.*, 1978).

All of these findings can be incorporated into the model shown in Figure 9. One can hypothesize that the nucleotide derivatives of GlcNAc and Man, synthesized in the cytoplasm, react with dolichol phosphate in the rough endoplasmic reticulum to form the monoglycosyl lipids. Further reactions leading to formation of oligosaccharide-P-P-dolichol could occur within the hydrophobic environment of the membrane. Glycosylation would occur when a -Asn-X-Ser/Thr- tripeptide sequence has been translated and is oriented so that it can interact with both the other substrate, oligosaccharide-P-P-dolichol, and the membrane-associated oligosaccharide transferase. In view of the finding discussed earlier on the *in vitro* glycosylation of both membranes and secretory proteins via lipid-linked oligosaccharide, it seems likely that the same mechanism could be utilized for both classes of proteins. Thus, the only difference could be that the secretory proteins upon completion of translation would no longer remain associated with the membrane, whereas the membrane protein, because of the presence of sequences of hydrophobic amino acids, would remain attached to the rough endoplasmic reticulum. In short, in this model the only distinguishing feature between membrane glycoprotein and secretory glycoprotein would be their primary structure.

In the second phase of this model (Figure 9) the secretory glycoproteins would move into the smooth endoplasmic reticulum–Golgi complex. Similarly, the membrane glycoproteins, perhaps by lateral diffusion in the plane of the membrane system, would move into this compartment. Here particular secretory or membrane glycoproteins might be modified by addition of Fuc and the distal sugars, GlcNAc, Gal, and SA to the basic carbohydrate chain. Although it is clear that the nucleotide derivatives of these sugars are involved in this process (see Chapter 3), it is not clear how these compounds enter the smooth endoplasmic reticulum–Golgi complex. One possibility that has been suggested, at least for UDP-Gal (Kuhn and White, 1976), is that a specific transport system is present. Another is that the activated monosaccharides are carried through the membrane via lipid-linked intermediates, but there is no evidence to support this idea.

In the third and final stage secretory vesicles are formed, and these fuse with the plasma membrane. If this process occurs as depicted in Figure 9, the secretory glycoproteins would be deposited outside the cell and the membrane glycoproteins would be oriented on the plasma membrane with the carbohydrate chain on the external surface. Indeed, such a surface orientation has been established in a number of systems (see review by Singer, 1974).

Clearly this model, which is merely an extension of earlier ones (Redman and Sabatini, 1966; Blobel and Dobberstein, 1975a,b) to include the glycosylation process, is grossly oversimplified, and it does not deal with many important unresolved questions (e.g., the energetics of the passage of glycosyl lipids through membranes; the relative rates of translation, glycosylation, and folding; and the mode of lateral movement of glycoproteins through each subcellular compartment). Its main feature is that it suggests that the same mechanism of glycosylation functions in synthesis of secretory and membrane glycoproteins and that the factor determining if either type of protein becomes a glycoprotein is the presence of an -Asn-X-Ser/Thr- sequence situated on the polypeptide so that it is accessible to the oligosaccharide transferase. The importance of the tripeptide sequence, -Asn-X-Ser/Thr-, in the formation of the N-glycosidic bond *in vivo* is supported by statistical analysis of its occurrence in proteins of known amino acid sequence (Hunt and Dayhoff, 1979; Sinohara and Marugama, 1973). In prokaryotes, which have few if any glycoproteins with N-glycosidically linked carbohydrates, the frequency of the tripeptide sequences is essentially that expected for random occurrence. However, in eukaryotes the frequency of such tripeptides is much lower than that calculated from the frequencies of occurrence of the individual amino acids. Most striking is the observation that this restriction applied only to secreted proteins: Intracellular, soluble eukaryotic proteins appear to have the tripeptide sequences at near normal frequencies (Sinohara and Marugama, 1973). This is consistent with the fact that intracellular, soluble proteins, which are rarely if ever glycosylated (Dayhoff, 1973), are synthesized on free cytoplasmic polyribosomes (Ganoza and Williams, 1969; Hicks *et al.*, 1969) at sites removed from the membrane-bound glycosyltransferases of the lipid-linked path-

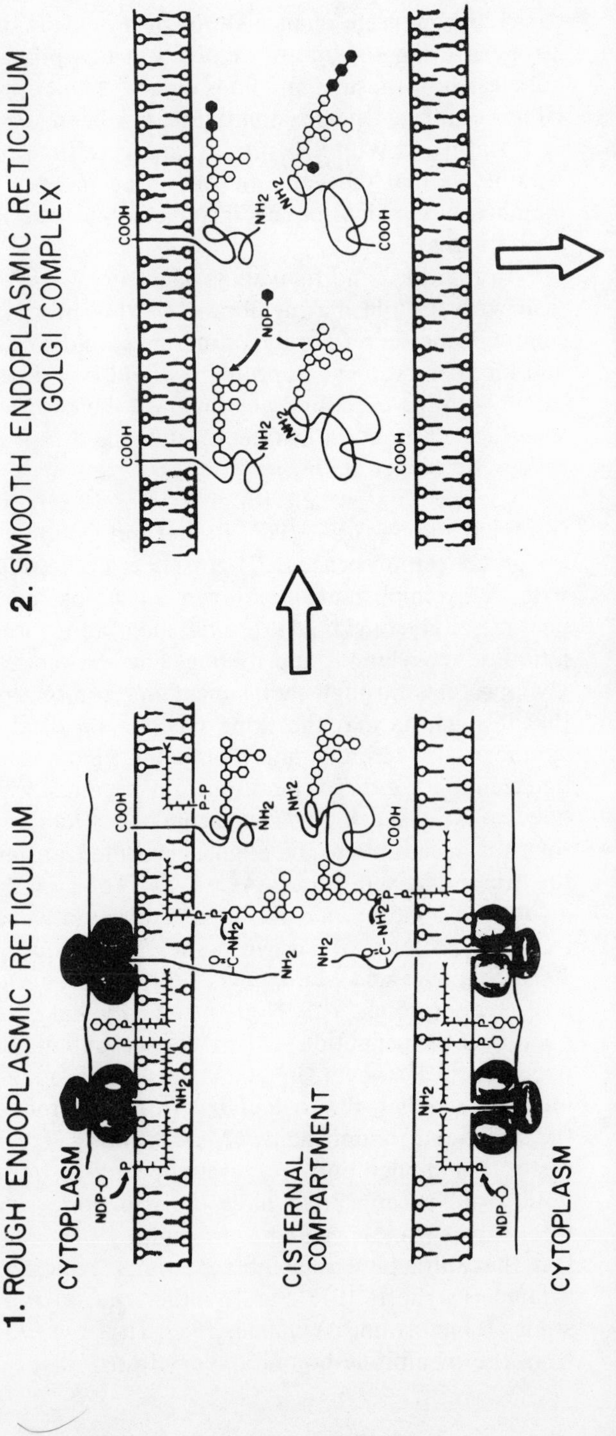
1. ROUGH ENDOPLASMIC RETICULUM
CYTOPLASM
CISTERNAL
COMPARTMENT
CYTOPLASM
2. SMOOTH ENDOPLASMIC RETICULUM
GOLGI COMPLEX

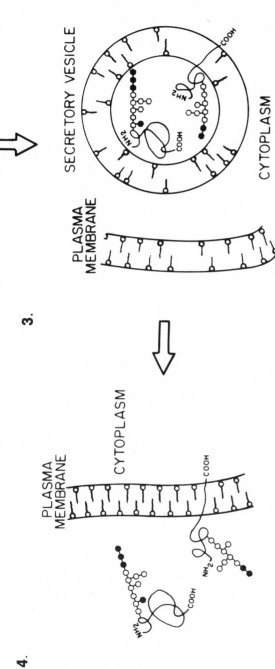

Figure 9. Model for the concerted translation and glycosylation of secretory and membrane glycoproteins. In stage 1 the transfer of the oligosaccharide chain from the oligosaccharide–lipid to the protein is arbitrarily depicted as occurring at the cisternal face of the RER; the precise site of transfer remains to be established.

way. Thus, the restricted occurrence of tripeptide sequences in eukaryotic secretory proteins, and perhaps also membrane-bound proteins,
would serve to limit glycosylation in the rough endoplasmic reticulum to
those sites which would not interfere with subsequent folding of the
protein into its normal three-dimensional structure.

8. REGULATION OF GLYCOPROTEIN SYNTHESIS

There is great interest in the biological significance of alterations in
cell surface glycoproteins (see Chapter 5). However, work concerned
with the factors that regulate the synthesis of N-glycosidically linked
glycoproteins has only begun in the last few years, and at this time only
fragmentary information is available. Obviously, there are a number of
potential sites of regulation. These include regulation of the level of: (1)
initial glycosyl donors, sugar nucleotides; (2) glycosyl acceptor, dolichol
phosphate; (3) transferases that catalyze these reactions and hence the
overall synthesis of oligosaccharide-P-P-dolichol; (4) oligosaccharide
transferase; and (5) acceptor apoprotein.

Specific control of N-glycosidically linked glycoproteins by regulation of the level of UDP-GlcNAc seems unlikely since this sugar is found
in other types of glycoproteins and in proteoglycans. In contrast, in higher
eukaryotes Man is uniquely found in N-glycosidically linked glycoproteins and in keratan sulfate type I (see below), so that it is possible that
control of the level of GDP-Man might affect the level of glycoprotein
synthesis. In any case, little is known about alterations in the level of
sugar nucleotides. Nor is there any information about regulation of the
glycosyltransferases involved in assembly of oligosaccharide–lipid. It is
of interest, however, that a concanavalin-A-resistant mutant of Chinese
hamster ovary cells has been shown to be defective in synthesis of
oligosaccharide-P-P-dolichol (Krag et al., 1977). As discussed in Chapter
4, it is clear that further studies using the genetic approach will provide
insight into the regulatory steps involved in synthesis of oligosaccharide–
lipid.

Two potential specific sites of regulation of protein glycosylation
would be those involving dolichol phosphate synthesis and regeneration
of dolichol phosphate from dolichol pyrophosphate. Based on a number
of studies it has been proposed (Hemming, 1974) that the initial steps in
dolichol phosphate synthesis are identical to those in sterol synthesis and
involve formation of isopentenyl-P-P and dimethylallyl-P-P from mevalonic acid, followed by condensation of these units to form *trans*-farnesyl-
P-P. As outlined in Figure 10, *trans*-farnesyl-P-P serves as the acceptor
for isoprene units from isopentenyl-P-P. The end-product formed *in vitro*,
using either a soluble pea seedling preparation (Daleo and Lezica, 1977)

or a liver preparation (Daleo *et al.*, 1977), has been shown to be dolichol phosphate, while that formed in oviduct membranes (Grange and Adair, 1977) is 2,3-dehydrodolichol phosphate. Thus, apparently the preparations from peas and from liver contained all the enzymes to complete synthesis of dolichol phosphate, whereas that from oviduct lacked the ability to reduce the double bond of the first isoprene unit. Because early studies (Hemming, 1974) showed that mevalonic acid serves as the precursor of dolichol phosphate, it seems unlikely that specific regulation of dolichol phosphate synthesis occurs at steps before the condensation between farnesyl-P-P and isopentenyl-P-P, since alterations in the activity or levels of any of these enzymes would concomitantly alter the level of cholesterol synthesis. Indeed, Mills and Adamany (1978) have recently shown that 25-hydroxycholesterol, a specific inhibitor of 3-hydroxy-3-methylglutaryl-CoA reductase, not only inhibits cholesterol synthesis in muscle cells, but also impairs synthesis of dolichol, dolichol-linked oligosaccharides, and glycoproteins. Thus, in the absence of further information it appears that possible sites for specific regulation of synthesis of dolichol phosphate are either those involving condensation of isopentenyl units with farnesyl pyrophosphate or the two reactions involved in final structural alterations to the complete chain, i.e., reduction of the double bond in the first isoprene unit or conversion of the pyrophosphate to the phosphate (see Figure 10). Regarding the dephosphorylation reaction, it should be noted that one of the end products of the reaction between oligosaccharide-P-P-dolichol and acceptor protein is dolichol-P-P. Presumably for this compound to "recycle" and serve again as an acceptor for glycose units it must be enzymatically converted to dolichol-P. Although an enzyme that converts undecaprenol-P-P to undecaprenol-P has been shown to exist in bacterial systems (Storm and Strominger,

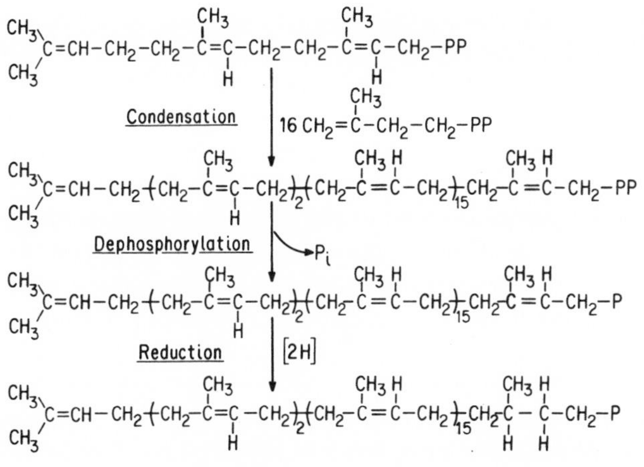

Figure 10. Proposed pathway for the biosynthesis of dolichol phosphate.

1973), an analogous enzyme that acts on dolichol-P-P has not yet been found in eukaryotes.

Although the mechanism of regulation of dolichol phosphate synthesis is completely unknown, two studies indicate that the level of dolichol phosphate does change during development. A study of pig brain white matter indicates that the level of dolichol phosphate, as assessed by the level of synthesis of Man-P-Dol, is lower in membranes from adult brain then in membranes from the actively myelinating brain of newborn pigs (Harford *et al.*, 1977). A study of the level of Man-P-Dol synthesis membranes from the oviducts of chicks before and after estrogen-induced differentiation indicates that the primary factor in enhanced synthesis is an increase in the level of dolichol phosphate, rather than the mannosyl-transferase (Lucas and Levin, 1977). Finally, in the context of development, studies on reticulocytes and erythrocytes should be mentioned. Martin-Barrientos and Parodi (1977) as well as Lucas and Nevar (1978) have shown that the disappearance of Man-P-Dol synthesis upon conversion of reticulocytes to erythrocytes is due to the loss of the sugar acceptor Dol-P. In the presence of exogenous Dol-P saccharide–lipid synthesis can be detected in membrane preparations from both cell types; however, glycosylation of proteins is only detected in the reticulocytes (Martin-Barrientos and Parodi, 1977). This is consistent with the fact that this cell, in contrast to the erythrocyte, is still actively synthesizing proteins and that there is a progressive decrease in the insertion of membrane proteins as the reticulocyte matures (Light and Tanner, 1978).

Finally, it should be mentioned that another possible mode of regulation is the interconversion of dolichol and dolichol phosphate. Although a phosphatase that acts on dolichol phosphate has not been reported, a kinase that converts dolichol to dolichol phosphate has been detected (Allen *et al.*, 1979; Burton *et al.*, 1979).

9. INHIBITORS OF GLYCOSYLATION

Inhibitors of glycosylation have become important tools for studying the biological role of the carbohydrate moiety of glycoproteins as well as the process of glycosylation itself. The most commonly used general inhibitors have been glucosamine and 2-deoxy-D-glucose (see review by Scholtissek, 1975). In addition, the related compounds 2-fluoro-2-deoxy-D-glucose and 2-fluoro-2-deoxy-D-mannose have also been shown to block glycosylation in eukaryotic cells (Biely *et al.*, 1973; Schmidt *et al.*, 1976). More recently, it has been demonstrated that the antibiotics tunicamycin and bacitracin inhibit the lipid-mediated glycosylation process. These antibiotics may be of greater value than the sugar analogues since their effects appear to be more specific.

9.1. In Vivo Effects of Deoxy Sugars on Protein Glycosylation

It has been shown that 2-deoxy-D-glucose interferes with the *in vivo* glycosylation of extracellular glycoproteins (Farkas *et al.*, 1970; Liras and Gascon, 1971; Kuo and Lampen, 1972), glucans (Johnson, 1968; Farkas *et al.*, 1969), and mannans (Farkas *et al.*, 1969) in yeast, viral glycoproteins in cultured mammalian cells (see review by Scholtissek, 1975), immunoglobulins in a myeloma tumor cell line (Melchers, 1970, 1973; Eagon and Heath, 1977), human interferon in cultured fibroblasts (Havell *et al.*, 1977), and plasma membrane proteins in hen oviduct cell suspensions (Struck and Lennarz, 1976). In most studies, the inhibition of glycosylation was determined by comparing the incorporation of iso-topically labeled sugars into particular glycoproteins in the presence and absence of the inhibitor. However, with the exception of the report by Eagon and Heath (1977), the protein products synthesized in the presence of the inhibitor have not been characterized in detail. Thus, it is not clear whether 2-deoxy-D-glucose completely blocks glycosylation or merely causes premature termination of the growing oligosaccharide chain after its incorporation into the chain as an analogue of glucose or mannose. This latter possibility is suggested by reports that 2-deoxy-D-glucose is incorporated into viral (Kaluza *et al.*, 1973) and cellular glycoproteins (Steiner *et al.*, 1973), yeast mannans and glucans (Biely *et al.*, 1972; Kratky *et al.*, 1975), and cellular glycolipids (Steiner and Steiner, 1973). However, using [³H]-2-deoxy-D-glucose at concentrations inhibitory to protein glycosylation, the normally glycosylated K-46 chain produced by the mouse myeloma MOPC-46 was synthesized and secreted devoid of all detectable carbohydrate (Eagon and Heath, 1977). Nevertheless, the fact that 2-deoxy-D-glucose may be incorporated into glycoproteins as a sugar analogue places a severe limitation on its use as a tool to study functional role of carbohydrate in such proteins.

Since 2-deoxy-D-glucose is an analogue of glucose, it might also effect the synthesis of macromolecules through its metabolites or by interfering with normal sugar metabolism. Thus, for proteins whose synthesis is controlled by the carbon source in the medium, decreased synthesis seen in the presence of 2-deoxy-D-glucose may be unrelated to its effects on glycosylation. Although this possibility was originally suggested for simple eukaryotes such as yeast, recent reports suggest that analogous situations can also occur in avian and mammalian cells (Banjo and Perdue, 1976; Pouyssegur *et al.*, 1977; Pouyssegur and Yamada, 1978). As already suggested (Kratky *et al.*, 1975), the overall effect of deoxysugars on glycoprotein synthesis is probably due to a combination of a general inhibition of protein synthesis, catabolite repression or induction, and an inhibition of glycosylation.

9.2. Effect of Deoxy Sugars on the Secretion of Glycoproteins

Of particular interest has been the effect of 2-deoxy-D-glucose on the secretion of glycoproteins, the migration of membrane-associated glycoproteins from their site of synthesis to the plasma membrane, and the assembly and maturation of infectious virus particles. In yeast, it appears that this analogue prevents the secretion of newly synthesized glucans (Farkas *et al.*, 1969), mannans (Farkas *et al.*, 1969), and extracellular, mannan-associated enzymes (Farkas *et al.*, 1970; Liras and Gascon, 1971; Kuo and Lampen, 1972). However, a more recent study (Kratky *et al.*, 1975) suggests that the decreased rate of secretion of mannan-associated enzymes is due to a decreased rate of synthesis of these enzymes and not to a direct involvement of carbohydrate in the secretion process. In fact, the biosynthesis of several intracellular enzymes and the well-characterized extracellular enzyme, invertase, is repressed by growing yeast in media containing 2-deoxy-D-glucose. Although the mechanism by which 2-deoxy-D-glucose depressed the synthesis of invertase was not investigated, it was suggested that this effect might be related to the known catabolite repression of invertase synthesis caused by high glucose concentrations (Montencourt *et al.*, 1973). Thus, in yeast, the decreased secretion of extracellular enzymes observed with yeast protoplasts incubated with 2-deoxy-D-glucose may reflect decreased enzyme synthesis rather than a requirement of bound carbohydrate for the secretion of glycoprotein enzymes.

Furthermore, it has been shown that the level of extracellular enzymes by *Micrococcus sodonensis* is decreased by 2-deoxy-D-glucose because of inhibition in the synthesis of a specific polysaccharide which protects these enzymes from proteolysis (Braatz and Heath, 1974). Related to this finding is the fact that viral glycoproteins synthesized in cells incubated with 2-deoxy-D-glucose exhibit increased sensitivities to proteolysis by host cell proteases while other virus-specific proteins are not affected (Kaluza *et al.*, 1973; Schwarz and Klenk, 1974). Similar effects have been observed with virus-infected cells treated with tunicamycin (Schwarz *et al.*, 1976; see below). Thus, if carbohydrate prosthetic groups protect certain glycoproteins from proteolysis, inhibitors of glycosylation might increase their susceptibility to intracellular degradation, resulting in an apparent decrease in secretion. Consistent with this proposal is the report that the glycoprotein, bovine pancreatic ribonuclease B, is less sensitive to tryptic digestion than unglycosylated bovine pancreatic ribonuclease A (Birkeland and Christensen, 1975). Since the two proteins are identical in amino acid sequence (Plummer and Hirs, 1964) and have similar, if not identical, three-dimensional configuration (Puett, 1973), the carbohydrate portion of ribonuclease B might function as a block to proteolysis.

9.3. Effect of Deoxy Sugars on Virus Replication

From the preceding discussion it is clear that the effect of 2-deoxy-D-glucose on protein secretion is more complex than once thought. The available data using this sugar analogue do not provide definitive evidence supporting the hypothesis that the carbohydrate moieties of glycoproteins are required for their secretion. Similarly, there exists a certain amount of confusion concerning the effect of 2-deoxy-D-glucose (as well as other inhibitors of glycosylation) on the assembly and maturation of infective virus particles in cultured cells. However, it is now clear that this confusion arises from the fact that the effects of deoxy sugars depends upon the virus strain and cell type employed (Nakamura and Compans, 1978). It is also apparent that, since 2-deoxy-D-glucose can be incorporated into viral glycoproteins, results obtained with this compound may not be comparable to those obtained using inhibitors, such as tunicamycin, that are not incorporated (see below). In the case of influenza virus, it has been shown that a variety of inhibitors of glycosylation, including 2-deoxy-D-glucose, have no effect on the migration of the influenza virus hemagglutinin glycoprotein from the rough endoplasmic reticulum to the plasma membrane (Klenk et al., 1974; Nakamura and Compans, 1978). While inhibitors of glycosylation did prevent the expression of hemagglutinin activity, they had a relatively lesser effect on the appearance of the unglycosylated hemagglutinin polypeptide at the cell surface, or on the production of virions containing the unglycosylated form of this protein. The major reasons for decreased virion production in the presence of 2-deoxy-D-glucose seems to be an inhibition of protein synthesis and interference with cellular metabolism (Nakamura and Compans, 1978). However, it is likely that in these studies 2-deoxy-D-glucose was incorporated into the hemagglutinin polypeptide, so that the behavior of an abnormally glycosylated protein rather than an unglycosylated protein was actually studied. As will be mentioned below, experiments using the inhibitor tunicamycin gave qualitatively similar results under conditions in which glycosylation of viral proteins was completely blocked. Consequently, normal glycosylation of the influenza virus hemagglutinin polypeptide is not required for virion production. The carbohydrate moiety of this protein appears to be required only for the expression of virus-specific hemagglutinin and neuraminidase activities (Klenk et al., 1972).

9.4. In Vitro Effect of Deoxy Sugars on Protein Glycosylation

Although a great deal of effort has been directed toward investigating the *in vivo* effects of the deoxy sugar analogues, the precise mechanism by which they inhibit glycosylation has been elusive. These compounds might act as inhibitors of phosphohexoseisomerases (Kuo and Lampen,

1972), as uridine traps (Keppler *et al.*, 1970), or by more direct effects mediated by their UDP and GDP derivatives (see review by Scholtissek, 1975).

The finding that lipid carriers are involved in the glycosylation of eukaryotic proteins (see above) has prompted investigations concerning the effect of deoxy sugars on this mechanism of protein glycosylation. Using suspensions of hen oviduct cells, it has been shown that 2-deoxy-D-glucose inhibits the synthesis of a mannose-containing oligosaccharide-lipid intermediate (Struck and Lennarz, 1976). Using membranes of *Saccharomyces cerevisiae*, the enzymatic transfer of 2-deoxy-D-glucose from GDP-2-deoxy-D-glucose (an analogue of GDP-mannose) to a lipid acceptor has been demonstrated (Lehle and Schwarz, 1976). The product which had the chromatographic properties of dolicholphosphoryl-2-deoxy-D-glucose, could then be transferred to endogenous protein acceptors in the membrane preparations. After this transfer, further elongation of the carbohydrate chains bound to acceptor mannoproteins was prevented, although such elongation was observed in controls. These observations are consistent with the earlier findings that while 2-deoxy-D-glucose inhibits (to varying degrees) sugar incorporation in proteins, glycoproteins synthesized in the presence of this analogue often have incomplete or defective oligosaccharide chains which contain 2-deoxy-D-glucose.

More recent work with virus-infected cells indicates that 2-deoxy-D-glucose can interfere with lipid-mediated glycosylation in a different manner (Schwarz *et al.*, 1978). Using chick embryo or BHK-21 fibroblast host cells, viral glycoproteins synthesized in the presence of 2-deoxy-D-glucose appear to be completely devoid of carbohydrate (Schwarz *et al.*, 1977). *In vitro* experiments utilizing particulate enzyme preparations from these virus-infected cells suggest that deoxy sugars interfere with the synthesis of the oligosaccharide–lipid intermediates which serve as carbohydrate donors to viral proteins (Schwarz *et al.*, 1978). This interference appears to involve dolicholphosphoryl-2-deoxy-D-glucose. In any case, it is clear that the entry of deoxy sugars, and perhaps fluorodeoxy sugars (see below), into the lipid-linked pathway via dolicholphosphoryl derivatives might explain their inhibitory activity towards protein glycosylation. The fact that GDP-2-deoxy-D-glucose inhibits the lipid mediated pathway *in vitro*, while UDP-2-deoxy-D-glucose does not (Schwarz *et al.*, 1978), could account for the fact that, with respect to protein glycosylation, 2-deoxy-D-glucose appears to be an antimetabolite of mannose rather than glucose (Kaluza *et al.*, 1973; Schmidt *et al.*, 1976).

9.5. Fluorodeoxy Sugars

Less information is currently available concerning the effect of fluorodeoxy sugars on protein glycosylation. 2-Fluoro-2-deoxy-glucose has been reported to block glucan synthesis in yeast (Biely *et al.*, 1973), whereas 6-fluoro-6-deoxyglucose prevented growth of mouse lymphoma cells in culture (Bessel *et al.*, 1973). 2-Fluoro-2-deoxy-D-glucose and 2-fluoro-2-deoxy-D-mannose are known to prevent the formation of infectious Semliki forest virus and fowl plaque virus in rabbit kidney cells by interfering with glycosylation of viral proteins (Schmidt *et al.*, 1976). The fact that 2-fluoro-2-deoxy-D-mannose is the more potent inhibitor of virus replication is consistent with the hypothesis that deoxy sugars act as antimetabolites of mannose rather than glucose. Unlike 2-deoxy-D-glucose, fluorodeoxy sugars are incorporated into cellular macromolecules at low, nearly undetectable rates even though they are readily activated to their nucleotide sugar derivatives by cellular enzymes (Schmidt *et al.*, 1978). Consequently, these analogues may inhibit glycosylation by a mechanism distinct from that of 2-deoxy-D-glucose. However, the fact that fluorodeoxy sugars are not readily incorporated into cellular macromolecules may make these analogues more suitable in studies concerning the biological function of protein-bound carbohydrate.

9.6. Amino Sugars

Glucosamine, galactosamine, and mannosamine are known to prevent virus replication (see review by Scholtissek, 1975), growth of transplantable tumors (Quastel and Cantero, 1953; Bekesi *et al.*, 1969; Bekesi and Winzler, 1970), and to be hepatotoxic in rats (Keppler *et al.*, 1970). Although some of the effects of amino sugars can be reversed by uridine, their effects on viral protein glycosylation are not reversible (Scholtissek *et al.*, 1975). Hepatotoxicity, by galactosamine in particular, may be due to the substitution of UDP-hexosamine for UDP-hexoses in various reactions utilizing sugar nucleotides (Tarentino and Maley, 1976). This would explain the incorporation of amino sugars into glycogen after administration of high levels of galactosamine to rats (Maley *et al.*, 1968). With regard to viral protein glycosylation, it appears that high levels of glucosamine do not completely block carbohydrate incorporation into viral glycoproteins (Nakamura and Compans, 1978), although completeness of the block may depend on host cell type (Schwarz and Klenk, 1974). Thus, glycoproteins synthesized in the presence of high concentrations of glucosamine may contain incomplete or abnormal carbohydrate moieties as is often seen with deoxy sugars (see above). However, there remains no clear explanation for the inhibition of glycosylation seen with amino sugars.

9.7. Bacitracin

This polypeptide antibiotic has long been known as an inhibitor of lipid-mediated glycosylation in prokaryotes (Siewert and Strominger,

1967) but little has been reported on its effects in eukaryotes. In bacteria, it has been suggested that bacitracin acts by blocking the dephosphorylation of undecaprenol pyrophosphate, thus preventing the recycling of the sugar carrier, undecaprenol phosphate (Stone and Strominger, 1971; Storm and Strominger, 1973). However, bacitracin has also been shown to affect membrane stability *in vitro* (MacDonald *et al.*, 1974). Similar effects *in vivo* could result in cell lysis, thus offering an alternative explanation for the bacteriocidal action of this antibiotic.

In eukaryotes, the effect of bacitracin on the lipid-mediated pathway appears to differ from one system to another. Using hen oviduct membranes, bacitracin causes an *in vitro* accumulation of a trisaccharide–lipid intermediate involved in the glycosylation of proteins endogenous to the membrane preparations (Chen and Lennarz, 1976). In contrast, studies using calf pancreas microsomes show that bacitracin blocks the synthesis of N-acetylglucosaminylpyrophosphoryldolichol (Herscovics *et al.*, 1977c). A more recent report demonstrates the ability of this antibiotic to block the formation of a di-N-acetylchitibiosylpyrophosphoryl lipid intermediate in yeast membranes (Reuvers *et al.*, 1978b). In light of the fact that bacitracin inhibits sterol and squalene synthesis in cell-free preparations of rat liver, it is possible that it interferes with all reactions involving polyisoprenoid pyrophosphates (Stone and Strominger, 1972). This effect could be mediated by complex formation of the polyisoprenoid pyrophosphate derivatives with bacitracin as already suggested (Stone and Strominger, 1972; Chen and Lennarz, 1976; Reuvers *et al.*, 1978b). Without further investigation, the usefulness of bacitracin for studying eukaryotic protein glycosylation remains an open question.

9.8. Tunicamycin

Tunicamycin, a potent inhibitor of protein glycosylation produced by *Streptomyces lysosuperficus*, was discovered by Tamura in 1970 (Takatsuki *et al.*, 1971). Its complete structure, which has been recently elucidated (Takatsuki *et al.*, 1977a; Ito *et al.*, 1977), is shown in Figure 11. Although first noticed for its activity against gram-positive bacteria (Takatsuki *et al.*, 1971), its ability to inhibit the replication of fungi, yeast, and viruses was discovered almost simultaneously (Takatsuki and Tamura, 1971a,b; Takatsuki *et al.*, 1972). Its composition and lipophilic nature, coupled with its known biological activity, have prompted *in vitro* investigations on its effects on the lipid-linked pathway for protein glycosylation. Studies using calf liver (Tkacz and Lampen, 1975) or chick embryo (Takatsuki *et al.*, 1975) microsomes indicated that tunicamycin blocks the synthesis of N-acetylglucosaminylpyrophosphorylpolyisoprenol. Further work using a yeast membrane system (Lehle and

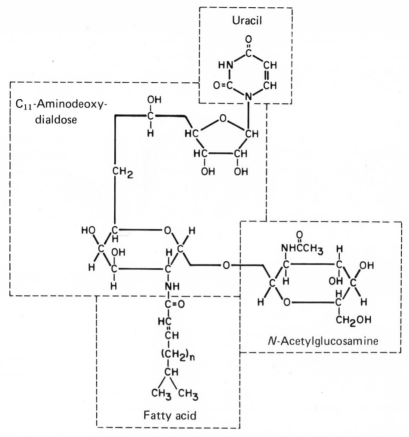

Figure 11. Structure of tunicamycin.

Tanner, 1976) confirmed this observation and also showed that N-acetylglucosaminylpyrophosphoryldolichol could be elongated to disaccharide- and trisaccharide-lipids in the presence of this drug (Figure 12). A later study demonstrated the preformed saccharide-lipid intermediates could serve as carbohydrate donors to protein in the presence of tunicamycin (Struck and Lennarz, 1977). From these *in vitro* studies, it can be concluded that tunicamycin blocks only the first step in the lipid-linked pathway of eukaryotes, the synthesis of N-acetylglucosaminyl-pyrophosphoryldolichol from UDP-N-acetylglucosamine and dolichol phosphate (Figure 12). Subsequent steps in the pathway and the final transfer of an oligosaccharide to protein are not affected.

In vivo studies using yeast (Kuo and Lampen, 1974), virus-infected cells (Takatsuki and Tamura, 1971b; Schwarz *et al.*, 1976; Leavitt *et al.*, 1977; Gibson *et al.*, 1978; Krag and Robbins, 1977), cultured fibroblasts

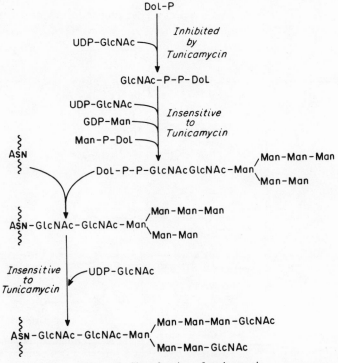

Figure 12. Site of action of tunicamycin.

(Duksin and Bornstein, 1977a; Olden *et al.*, 1978), cultured ovary cells (Krag *et al.*, 1977), myeloma tumor lines (Hickman *et al.*, 1977), oviduct tissue slices (Struck and Lennarz, 1977), primary hepatocytes (Struck *et al.*, 1978b), and chick embryo cornea (Hart and Lennarz, 1978) have documented the fact that tunicamycin prevents the glycosylation of proteins containing N-glycosidically linked carbohydrate. Since tunicamycin is not metabolized *in vivo* or *in vitro* (Kuo and Lampen, 1976), it appears unlikely that the inhibition of glycosylation observed in these investigations was caused by the incorporation of this drug into the growing oligosaccharide chains, a possibility discussed above with respect to 2-deoxy-D-glucose. Furthermore, studies in yeast have indicated that tunicamycin does not exert its effect by blocking sugar uptake or by interfering with the synthesis of UDP-*N*-acetylglucosamine or glucosamine-6-phosphate (Kuo and Lampen, 1974, 1976; Tkacz and Lampen, 1975).

Thus, unlike sugar analogues or the amino sugars, tunicamycin does not interfere with either sugar or nucleotide metabolism. Further evidence supporting the assumption that tunicamycin blocks glycosylation *in vivo* by a mechanism identical to that demonstrated *in vitro* has been provided by studies using hen oviduct tissue slices (Struck and Lennarz, 1977). Although this antibiotic was found to prevent the incorporation of both mannose and N-acetylglucosamine into the core oligosaccharide of ovalbumin, the addition of terminal N-acetylglucosaminyl residues was not affected (Figure 12). It is known that these terminal sugars are added directly from UDP-N-acetylglucosamine by reactions not mediated by lipid carriers (see Chapter 3). Thus, in the presence of tunicamycin, incorporation of N-acetylglucosamine (and presumably other sugars, i.e., galactose or sialic acid) occurs by terminal glycosylation of a core oligosaccharide already attached to a polypeptide chain and not by the *de novo* synthesis of the core oligosaccharide itself. This possibility is consistent with *in vitro* studies showing that not all N-acetylglucosaminyl-transferases are inhibited by tunicamycin (Lehle and Tanner, 1976; Kuo and Lampen, 1976; Struck and Lennarz, 1977; Waechter and Harford, 1977). All of these results clearly indicate that tunicamycin inhibits protein glycosylation by blocking the transfer of the N-acetylglucosamine-1-phosphate moiety from UDP-N-acetylglucosamine to dolichol phosphate. The other reactions involving the transfer of the N-acetylglucosamine moiety *per se* are not affected (Figure 12).

However, tunicamycin is not without cytotoxic effects (Takatsuki *et al.*, 1971, 1972; Takatsuki and Tamura, 1971*b*; Duksin and Bornstein, 1977*b*). Impairment of DNA, RNA, and protein synthesis as well as induction of RNA and DNA degradation have been shown to occur in its presence (Takatsuki and Tamura, 1971*b*; Takatsuki *et al.*, 1972). Although the mechanisms for this cytotoxicity are not known, it has been reported that tunicamycin has no effect on protein synthesis *in vitro* (Hickman *et al.*, 1977). It is possible that tunicamycin cytotoxicity is due to secondary effects reflecting the involvement of glycoproteins in a variety of cellular processes, rather than direct effects on events unrelated to protein glycosylation. Despite these problems, tunicamycin has been shown to inhibit protein glycosylation by greater than 80% under conditions in which protein synthesis is depressed by no more than 15-25% (Struck and Lennarz, 1977; Hickman *et al.*, 1977; Duksin and Bornstein, 1977*a*). Moreover, Mahoney and Duksin (1979) have recently shown that tunicamycin consists of two major components and that one of these inhibits glycosylation without any effect on protein synthesis. Thus, in future studies it should be possible to avoid secondary effects of tunicamycin.

As with the deoxy sugars, tunicamycin has been used to investigate the role of carbohydrate in the secretion of glycoproteins. Using yeast protoplasts, this antibiotic had no effect on the secretion of the glycoprotein enzymes invertase and acid phosphatase, although a marked reduction in the synthesis of these enzymes was observed (Kuo and

Lampen, 1974). In cultured fibroblasts, the secretion of procollagen was unimpaired by the presence of this antibiotic (Duksin and Bornstein, 1977a). In a separate study using chick embryo fibroblasts, tunicamycin did not alter the percentage of cellular CSP (the major cell surface glycoprotein) appearing at the cell surface or in the medium (Olden et al., 1978). However, it was found that tunicamycin treatment resulted in a reduction in the amount of immunoprecipitable CSP present in fibroblast cultures. This decrease in total CSP was caused by an increased intracellular degradation of CSP by cells treated with tunicamycin. Consequently, it appears that unglycosylated CSP has a higher susceptibility to proteolytic digestion than does normally glycosylated CSP. Tunicamycin has also been shown to have little effect on the secretion of rat liver transferrin or the apoproteins of chicken liver VLDL by cultures of primary hepatocytes (Struck et al., 1978b). Furthermore, very recently it has been shown that tunicamycin does not inhibit secretion of ovalbumin (Keller and Swank, 1978). Taken together, these results do not support the hypothesis that the oligosaccharide units of glycoproteins are required for their secretion (Eylar, 1965; Melchers, 1973). Also arguing against this suggestion is the fact that some proteins (α-lactalbumin and ribonuclease) are normally secreted in both glycosylated and unglycosylated forms (Plummer and Hirs, 1964; Hill and Brew, 1975). In fact, it has been suggested that, in the case of ribonuclease, glycosylation may merely reflect the synthetic capabilities of the membranes with which the protein is associated during the process of its synthesis and secretion (Jackson and Hirs, 1970).

In contrast, it has been reported that tunicamycin blocks not only the glycosylation of μ- and α-immunoglobulin heavy chains, but also their secretion from myeloma tumor cell lines (Hickman et al., 1977). Electron microscopy and direct immunofluorescence revealed the presence of distended rough endoplasmic reticulum containing a granular, electron-dense precipitate identified as immunoglobulin. These authors pointed out that the lack of carbohydrate may alter the physical properties of these particular immunoglobulins, resulting in aggregation or inadequate solubility which prevents passage of the proteins from the endoplasmic reticulum to the extracellular environment. This possibility, as well as the observations on fibroblast CSP, indicate that tunicamycin treatment can cause a decrease in the secreted amounts of a particular glycoprotein via mechanisms which do not involve the participation of the oligosaccharide units in the secretory process per se.

Investigations using glycosylation-deficient mutant cell lines underscore the observation made with inhibitors of glycosylation, namely that it is too early to generalize about the role of bound carbohydrate in the intracellular migration and secretion of glycoproteins. The ADG mutant of Balb/c 3T3 has been shown to produce abnormal glycoproteins because of a defect on its ability to synthesize N-acetylglucosamine (Pouyssegur and Pastan, 1976, 1977). This mutant releases normal amounts of the glycoproteins CSP and procollagen into the medium (Olden et al., 1978). Furthermore, mutant M3.11 of the mouse myeloma MPC-11 has been shown to secrete unglycosylated IgG heavy chain at a rate similar to the parent cell line (Weitzman and Scharf, 1976). In contrast, it has been suggested that, in certain cases, clinical agammaglobulinemia may be associated with impaired glycosylation of IgG heavy chains (Ciccimarra et al., 1976). Investigations with a number of glycosylation-deficient lymphoma cell lines have shown that normal glycosylation of the T25 glycoprotein is necessary for its incorporation and expression at the cell surface (Trowbridge et al., 1978). However, these studies also indicated that, in the same mutant cells, other glycoproteins with altered oligosaccharide side chains were present at the cell surface in normal amounts. A more detailed discussion of glycosylation-deficient mutants is presented in Chapter 4.

Although the antiviral activity of tunicamycin has long been known, its effects seem to depend on the virus strain and cell type used. In baby hamster kidney (BHK) cells, tunicamycin causes a marked decrease in the yield of infective vesicular stomatitis virions (Leavitt et al., 1977). This effect is not due to lowered specific infectivity of virions containing unglycosylated viral proteins, but to a block in virion assembly (Gibson et al., 1978). This block appears to be due to impaired intracellular migration of the unglycosylated viral proteins to the plasma membrane, which is assumed to be an obligatory step in the budding process (Leavitt et al., 1977). In a different system, Nakamura and Compans (1978) have shown that tunicamycin has relatively little effect on the production of influenza virus. The virions produced in the presence of the drug were found to lack the "spiked" appearance of normal virus particles. However, it is not clear whether these defective particles had a specific infectivity similar to normal influenza virions. Finally, it is also clear that the effect of tunicamycin on virus replication can also reflect changes in the metabolic stability of unglycosylated viral glycoproteins in the host cell. Schwarz et al. (1976) have shown that the carbohydrate-free hemagglutinin peptide of fowl plaque virus is rapidly degraded by the action of cellular proteases in chick embryo fibroblasts. In fact, the unglycosylated hemagglutinin protein could only be detected when a protease inhibitor was included in the medium. The fact that abnormal proteins are often rapidly degraded by intracellular proteases (Goldberg and Dice, 1974) should always be considered when interpreting experiments utilizing inhibitors of glycosylation.

Another use of drugs such as tunicamycin that specifically inhibit the assembly of oligosaccharide units linked N-glycosidically to protein is in studies on the biosynthesis of macromolecules that are not normally viewed as glycoproteins but which contain such carbohydrate moieties. For example, unlike other mucopolysaccharides, the glycosaminoglycan moiety of keratan sulfate type I is linked to protein via a mannose- and N-acetylglucosamine-containing core oligosaccharide by an N-glycosidic bond (see Chapter 7). On the basis of this structural feature, one might expect that the synthesis of this mucopolysaccharide would be inhibited by tunicamycin. Indeed, experiments concerning the synthesis of keratan sulfate type I in chick embryo cornea have shown that this process is markedly inhibited by tunicamycin (Hart and Lennarz, 1978). Furthermore, studies on the possible participation of saccharide–lipid intermediates in the synthesis of the more classical, O-glycosidically linked proteoglycans seem warranted because (1) glucuronic acid containing saccharide–lipid intermediates have been found in a variety of fibroblastic cell lines (Turco and Heath, 1977; Hopwood and Dorfman, 1977), (2) mannosylphosphoryldolichol serves as a mannosyl donor to yeast cell wall mannoprotein acceptor to form a product susceptible to β-elimination (Lehle and Tanner, 1978), and (3) tunicamycin blocks the synthesis of secreted mucopolysaccharides by chick embryo fibroblasts (Takatsuki *et al.*, 1977*b*).

Tunicamycin may also prove useful in studies involving the role of glycoproteins in complex biological processes such as growth, cellular recognition, and differentiation. Tunicamycin has been found to have profound effects on the morphology and surface properties of prokaryotes and eukaryotes. This drug was found to alter the lectin agglutinability of virally transformed mouse 3T3 or human WI38 cells (Duksin and Bornstein, 1977*b*). However, no significant changes in lectin binding was observed, so that tunicamycin may have affected a subset of lectin binding sites or caused changes in membrane organization. Regarding differentiation, tunicamycin has been shown to specifically block gastrulation of sea urchin embryos without impairing earlier developmental steps (Schneider *et al.*, 1978; Heifetz and Lennarz, 1979).

At this time, tunicamycin appears to be the most useful of the many inhibitors used to study protein glycosylation. With this drug, the effects on cellular metabolism caused by deoxy and fluorodeoxy sugars are avoided, although some cytotoxicity still occurs. Unlike sugar analogues, the mechanism of action of tunicamycin is known from *in vitro* studies and in some instances its effects *in vivo* on protein glycosylation can be clearly attributed to inhibition of the lipid-linked pathway. As screening techniques are developed, it is hoped that other natural inhibitors of protein glycosylation can be discovered. Such inhibitors, in conjunction with mutants in protein glycosylation (see Chapter 4), will be of great value in studying both the biosynthesis and the function of glycoproteins.

10. REFERENCES

Allen, C. M., Kalin, J. R., Sack, J., and Verizzo, D., 1978, CTP-dependent dolichol phosphorylation by mammalian cell homogenates, *Biochemistry* **17**:5020.

Anttinen, H., Oikarinen, A., Ryhanen, L., and Kivirikko, K. I., 1978, Evidence for the transfer of mannose to the extension peptides of procollagen within the cisternae of the rough endoplasmic reticulum, *FEBS Lett.* **87**:222.

Aubert, J.-P., Biserte, G., and Loucheux-LeFebvre, M.-H., 1976, Carbohydrate–peptide linkage in glycoproteins, *Arch. Biochem. Biophys.* **175**:410.

Banjo, B., and Perdue, J., 1976, Increased synthesis of selected membrane polypeptides correlated with increased sugar transport sites in glucose-starved chick embryo fibroblasts, *J. Cell Biol.* **70**:270a.

Baynes, J. W., and Wold, F., 1976, Effect of glycosylation on the *in vivo* circulating halflife of ribonuclease, *J. Biol. Chem.* **251**:6016.

Baynes, J. W., Hsu, A.-F., and Heath, E. C., 1973, The role of mannosylphosphoryldihydropolyisoprenol in the synthesis of mammalian glycoproteins, *J. Biol. Chem.* **248**:5693.

Behrens, N. H., and Leloir, L. F., 1970, Dolichol monophosphate glucose: An intermediate in glucose transfer in liver, *Proc. Natl. Acad. Sci. U.S.A.* **66**:153.

Behrens, N. H., Parodi, A. J., and Leloir, L. F., 1971, Glucose transfer from dolichol monophosphate glucose: The product formed with endogenous microsomal acceptor, *Proc. Natl. Acad. Sci. U.S.A.* **68**:2857.

Behrens, N. H., Carminatti, H., Staneloni, R. J., Leloir, L. F., and Cantarella, A. I., 1973, Formation of lipid-bound oligosaccharides containing mannose. Their role in glycoprotein synthesis, *Proc. Natl. Acad. Sci. U.S.A.* **70**:3390.

Bekesi, J. G., and Winzler, R. J., 1970, Inhibitory effects of D-glucosamine on the growth of Walker 256 carcinosarcoma and on protein, RNA, and DNA synthesis, *Cancer Res.* **30**:2905.

Bekesi, J. G., Molnar, Z., and Winzler, R. J., 1969, Inhibitory effect of D-glucosamine and other sugar analogs on the viability and transplantability of ascites tumor cells, *Cancer Res.* **29**:353.

Bergman, L. W., and Kuehl, W. M., 1977, Addition of glucosamine and mannose to nascent immunoglobulin heavy chains, *Biochemistry* **16**:4490.

Bessell, E. M., Courtenay, V. D., Foster, A. B., Jones, M., and Westwood, J. H., 1973, Some *in vivo* and *in vitro* antitumour effects of the deoxyfluoro-D-glucopyranoses, *Eur. J. Cancer* **9**:463.

Biely, P., Kratky, Z., and Bauer, S., 1972, Metabolism of 2-deoxy-D-glucose by baker's yeast, *Biochim. Biophys. Acta* **255**:631.

Biely, P., Kovarik, J., and Bauer, S., 1973, Lysis of *Saccharomyces cerevisiae* with 2-deoxy-2-fluoro-D-glucose, an inhibitor of the cell wall glucan synthesis, *J. Bacteriol.* **115**:1108.

Birkeland, A. J., and Christensen, T. B., 1975, Resistance of glycoproteins to proteolysis ribonuclease-A and -B compared, *J. Carbohydr., Nucleotides, Nucleosides* **2**:83.

Birshtein, B. K., Preud'homme, J.-L., and Scharff, M. D., 1974, Variants of mouse myeloma cells that produce short immunoglobulin chains, *Proc. Natl. Acad. Sci. U.S.A.* **71**:3478.

Blobel, G., and Dobberstein, B., 1975a, Transfer of proteins across membranes. I. Presence of proteolytically processed and unprocessed nascent immunoglobulin light chains on membrane-bound ribosomes of murine myeloma, *J. Cell Biol.* **67**:835.

Blobel, G., and Dobberstein, B., 1975b, Transfer of proteins across membranes. II. Reconstitution of functional rough microsomes from heterologous components, *J. Cell Biol.* **67**:852.

Braatz, J. A., and Heath, E. C., 1974, The role of polysaccharide in the secretion of protein by *Micrococcus sodonensis, J. Biol. Chem.* **249:**2536.

Burton, W. A., Scher, M. G., and Waechter, C. J., 1979, Enzymatic phosphorylation of dolichol in central nervous tissue, *J. Biol. Chem.* **254:**7129.

Chambers, J., Forsee, W. T., and Elbein, A. D., 1977, Enzymatic transfer of mannose from mannosyl-phosphoryl-polyprenol to lipid-linked oligosaccharides by pig aorta, *J. Biol. Chem.* **252:**2498.

Chen, W. W., 1978, The role of glucose-containing oligosaccharide–lipid in glycosylation of protein acceptors, *Fed. Proc.* **37:**2288.

Chen, W. W., and Lennarz, W. J., 1976, Participation of a trisaccharide-lipid in glycosylation of oviduct membrane glycoproteins, *J. Biol. Chem.* **251:**7802.

Chen, W. W., and Lennarz, W. J., 1977, Metabolism of lipid-linked *N*-acetylglucosamine intermediates, *J. Biol. Chem.* **252:**3472.

Chen, W. W., and Lennarz, W. J., 1978*a*, Enzymatic synthesis of a glucose-containing oligosaccharide–lipid involved in glycosylation of proteins, *J. Biol. Chem.* **253:**5774.

Chen, W. W., and Lennarz, W. J., 1978*b*, Enzymatic excision of glucosyl units linked to the oligosaccharide chains of glycoproteins, *J. Biol. Chem.* **253:**5780.

Chen, W. W., Lennarz, W. J., Tarentino, A. L., and Maley, F., 1975, A lipid-linked oligosaccharide intermediate in glycoprotein synthesis in oviduct. Structural studies on the oligosaccharide chain, *J. Biol. Chem.* **250:**7006.

Chou, P. Y., and Fasman, G. D., 1974*a*, Conformational parameters for amino acids in helical, β-sheet, and random coil regions calculated from proteins, *Biochemistry* **13:**211.

Chou, P. Y., and Fasman, G. D., 1974*b*, Prediction of protein conformation, *Biochemistry* **13:**222.

Ciccimarra, F., Rosen, F. S., Schnesberger, E., and Merler, E., 1976, Failure of heavy chain glycosylation of IgG in some patients with common, variable agammaglobulinemia, *J. Clin. Invest.* **57:**1386.

Czichi, U., and Lennarz, W. J., 1977, Localization of the enzyme system for glycosylation of proteins via the lipid-linked pathway in rough endoplasmic reticulum, *J. Biol. Chem.* **252:**7901.

Daleo, G. R., and Lezica, R. P., 1977, Synthesis of dolichol phosphate by a cell-free extract from pea, *FEBS Lett.* **74:**247.

Daleo, G. R., Hopp, H. E., Romero, P. A., and Lezica, R. P., 1977, Biosynthesis of dolichol phosphate by subcellular fraction from liver, *FEBS Lett.* **81:**411.

Dayhoff, M. O. (ed.), 1973, *Atlas of Protein Sequence and Structure,* Vol. 5, Suppl. 1, National Biomedical Research Foundation, Washington, D.C.

DeLuca, L., Maestri, N., Rosso, G.C., and Wolf, G., 1973, Retinal glycolipids, *J. Biol. Chem.* **248:**641.

DeLuca, L. M., Silverman-Jones, C. S., and Barr, R. M., 1975, Biosynthetic studies on mannolipids and mannoproteins of normal and vitamin A-depleted hamster livers, *Biochim. Biophys. Acta* **409:**342.

Devillers-Thiery, A. Kindt, T., Scheele, G., and Blobel, G., 1975, Homology in amino-terminal sequence of precursors to pancreatic secretory proteins, *Proc. Natl. Acad. Sci. U.S.A.* **72:**5016.

Duksin, D., and Bornstein, P., 1977*a*, Impaired conversion of procollagen to collagen by fibroblasts and bone treated with tunicamycin, and inhibitor of protein glycosylation, *J. Biol. Chem.* **252:**955.

Duksin, D., and Bornstein, P., 1977*b*, Changes in surface properties of normal and transformed cells caused by tunicamycin, an inhibitor of protein glycosylation, *Proc. Natl. Acad. Sci. U.S.A.* **74:**3433.

Eagon, P. K., and Heath, E. C., 1977, Glycoprotein biosynthesis in myeloma cells. Characterization of nonglycosylated immunoglobulin light chain secreted in presence of 2-deoxy-D-glucose, *J. Biol. Chem.* **252:**2372.

Eagon, P. K., Hsu, A. F., and Heath, E. C., 1975, Role of dolichol intermediates in the glycosylation of endogenous and exogenous proteins, *Fed. Proc.* **34:**2609.

Evans, P. J., and Hemming, F. W., 1973, The unambiguous characterization of dolichol phosphate mannose as a product of mannosyl transferase in pig liver endoplasmic reticulum, *FEBS Lett.* **31:**335.

Eylar, E. H., 1965, On the biological role of glycoproteins, *J. Theor. Biol.* **10:**89.

Farkas, V., Svoboda, A., and Bauer, S., 1969, Inhibitory effect of 2-deoxy-D-glucose on the formation of the cell wall in yeast protoplasts, *J. Bacteriol.* **98:**744.

Farkas, V., Svoboda, A., and Bauer, S., 1970, Secretion of cell-wall glycoproteins by yeast protoplasts, *Biochem. J.* **118:**755.

Forsee, W. T., Griffin, J. A., and Schutzbach, J. S., 1977, Mannosyltransfer from GDP-mannose to oligosaccharide–lipids, *Biochem. Biophys. Res. Commun.* **75:**799.

Ganoza, M., and Williams, C., 1969, *In vitro* synthesis of different categories of specific proteins by membrane bound and free ribosomes, *Proc. Natl. Acad. Sci. U.S.A.* **63:**1370.

Ghalambor, M. A., Warren, C. D., and Jeanloz, R. W., 1974, Biosynthesis of a P^1-2-acetamido-2-deoxy-D-glucosyl P^2-polyisoprenyl pyrophosphate by calf pancreas microsomes, *Biochem. Biophys. Res. Commun.* **56:**407.

Gibson, R., Leavitt, R., Kornfeld, S., and Schlesinger, S., 1978, Synthesis and infectivity of vesicular stomatitis virus containing nonglycosylated G protein, *Cell* **13:**671.

Goldberg, A. L., and Dice, J. F., 1974, Intracellular protein degradation in mammalian and bacterial cells, *Annu. Rev. Biochem.* **43:**835.

Grange, D. K., and Adair, W. L., Jr., 1977, Studies on the biosynthesis of dolichyl phosphate: Evidence for the *in vitro* formation of 2,3-dehydrodolichyl phosphate, *Biochem. Biophys. Res. Commun.* **79:**734.

Harford, J. B., Waechter, C. J., and Earl, F. L., 1977, Effect of exogenous dolichyl monophosphate on a developmental change in mannosylphosphoryldolichol biosynthesis, *Biochem. Biophys. Res. Commun.* **76:**1036.

Hart, G. W., and Lennarz, W. J., 1978, Effects of tunicamycin on the biosynthesis of glycosaminoglycans by embryonic chick cornea, *J. Biol. Chem.* **253:**5795.

Hart, G. W., Brew, K., Grant, G. A., Bradshaw, R. A., and Lennarz, W. J., 1979, Primary structural requirements for the enzymatic formation of the N-glycosidic bond in glycoproteins. Studies with natural and synthetic peptides, *J. Biol. Chem.* **254:**9747.

Hassell, J. R., Silverman-Jones, C. S., and DeLuca, L. M., 1978, The *in vivo* stimulation of mannose incorporation into mannosylretinylphosphate, dolichylmannosylphosphate, and specific glycopeptides of rat liver by high doses of retinylpalmitate, *J. Biol. Chem.* **253:**1627.

Havell, E. A., Yamazaki, S., and Vilcek, J., 1977, Altered molecular species of human interferon produced in the presence of inhibitors of glycosylation, *J. Biol. Chem.* **252:**4425.

Heifetz, A., and Elbein, A. D., 1977a, Solubilization and properties of mannose and N-acetylglucosamine transferase involved in formation of polyprenol-sugar intermediates, *J. Biol. Chem.* **252:**3057.

Heifetz, A., and Elbein, A. D., 1977b, Biosynthesis of Man-β-GlcNAc-GlcNAc-pyrophosphoryl-polyprenol by a solubilized enzyme from aorta, *Biochem. Biophys. Res. Commun.* **75:**20.

Heifetz, A., and Lennarz, W. J., 1979, Biosynthesis of N-glycosidically linked glycoproteins during gastrulation of sea urchin embryos, *J. Biol. Chem.* **254:**6119.

Hemming, F. W., 1974, Lipids in glycan biosynthesis, in: *Biochemistry of Lipids* (T. W. Goodwin, ed.), pp. 39–97, Butterworth, London.

Herscovics, A., 1969, Biosynthesis of thyroglobulin. Incorporation of [1-^{14}C]galactose, [1-^{14}C]mannose and [4,5-^{3}H$_2$]leucine into soluble proteins by rat thyroids *in vitro*, *Biochem. J.* **112:**709.

Herscovics, A., 1970, Biosynthesis of thyroglobulin: Incorporation of [³H]fucose into proteins by rat thyroids *in vitro, Biochem. J.* **117:**411.

Herscovics, A., Warren, C. D., and Jeanloz, R. W., 1975, Anomeric configuration of the dolichyl-D-mannosyl phosphate formed in calf pancreas microsomes, *J. Biol. Chem.* **250:**8079.

Herscovics, A., Golovtchenko, A. M., Warren, C. D., Bugge, B., and Jeanloz, R. W., 1977a, Mannosyltransferase activity in calf pancreas microsomes. Formation of ¹⁴C-labeled lipid-linked oligosaccharides from GDP-D-[¹⁴C]mannose and pancreatic dolichyl β-D-[¹⁴C]mannopyranosyl phosphate, *J. Biol. Chem.* **252:**224.

Herscovics, A., Bugge, B., and Jeanloz, R. W., 1977b, Glucosyltransferase activity in calf pancreas microsomes. Formation of dolichyl D-[¹⁴C]glucosyl phosphate and ¹⁴C-labeled lipid-linked oligosaccharides from UDP-D-[¹⁴C]glucose, *J. Biol. Chem.* **252:**2271.

Herscovics, A., Bugge, B., and Jeanloz, R. W., 1977c, Effect of bacitracin on the biosynthesis of dolichol derivatives in calf pancreas microsomes, *FEBS Lett.* **82:**215.

Herscovics, A., Warren, C. D., Bugge, B., and Jeanloz, R. W., 1978, Biosynthesis of P^1-di-N-acetyl-α-chitobiosyl P^2-dolichyl pyrophosphate in calf pancreas microsomes, *J. Biol. Chem.* **253:**160.

Hickman, S., Kulczcki, A., Jr., Lynch, R. G., and Kornfeld, S., 1977, Studies of the mechanism of tunicamycin inhibition of IgA and IgE secretion by plasma cells, *J. Biol. Chem.* **252:**4402.

Hicks, S. J., Drysdale, J. W., and Munro, H., 1969, Preferential synthesis of ferritin and albumin by different populations of liver polysomes, *Science* **164:**584.

Hill, R. L., and Brew, K., 1975, Lactose synthetase, *Adv. Enzymol.* **43:**411.

Hopwood, J. J., and Dorfman, A., 1977, Isolation of lipid glucuronic acid and N-acetylglucosamine derivatives from a rat fibrosarcoma, *Biochem. Biophys. Res. Commun.* **75:**472.

Hubbard, S. C., Turco, S. J., Liu, T., Sultzman, L., and Robbins, P. W., 1978, Processing of asparagine-linked oligosaccharides *in vivo* and *in vitro, Fed. Proc.* **37:**2287.

Hunt, L. T., and Dayhoff, M. O., 1970, The occurrence in proteins of the tripeptides Asn-X-Ser and Asn-X-Thr and of bound carbohydrate, *Biochem. Biophys. Res. Commun.* **39:**757.

Hunt, L. T., Etchison, J. R., and Summers, D. F., 1978, Oligosaccharide chains are trimmed during synthesis of the envelope glycoprotein of vesicular stomatitis virus, *Proc. Natl. Acad. Sci. U.S.A.* **75:**754.

Ito, T., Kodama, Y., Kawamura, K., Suzuki, K., Takatsak, A., and Tamura, G., 1977, The structure of tunicaminyl uracil, a degradation product of tunicamycin, *Agric. Biol. Chem.* **41:**2303.

Jackson, R. L., and Hirs, C. H. W., 1970, The primary structure of porcine pancreatic ribonuclease: The distribution and sites of carbohydrate attachment, *J. Biol. Chem.* **245:**624.

Jilka, R. L., Cavalieri, L. Y., and Pestka, S., 1977, Synthesis and glycosylation of the MOPC-46B immunoglobulin kappa chain in *Xenopus laevis* oocytes, *Biochem. Biophys. Res. Commun.* **79:**625.

Johnson, B. B., 1968, Lysis of yeast cell walls induced by 2-deoxyglucose at their sites of glucan synthesis, *J. Bacteriol.* **95:**1169.

Kaluza, G., Schmidt, M. F. G., and Scholtissek, C., 1973, Effect of 2-deoxy-D-glucose on the multiplication of Semliki forest virus and the reversal of the block by mannose, *Virology* **54:**179.

Katz, F. N., Rothman, J. E., Lingappa, V. R., Blobel, G., and Lodish, H. F., 1977, Membrane assembly *in vitro*: Synthesis, glycosylation, and asymmetric insertion of a transmembrane protein, *Proc. Natl. Acad. Sci. U.S.A.* **74:**3278.

Kean, E. L., 1977, GDP-mannose-polyprenyl phosphate mannosyltransferases of the retina, *J. Biol. Chem.* **252:**5622.

Keenan, R. W., Fischer, J. B., and Kruczek, M. E., 1977, The tissue and subcellular distribution of [³H]dolichol in the rat, *Arch. Biochem. Biophys.* **179:**634.

Keller, R. K., and Adair, W. L., 1977, Microdetermination of dolichol in tissues, *Biochem. Biophys. Acta* **489:**330.

Keller, R. K., and Swank, G. B., 1978, Tunicamycin does not block ovalbumin secretion in the oviduct, *Biochem. Biophys. Res. Commun.* **85:**762.

Keppler, D. O. R., Rudigier, J. F. M., Bischoff, E., and Decker, K. F. A., 1970, The trapping of uridine phosphates by D-galactosamine, D-glucosamine, and 2-deoxy-D-galactose. A study of the mechanism of galactosamine hepatitis, *Eur. J. Biochem.* **17:**246.

Khalkhali, Z., and Marshall, R. D., 1975, Glycosylation of ribonuclease A catalysis by rabbit liver extracts, *Biochem. J.* **146:**299.

Khalkhali, Z., and Marshall, R. D., 1976, URP-*N*-acetyl-D-glucosamine-asparagine sequon *N*-acetyl-β-D-glucosaminyl-transfease-activity in human serum, *Carbohydr. Res.* **49:**455.

Khalkhali, Z., Marshall, R. D., Reuvers, F., Habets-Willems, C., and Boer, P., 1976, Glycosylation *in vitro* of an asparagine sequon catalysed by preparation of yeast cell membranes, *Biochem. J.* **160:**37.

Khalkhali, Z., Serafini-Cessi, F., and Marshall, R. D., 1977, The UDP-*N*-acetylglucosamine-asparagine sequon *N*-acetyl-β-D-glucosaminyltransferase activity in preparation of rough endoplasmic reticulum from regenerating rat liver, *Biochem. J.* **164:**257.

Kiely, M. L., McKnight, G. S., and Schimke, R. T., 1976, Studies on the attachment of carbohydrate to ovalbumin nascent chains in hen oviduct, *J. Biol. Chem.* **251:**5490.

Klenk, H.-D., Rott, R., and Scholtissek, C., 1972, Inhibition of glycoprotein biosynthesis of influenza virus by D-glucosamine and 2-deoxy-D-glucose, *Virology* **49:**723.

Klenk, H.-D., Wollert, W., Rott, R., and Scholtissek, C., 1974, Association of influenza virus proteins with cytoplasmic fractions, *Virology* **57:**28.

Krag, S. S., and Robbins, P. W., 1977, Sindbis envelope proteins as endogenous acceptors in reactors of guanosine diphosphate-[¹⁴C]mannose with preparations of infected chicken embryo fibroblasts, *J. Biol. Chem.* **252:**2621.

Krag, S. S., Cifone, M., Robbins, P. W., and Baker, R. M., 1977, Reduced synthesis of [¹⁴C]mannosyl oligosaccharide–lipid by membranes prepared from concanavalin A-resistant Chinese hamster ovary cells, *J. Biol. Chem.* **252:**3561.

Kratky, Z., Biely, P., and Bauer, S., 1975, Mechanism of 2-deoxy-D-glucose inhibition of cell-wall polysaccharide and glycoprotein biosynthesis in *Saccharomyces cerevisiae,* *Eur. J. Biochem.* **54:**459.

Kronquist, K. E., and Lennarz, W. J., 1978, Enzymatic conversion of proteins to glycoproteins by lipid-linked saccharides. A study of potential exogenous acceptor proteins, *J. Supramol. Struct.* **8:**51.

Kuhn, N. J., and White, A., 1976, Evidence for specific transport of UDP-galactose across the Golgi membrane of rat mammary gland, *Biochem. J.* **154:**243.

Kuo, S.-C., and Lampen, J. O., 1972, Inhibition by 2-deoxy-D-glucose of synthesis of glycoprotein enzymes by protoplasts of *Saccharomyces*: Relation of inhibition of sugar uptake and metabolism, *J. Bacteriol.* **111:**419.

Kuo, S.-C., and Lampen, J. O., 1974, Tunicamycin—An inhibitor of yeast glycoprotein synthesis, *Biochem. Biophys. Res. Commun.* **58:**287.

Kuo, S.-C., and Lampen, J. O., 1976, Tunicamycin inhibition of [³H]glucosamine incorporation into yeast glycoproteins: Binding of tunicamycin and interaction with phospholipids, *Arch. Biochem. Biophys.* **172:**574.

Leavitt, R., Schlesinger, S., and Kornfeld, S., 1977, Tunicamycin inhibits glycosylation and multiplication of Sindbis and vesicular stomatitis virus, *J. Virol.* **21**:375.

Lehle, L., and Schwarz, R. T., 1976, Formation of dolichol monophosphate 2-deoxy-D-glucose and its interference with the glycosylation of mannoproteins in yeast, *Eur. J. Biochem.* **67**:239.

Lehle, L., and Tanner, W., 1976, The specific site of tunicamycin inhibition in the formation of dolichol-bound-*N*-acetylglucosamine derivatives, *FEBS Lett.* **71**:167.

Lehle, L., and Tanner, W., 1978, Glycosyl transfer from dolichyl phosphate sugars to endogenous and exogenous glycoprotein acceptors in yeast, *Eur. J. Biochem.* **83**:563.

Lennarz, W. J., 1975, Lipid-linked sugars in glycoprotein synthesis, *Science* **188**:986.

Lenstra, J. A., Hofsteenge, J., and Beintema, J. J., 1977, Invariant features of the structure of pancreatic ribonuclease. A test of different predictive models, *J. Mol. Biol.* **109**:185.

Levy, J. A., Carminatti, H., Cantarella, A. I., Behrens, N. H., Leloir, L. F., and Tabora, E., 1974, Mannose transfer to lipid linked di-*N*-acetylchitobiose, *Biochem. Biophys. Res. Commun.* **60**:118.

Li, E., Tabas, I., and Kornfeld, S., 1978, The structure of oligosaccharide–lipid of vesicular stomatitis virus (VSV)-infected Chinese hamster ovary (CHO) cells, *Fed. Proc.* **37**:169.

Light, N. D., and Tanner, M. J. A., 1978, Erythrocyte membrane proteins. Sequential accumulation in the membrane during reticulocyte maturation, *Biochim. Biophys. Acta* **508**:571.

Lingappa, V. R., Lingappa, J. R., Prasad, R., Ebner, K. E., and Blobel, G., 1978, Coupled cell-free synthesis, segregation, and core glycosylation of a secretory protein, *Proc. Natl. Acad. Sci. U.S.A.* **75**:2338.

Liras, P., and Gascon, S., 1971, Biosynthesis and secretion of yeast invertase. Effect of cycloheximide and 2-deoxy-D-glucose, *Eur. J. Biochem.* **23**:160.

Lucas, J. J., and Levin, E., 1977, Increase in the lipid intermediate pathway of protein glycosylation during hen oviduct differentiation, *J. Biol. Chem.* **252**:4330.

Lucas, J. J., and Nevar, C., 1978, Loss of mannosyl phosphoryl polyisoprenol synthesis upon conversion of reticulocytes to erythrocytes, *Biochim. Biophys. Acta* **528**:475.

Lucas, J. J., and Waechter, C. J., 1976, Polyisoprenoid glycolipids involved in glycoprotein biosynthesis, *Mol. Cell. Biochem.* **11**:67.

MacDonald, R. I., MacDonald, R. C., and Cornell, N. W., 1974, Perturbation of liposomal and planar lipid bilayer membranes by bacitracin-cation complex, *Biochemistry* **13**:4018.

Mahoney, W. C., and Duksin, D., 1979, Biological activities of the two major components of tunicamycin, *J. Biol. Chem.* **254**:6572.

Maley, F., Tarentino, A. L., McGarrahan, J. F., and Del Giacco, R., 1968, The metabolism of D-galactosamine and *N*-acetyl-D-galactosamein in rat liver, *Biochem. J.* **107**:637.

Mankowski, T., Sasak, W., and Chojnacki, T., 1975, Hydrogenated polyprenol phosphates—Exogenous lipid acceptors of glucose from UDP glucose in rat liver microsomes, *Biochem. Biophys. Res. Commun.* **65**:1292.

Mankowski, T., Sasak, W., Janczura, E., and Chojnacki, T., 1977, Specificity of polyprenyl phosphates in the *in vitro* formation of lipid-linked sugars, *Arch. Biochem. Biophys.* **181**:393.

Marshall, R. D., 1974, The nature and metabolism of the carbohydrate–peptide linkage of glycoproteins, *Biochem. Soc. Symp.* **40**:17.

Martin-Barrientos, J., and Parodi, A. J., 1977, Synthesis of dolichol derivatives in human erythrocyte membranes, *Mol. Cell. Biochem.* **16**:111.

McEvoy, F. A., Ellis, D. E., and Shall, S., 1977, Effects of dolichol monophosphate on galactose incorporation into glycoconjugates of cell cultures, *Biochem. J.* **164**:273.

Melchers, F., 1970, Biosynthesis of the carbohydrate portion of immunoglobulins. Kinetics of synthesis and secretion of [³H]leucine-, [³H]galactose- and [³H]mannose-labeled myeloma protein by two plasma-cell tumours, *Biochem. J.* 119:765.

Melchers, F., 1973, Biosynthesis, intracellular transport, and secretion of immunoglobulins. Effect of 2-deoxy-D-glucose in tumor plasma cells producing and secreting immunoglobulin G₁, *Biochemistry* 12:1471.

Mills, J. T., and Adamany, A. M., 1978, Impairment of dolichyl saccharide synthesis and dolichol-mediated glycoprotein assembly in the aortic smooth muscle cell in culture by inhibitors of cholesterol biosynthesis, *J. Biol. Chem.* 253:5270.

Montenecourt, B. S., Kuo, S.-C., and Lampen, J. O., 1973, *Saccharomyces* mutants with invertase formation resistant to repression by hexoses, *J. Bacteriol.* 114:233.

Nakajima, T., and Ballou, C. E., 1974, Structure of the linkage region between the polysaccharide and protein parts of *Saccharomyces cerevisiae* mannan, *J. Biol. Chem.* 249:7685.

Nakamura, K., and Compans, R. W., 1978, Effects of glucosamine, 2-deoxyglucose, and tunicamycin on glycosylation, sulfation and assembly of influenza viral proteins, *Virology* 84:303.

Nakayama, K., Araki, Y., and Ito, E., 1976, The formation of a mannose-containing trisaccharide on a lipid and its transfer to proteins in yeast, *FEBS Lett.* 72:287.

Olden, K., Pratt, R. M., and Yamada, K. M., 1978, Role of carbohydrates in protein secretion and turnover: Effects of tunicamycin on the major cell surface glycoprotein of chick embryo fibroblasts, *Cell* 13:461.

Parodi, A. J., 1976, Protein glycosylation through dolichol derivatives in baker's yeast, *FEBS Lett.* 71:283.

Pennock, J. F., Hemming, F. W., and Morton, R. A., 1960, Dolichol: A naturally occurring isoprenoid alcohol, *Nature* 186:470.

Peterson, P. A., Rask, L., Helting, T., Östberg, L., and Fernstedt, Y., 1976, Formation and properties of retinylphosphate galactose, *J. Biol. Chem.* 251:4986.

Pless, D. D., and Lennarz, W. J., 1975, A lipid-linked oligosaccharide intermediate in protein glycosylation, *J. Biol. Chem.* 250:7014.

Pless, D. D., and Lennarz, W. J., 1977, Enzymatic conversion of proteins to glycoproteins, *Proc. Natl. Acad. Sci. U.S.A.* 74:134

Pless, D. D., and Palamarczyk, G., 1978, Comparison of polyprenyl derivatives in yeast glycosyl transfer reactions, *Biochim. Biophys. Acta* 529:21.

Plummer, T. H., Jr., and Hirs, C. H. W., 1964, On the structure of bovine pancreatic ribonuclease B, *J. Biol. Chem.* 239:2530.

Pouyssegur, J. M., and Pastan, I., 1976, Mutants of Balb/c 3T3 fibroblasts defective in adhesiveness to substratum: Evidence for alteration in cell surface proteins, *Proc. Natl. Acad. Sci. U.S.A.* 73:544.

Pouyssegur, J. M., and Pastan, I., 1977, Mutants of mouse fibroblasts altered in the synthesis of cell surface glycoproteins, *J. Cell Biol.* 252:1639.

Pouyssegur, J., and Yamada, K. M., 1978, Isolation and immunological characterization of a glucose-regulated fibroblast cell surface glycoprotein and its nonglycosylated precursor, *Cell* 13:139.

Pouyssegur, J., Shiu, R. P. C., and Pastan, I., 1977, Induction of two transformation-sensitive membrane polypeptides in normal fibroblasts by a block in glycoprotein synthesis or glucose deprivation, *Cell* 11:941.

Puett, D., 1973, Conformational studies on a glycosylated bovine pancreatic ribonuclease, *J. Biol. Chem.* 248:3566.

Quastel, J. H., and Cantero, A., 1953, Inhibition of tumor growth by D-glucosamine, *Nature* 171:252.

Redman, C. M., and Sabatini, D. D., 1966, Vectorial discharge of peptides released by puromycin from attached ribosomes, *Proc. Natl. Acad. Sci. U.S.A.* **56**:608.

Reuvers, F., Habets-Willems, C., Reinking, A., and Boer, P., 1977, Glycolipid intermediates involved in the transfer of *N*-acetylglucosamine to endogenous proteins in a yeast membrane preparation, *Biochim. Biophys. Acta* **486**:541.

Reuvers, F., Boer, P., and Hemming, F. W., 1978*a*, The presence of dolichol in a lipid diphosphate *N*-acetylglucosamine from *Saccharomyces cerevisiae* (baker's yeast), *Biochem. J.* **169**:505.

Reuvers, F., Boer, P., and Steyn-Parve, E. P., 1978*b*, The effect of bacitracin on the formation of polyprenol derivatives in yeast membrane vesicles, *Biochem. Biophys. Res. Commun.* **82**:800.

Richards, J. B., and Hemming, F. W., 1972, The transfer of mannose from guanosine disphosphate mannose to dolichol phosphate and protein by pig liver endoplasmic reticulum, *Biochem. J.* **130**:77.

Robbins, P. W., Krag, S. S., and Liu, T., 1977*a*, Effects of UDP-glucose addition on the synthesis of mannosyl lipid-linked oligosaccharides by cell-free fibroblast preparations, *J. Biol. Chem.* **252**:1780.

Robbins, P. W., Hubbard, S. C., Turco, S. J., and Wirth, D. F., 1977*b*, Proposal for a common oligosaccharide intermediate in the synthesis of membrane glycoproteins, *Cell* **12**:893.

Rosso, G. C., DeLuca, L., Warren, C. D., and Wolf, G., 1975, Enzymatic synthesis of mannosyl retinyl phosphate from retinyl phosphate and guanosine diphosphate mannose, *J. Lipid Res.* **16**:235.

Rosso, G. C., Masushige, S., Warren, C. D., Kiorpes, T. C., and Wolfe, G., 1976, The anomeric configuration of the D-mannosyl retinyl phosphate formed in rat liver microsomes, *J. Biol. Chem.* **251**:6465.

Rosso, G. C., Masushige, S., Quill, H., and Wolfe, G., 1977, Transfer of mannose from mannosyl retinyl phosphate to protein, *Proc. Natl. Acad. Sci. U.S.A.* **74**:3762.

Scher, M., and Lennarz, W. J., 1972, Metabolism and function of polyisoprenol sugar intermediates in membrane associated reactions, *Biochim. Biophys. Acta Rev.* **265**:417.

Scher, M., Lennarz, W. J., and Sweeley, C. C., 1968, The biosynthesis of mannosyl-1-phosphoryl-polyisoprenol in *Micrococcus lysodeikticus* and its role in mannan synthesis, *Proc. Natl. Acad. Sci. U.S.A.* **59**:1313.

Scher, M. G., Jochen, A., and Waechter, C. J., 1977, Biosynthesis of glucosylated derivatives of dolichol: Possible intermediates in the assembly of white matter glycoproteins, *Biochemistry* **16**:5037.

Schmidt, M. F. G., Schwarz, R. T., and Ludwig, H., 1976, Fluorosugars inhibit biological properties of different enveloped viruses, *J. Virol.* **18**:819.

Schmidt, M. F. G., Biely, P., Kratky, Z., and Schwarz, R. T., 1978, Metabolism of 2-deoxy-2-fluoro-D-[^{13}H]glucose and 2-deoxy-2-fluoro-D-[^{3}H]mannose in yeast and chick-embryo cells, *Eur. J. Biochem.* **87**:55.

Schneider, E. G., Nguyen, H., and Lennarz, W. J., 1978, The effect of tunicamycin, an inhibitor of protein glycosylation, on embryonic development of sea urchins, *J. Biol. Chem.* **253**:2348.

Scholtissek, C., 1975, Inhibition of the multiplication of enveloped viruses by glucose derivatives, *Curr. Top. Microbiol. Immunol.* **70**:101.

Scholtissek, C., Rott, R., and Klenk, H.-D., 1975, Two different mechanisms of the inhibition of the multiplication of enveloped viruses by glucosamine, *Virology* **63**:191.

Schwarz, R. T., and Klenk, H.-D., 1974, Inhibition of glycosylation of the influenza virus hemagglutinin, *J. Virol.* **14**:1023.

Schwarz, R. T., Rohrschneider, J. M., and Schmidt, M. F. G., 1976, Suppression of glycoprotein formation of Semliki forest, influenza, and avian sarcoma virus by tunicamycin, *J. Virol.* **19**:782.

Schwarz, R. T., Schmidt, M. F. G., Anwer, U., and Klenk, H.-D., 1977, Carbohydrates of influenza virus, *J. Virol.* **23:**217.

Schwarz, R. T., Schmidt, M. F. G., and Lehle, L., 1978, *In vitro* glycosylation of Semliki forest and influenza virus glycoproteins and its suppression by nucleotide-2-deoxy-sugar, *Eur. J. Biochem.* **88:**163.

Siewert, G., and Strominger, J. L., 1967, Bacitracin: An inhibitor of the dephosphorylation of lipid pyrophosphate an intermediate in biosynthesis of the peptidoglycan of bacterial cell wall, *Proc. Natl. Acad. Sci. U.S.A.* **57:**767.

Singer, S. J., 1974, The molecular organization of membranes, *Annu. Rev. Biochem.* **43:**805.

Sinohara, H., and Marugama, T., 1973, Evolution of glycoproteins as judged by the frequency of occurrence of the tripeptides Asn-X-Ser and Asn-X-Thr in proteins, *J. Mol. Evol.* **2:**117.

Spiro, R. G., and Spiro, M. J., 1966, Glycoprotein biosynthesis studies on thyroglobulin. Characterization of a particulate precursor and radioisotope incorporation by thyroid slices and particle systems, *J. Biol. Chem.* **211:**1271.

Spiro, M. J., Spiro, R. G., and Bhoyroo, V. D., 1976a, Lipid-saccharide intermediates in glycoprotein biosynthesis. III. Comparison of oligosaccharide–lipids formed by slices from several tissues, *J. Biol. Chem.* **251:**6420.

Spiro, R. G., Spiro, M. J., and Bhoyroo, V. D., 1976b, Lipid–saccharide intermediates in glycoprotein biosynthesis. II. Studies on the structure of an oligosaccharide–lipid from thyroid, *J. Biol. Chem.* **251:**6409.

Spiro, M. J., Spiro, R. G., and Bhoyroo, V. D., 1978a, Utilization of oligosaccharide-lipids in glycoprotein biosynthesis by thyroid enzyme, *Fed. Proc.* **37:**2285.

Spiro, R. G., Spiro, M. J., and Bhoyroo, V. D., 1978b, Processing of the carbohydrate units of glycoproteins: Action of a thyroid glucosidase, *Fed. Proc.* **37:**2286.

Staneloni, R. J., and Leloir, L. F., 1978, Oligosaccharides containing glucose and mannose in glycoproteins of the thyroid gland. *Proc. Natl. Acad. Sci. U.S.A.* **75:**1162.

Steiner, S., and Steiner, M. R., 1973, Incorporation of 2-deoxy-D-glucose into glycolipids of normal and SV_{40}-transformed hamster cells, *Biochim. Biophys. Acta* **296:**403.

Steiner, S., Courtney, R. J., and Melnick, J. L., 1973, Incorporation of 2-deoxy-D-glucose into glycoproteins of normal and Sinian virus 40-transformed hamster cells, *Cancer Res.* **33:**2402.

Stone, K. J., and Strominger, J. L., 1971, Mechanism of action of bacitracin: Complexation with metal ion and C_{55}-isoprenyl pyrophosphate, *Proc. Natl. Acad. Sci. U.S.A.* **68:**3223.

Stone, K. J., and Strominger, J. L., 1972, Inhibition of sterol biosynthesis by bacitracin, *Proc. Natl. Acad. Sci. U.S.A.* **69:**1287.

Storm, D. R., and Strominger, J. L., 1973, Complex formation between bacitracin peptides and isoprenyl pyrophosphates, *J. Biol. Chem.* **248:**3940.

Struck, D. K., and Lennarz, W. J., 1976, Utilization of exogenous GDP-mannose for the synthesis of mannose-containing lipids and glycoproteins by oviduct cells, *J. Biol. Chem.* **251:**2511.

Struck, D. K., and Lennarz, W. J., 1977, Evidence for the participation of saccharide-lipids in the synthesis of the oligosaccharide chain of ovalbumin, *J. Biol. Chem.* **252:**1007.

Struck, D. K., Lennarz, W. J., and Brew, K., 1978a, Primary sequence requirements for the formation of the N-glycosidic bond. Studies with α-Lactalbumin, *J. Biol. Chem.* **253:**5786.

Struck, D. K., Siuta, P. B., Lane, M. D., and Lennarz, W. J., 1978b, Effect of tunicamycin on the secretion of serum proteins by primary cultures of rat and chick hepatocytes. Studies on transferrin, very-low-density lipoprotein, and serum albumin, *J. Biol. Chem.* **253:**5332.

Tabas, I., and Kornfeld, S., 1978, The synthesis of complex-type oligosaccharides. III.

Identification of an α-D-mannosidase activity involved in a late stage of processing of complex-type oligosaccharides, *J. Biol. Chem.* **253**:7779.

Tabas, I., Schlesinger, S., and Kornfeld, S., 1978, Processing of high mannose oligosaccharides to form complex type oligosaccharides on the newly synthesized polypeptides of the vesicular stomatitis virus G protein and IgG heavy chain, *J. Biol. Chem.* **253**:716.

Tabora, E., and Behrens, N. H., 1977, A lipid-linked oligosaccharide which contains hexosamine, mannose and glucose, *Mol. Cell. Biochem.* **16**:193.

Takatsuki, A., and Tamura, G., 1971*a*, Tunicamycin, a new antibiotic. III. Reversal of the antiviral activity of tunicamycin by aminosugars and their derivatives, *J. Antibiot.* **24**:232.

Takatsuki, A., and Tamura, G., 1971*b*, Effect of tunicamycin on the synthesis of macromolecules in cultures of chick embryo fibroblasts infected with Newcastle disease virus, *J. Antibiot.* **24**:785.

Takatsuki, A., Arima, K., and Tamura, G., 1971, Tunicamycin, a new antibiotic. I. Isolation and characterization of tunicamycin, *J. Antibiot.* **24**:215.

Takatsuki, A., Shimizu, K.-I., and Tamura, G., 1972, Effect of tunicamycin on microorganisms: Morphological changes and degradation of RNA and DNA induced by tunicamycin, *J. Antibiot.* **25**:75.

Takatsuki, A., Kohno, K., and Tamura, G., 1975, Inhibition of biosynthesis of polyisoprenol sugars in chick-embryo microsomes by tunicamycin, *Agric. Biol. Chem.* **39**:2089.

Takatsuki, A., Fukui, Y., and Tamura, G., 1977*a*, Tunicamycin inhibits biosynthesis of acid mucopolysaccharides in cultures of chick embryo fibroblasts, *Agric. Biol. Chem.* **41**:425.

Takatsuki, A., Kawamura, K., Okina, M., Kodama, Y., Ito, T., and Tamura, G., 1977*b*, The structure of tunicamycin, *Agric. Biol. Chem.* **41**:2307.

Tarentino, A. L., and Maley, F., 1976, Direct evidence that D-galactosamine incorporation into glycogen occurs via UDP-glucosamine, *FEBS Lett.* **69**:175.

Tkacz, J. S., and Lampen, J. O., 1975, Tunicamycin inhibition of polyisoprenyl N-acetylglucosaminyl pyrophosphate formation in calf-liver microsomes, *Biochem. Biophys. Res. Commun.* **65**:248.

Toneguzzo, F., and Ghosh, H. P., 1978, *In vitro* synthesis of vesicular stomatitis virus membrane glycoprotein and insertion into membranes, *Proc. Natl. Acad. Sci. U.S.A.* **75**:715.

Trowbridge, I. S., Hyman, R., and Mazauskas, C., 1978, The synthesis and properties of T25 glycoprotein in Thy-1-negative mutant lymphoma cells, *Cell* **14**:21.

Turco, S. J., and Heath, E. C., 1977, Glucuronosyl-N-acetylglucosaminyl pyrophosphoryldilichol. Formation in SV$_{40}$-transformed human lung fibroblasts and biosynthesis in rat lung microsomal preparations, *J. Biol. Chem.* **252**:2918.

Turco, S. J., Stetson, B., and Robbins, P. W., 1977, Comparative rates of transfer of lipid-linked oligosaccharides to endogenous glycoprotein acceptors *in vitro*, *Proc. Natl. Acad. Sci. U.S.A.* **74**:4411.

Vargas, V. I., and Carminatti, H., 1977, Glycosylation of endogenous protein(s) of the rough and smooth microsomes by a lipid sugar intermediate, *Mol. Cell. Biochem.* **16**:171.

Waechter, C. J., and Harford, J. B., 1977, Evidence for the enzymatic transfer of N-acetylglucosamine from UDP-N-acetylglucosamine ino dolichol derivatives and glycoproteins by calf brain membranes, *Arch. Biochem. Biophys.* **181**:185.

Waechter, C. J., and Lennarz, W. J., 1976, The role of polyprenol-linked sugars in glycoprotein synthesis, *Annu. Rev. Biochem.* **45**:95.

Waechter, C. J., and Scher, M. G., 1978, Glucosylphosphoryldolichol: Role as a glucosyl donor in the biosynthesis of an oligosaccharide–lipid intermediate by calf brain membranes, *Arch. Biochem. Biophys.* **188**:385.

Waechter, C. J., Lucas, J. J., and Lennarz, W. J., 1973, Membrane glycoproteins. I. Enzymatic synthesis of mannosyl phosphoryl polyisoprenol and its role as a mannosyl donor in glycoprotein synthesis, *J. Biol. Chem.* **248**:7570.

Warren, C. D., and Jeanloz, R. W., 1974, Chemical synthesis of P'-2-acetamido-2-deoxy-D-glucopyranosyl P^2-dolichyl pyrophosphate, *Carbohydr. Res.* **37**:252.

Wedgwood, J. B., Warren, C. D., Jeanloz, R. W., and Strominger, J. L., 1974, Enzymatic utilization of P^1-di-N-acetylchitobiosyl P^2-dolichyl pyrophosphate and its chemical synthesis, *Proc. Natl. Acad. Sci. U.S.A.* **71**:5022.

Weitzman, S., and Scharf, M. D., 1976, Mouse myeloma mutants blocked in the assembly, glycosylation, and secretion of immunoglobulins, *J. Mol. Biol.* **102**:237.

White, D. A., 1978, The formation of lipid-linked sugars by cell-free preparations of lactating rabbit mammary gland, *Biochem. J.* **170**:479.

Mammalian Glycosyltransferases

Their Role in the Synthesis and Function of Complex Carbohydrates and Glycolipids

Harry Schachter and Saul Roseman

PART A

1. INTRODUCTION

1.1. Sugar Nucleotides and Transglycosylation

The general process for the synthesis of complex carbohydrates is that of transglycosylation:

$$G—O—R_1 + R_2—OH \rightleftharpoons G—O—R_2 + R_1—OH$$

$G—O—R_1$ is the glycosyl donor which is usually a sugar phosphate compound such as a sugar nucleotide. $R_2—OH$ is the acceptor, which can be an alcohol, a sugar phosphate, a monosaccharide, or an oligosaccharide. The discovery of glycogen synthetase by Leloir and Cardini (1957) established the role of sugar nucleotides in complex carbohydrate assembly *in vivo*; phosphorylase was seen to play a catabolic role under

Harry Schachter • Department of Biochemistry, Hospital for Sick Children, Toronto, Canada M5G 1X8.   *Saul Roseman* • Department of Biology and the McCollum Pratt Institute, The Johns Hopkins University, Baltimore, Maryland 21218.

physiologic conditions. Sugar nucleotides are, from a thermodynamic point of view, superior donors for complex carbohydrate assembly; these "high-energy" compounds are now known to be the glycosyl donors involved in glycoprotein and glycosphingolipid biosynthesis (see Part B of this chapter).

Sugar nucleotide synthesis usually involves the following reaction: nucleoside triphosphate + sugar 1-phosphate $\rightleftharpoons$ nucleoside diphosphate sugar + PP_i. Most sugar nucleotides are nucleoside diphosphate sugars (UDP-α-D-glucose, UDP-α-D-galactose, UDP-α-N-acetyl-D-glucosamine, UDP-α-N-acetyl-D-galactosamine, GDP-α-D-mannose and GDP-β-L-fucose). The exception, CMP-sialic acid, is synthesized by the following reaction:

$$\text{CTP + sialic acid} \rightarrow \text{CMP-sialic acid} + PP_i$$

The synthesis and interconversions of the various sugar nucleotides have been thoroughly reviewed (Hassid, 1967; Schachter and Rodén, 1973) and will not be discussed in detail in the present review.

The availability of sugar nucleotide is obviously essential for normal complex carbohydrate biosynthesis, and factors controlling the synthesis and catabolism of sugar nucleotides may play important roles in the control of glycosylation reactions. It is interesting, for example, that CMP-sialic acid synthetase has been reported to be a nuclear enzyme in several different mammalian tissues (Kean, 1970; Kean and Bruner, 1971; Van Dijk et al., 1973; Van den Eijnden et al., 1972; Van den Eijnden, 1973). Although these studies could not rule out the presence of some synthetase in other organelles, the marked enrichment in the nucleus is intriguing because of the localization of various sialyltransferases in the Golgi apparatus (see Section 2.6). It is conceivable that CMP-sialic acid may have to move through the cell's endomembrane system from nucleus to Golgi apparatus and that this process may play a role in the control of sialylation.

Sialic acids are N-acetyl and N-glycolyl derivatives of neuraminic acid, often with additional O-acetyl groups, and are widely distributed in the animal kingdom (Schauer et al., 1974). N-Acetylneuraminic acid is the precursor of N-glycolylneuraminic acid; the conversion is catalyzed by the enzyme N-acetylneuraminate ascorbate or NADPH:oxygen oxidoreductase (N-acetyl hydroxylating). N-Acetylneuraminic acid can also be converted to a series of N-acetylmono-O-acetylneuraminic acids and N-acetyloligo-O-acetylneuraminic acids by various acetyl CoA-requiring acetyltransferases. In the salivary gland, it appears that hydroxylation and O-acetylation can occur both with free N-acetylneuraminic acid and with protein-bound N-acetylneuraminic acid (Schauer et al., 1974; Corfield *et al.*, 1976); however, these enzymes cannot modify *N*-acetylneu-raminic acid while it is bound as CMP-*N*-acetylneuraminic acid. The modifying enzymes which act on protein-bound *N*-acetylneuraminic acid appear to be in the Golgi apparatus (Schauer *et al.*, 1974; Corfield *et al.*, 1976; Buscher *et al.*, 1977).

Crude CMP-sialic acid synthetase preparations can convert *N*-ace-tylneuraminic acid, *N*-glycolylneuraminic acid, *N*-acetyl-7-*O*-acetylneu-raminic acid, *N*-acetyl-9-*O*-acetylneuraminic acid, and *N*-acetyl-4-*O*-ace-tylneuraminic acid to the corresponding CMP-sialic acids (Kean, 1970; Schauer *et al.*, 1972; Schauer and Wember, 1973). Kinetic studies have indicated that a single enzyme may catalyze these various reactions (Schauer and Wember, 1973). Also, salivary gland sialyltransferases can utilize CMP-sialic acids other than CMP-*N*-acetylneuraminic acid as sialic acid donors in glycoprotein synthesis (Carlson *et al.*, 1966, 1973c; Schauer and Wember, 1973). It appears therefore that the type of sialic acid which is incorporated depends neither on the CMP-sialic acid syn-thetase nor on the sialyltransferase, but rather on the hydroxylating and *O*-acetylating enzymes which modify *N*-acetylneuraminic acid.

Thus, CMP-sialic acid is made in the nucleus, and incorporation of sialic acid into protein as well as the hydroxylation and *O*-acetylation of protein-bound *N*-acetylneuraminic acid all occur in the Golgi apparatus. The nucleoside diphosphate sugar pyrophosphorylases, in contrast to the nuclear location of CMP-sialic acid synthetase, appear to be cytosolic enzymes (Mendicino and Rao, 1975).

Studies have also been carried out on enzymes which degrade sugar nucleotides. Shoyab and Bachhawat (1967, 1969) reported the presence of CMP-sialic acid hydrolase in various rat tissues; it was later shown that the enzyme was enriched in plasma membrane fractions from sheep and rat liver (Kean and Bighouse, 1974) and calf kidney (Van Dijk *et al.*, 1976). An enzyme has been isolated from rat liver with nucleotide pyro-phosphatase activity (Evans *et al.*, 1973; Evans, 1974; Decker and Bis-choff, 1972; Bischoff *et al.*, 1975); this enzyme is present in the endo-plasmic reticulum but is also enriched in the plasma membrane (Evans, 1974; Bischoff *et al.*, 1976). It is not yet known whether CMP-sialic acid hydrolase and nucleotide pyrophosphatase are one and the same enzyme nor whether there is only a single nucleotide pyrophosphatase capable of hydrolyzing a variety of nucleoside diphosphate sugars. What is clear, however, is that enzymes capable of hydrolyzing both CMP-sialic acid and various nucleoside diphosphate sugars are present in many tissues and that these activities may be enriched in the plasma membrane (Kean and Bighouse, 1974; Van Dijk *et al.*, 1976; Evans, 1974; Munro *et al.*, 1975; Mookerjea and Yung, 1975; Bischoff *et al.*, 1976). Further work will have to be done to determine what roles in the control of glycosylation are played by this elaborate system of membrane-bound and cytosolic enzymes involved in the synthesis and hydrolysis of sugar nucleotides.

1.2. Glycosyltransferases: General Comments

The reactions to be discussed in this chapter take the following forms:

$$CMP\text{-sialic acid} + R\text{—}OH \rightarrow Sialic\ acid\text{—}O\text{—}R + CMP$$

or

$$Nucleoside\ diphosphate\ sugar + R\text{—}OH \rightarrow Sugar\text{—}O\text{—}R + nucleoside\ diphosphate$$

The glycosyl donor is a sugar nucleotide, and the acceptor may be a monosaccharide, oligosaccharide, glycopeptide, glycoprotein or, as discussed in Part B, a glycolipid. Reactions involving dolichol intermediates as glycosyl donors are discussed in Chapter 2. Some of the glycosyltransferases to be discussed transfer a sugar residue not to the hydroxyl group of a sugar but to the hydroxyl group of an amino acid such as serine, threonine, or hydroxylysine. The linkage between sugar and phosphate in all sugar nucleotides involves the anomeric carbon of the sugar (C-1 of D-glucose, D-galactose, D-mannose, L-fucose, N-acetyl-D-glucosamine, and N-acetyl-D-galactosamine and C-2 of sialic acid); this same anomeric carbon becomes glycosidically bonded to the acceptor hydroxyl group.

In mammalian tissues, the sugar nucleotides of D-glucose, D-galactose, D-mannose, N-acetyl-D-glucosamine, and N-acetyl-D-galactosamine are all α-linked whereas L-fucose is β-linked to GDP. Some sugars, for example, L-fucose and N-acetyl-D-glucosamine, appear to be transferred from sugar nucleotide to acceptor primarily by a single inversion of anomeric configuration, i.e., the anomeric configurations of L-fucose derived from GDP-β-L-fucose and of N-acetyl-D-glucosamine derived from UDP-α-N-acetyl-D-glucosamine are α and β, respectively, in most mammalian complex carbohydrates. It is probable that D-mannose transfer also occurs by a single anomeric inversion since β-linked D-mannose derives from GDP-α-D-mannose (Wedgwood et al., 1974; Levy et al., 1974) whereas α-linked mannose derives from dolichol monophosphate β-D-mannose (see Chapter 2). Other sugars, however, are found in complex carbohydrates in both α- and β-linkages. For example, as will be discussed in this chapter, D-galactose and N-acetyl-D-galactosamine are found both α- and β-linked and the α- and β-galactosyltransferases and α- and β-N-acetyl-D-galactosaminyltransferases carrying out these incorporations all use α-linked sugar nucleotides. The detailed enzymatic mechanisms by which this is accomplished are yet to be determined.

Although it is possible to assay lactose synthetase and other glycosyltransferases by spectrophotometric methods (Fitzgerald *et al.*, 1970*b*), the most accurate (especially with crude enzyme preparations) and, therefore, the most commonly used assay method involves measuring the transfer of radioactive sugar from sugar nucleotide to acceptor. Crude enzyme preparations may contain nucleotide pyrophosphatase, CMP-sialic acid hydrolase, glycosidases or nonspecific phosphatases and, therefore, a typical enzyme incubation may contain several radioactive components, i.e., sugar nucleotide, sugar phosphate, free sugar, and product. Further, the product may be due to the transfer of radioactive sugar to both exogenous and endogenous acceptors; the endogenous acceptors may be small or large oligosaccharides, glycoproteins, or glycolipids or even non-carbohydrate-containing contaminants. It is therefore imperative that proper assay conditions be established for every individual experimental situation.

Several techniques have been developed for the separation of product from the mixture described above. A very reliable but relatively tedious method is the use of high-voltage paper electrophoresis in 1% sodium tetraborate at pH 9.0 (Roseman *et al.*, 1966). Sugar nucleotide migrates rapidly towards the anode and is cleanly separated from product which usually remains at or near the origin. Sugar phosphate also migrates rapidly towards the anode, often at a slightly different rate than sugar nucleotide; a scan of the electrophoretogram may therefore conveniently indicate how much sugar nucleotide remains intact at the end of the incubation. Free sialic acid, galactose, mannose, and fucose migrate towards the anode either because of an inherent negative charge or because of borate complex formation. However, N-acetylhexosamines do not complex borate efficiently, and these materials must be removed from the origin of the paper by descending chromatography with 80% ethanol. Product remaining at or near the origin can then be counted directly on the paper by liquid scintillation techniques. It is important to count several (4 to 6) inches of paper near the origin since many products, even proteins, migrate slightly and tend to smear; counting a large segment of paper indicates whether there is efficient separation of product from other radioactive materials. Endogenous acceptor assays should be carried out routinely and appropriate corrections applied. If the endogenous acceptor is a protein or a lipid, the resulting product will remain at the origin and interfere with the exogenous acceptor assay. The endogenous acceptor activity can either be subtracted or, if a lipid, can be washed out of the paper with the appropriate organic solvent. The high-voltage electrophoresis assay is strongly recommended for work with crude enzyme preparations.

A more rapid assay involves the precipitation of product with trichloroacetic acid–phosphotungstic acid (Letts *et al.*, 1974a,b) followed by filtration on a glass fiber filter. The filter is then thoroughly washed with trichloroacetic acid–phosphotungstic acid followed by organic solvents, dried, and counted. All low-molecular-weight radioactive compounds are removed by this procedure. The method works only with high-molecular-weight acceptors. Endogenous lipid acceptors do not interfere since the resulting products are removed with the organic solvent wash. Caution is advised when sialyltransferases are assayed by this method due to the acid lability of bound sialic acid; the entire procedure should be carried out with ice-cold solutions. Crude enzyme preparations can also cause problems, especially when working with mucin acceptors, due to clogging of the filters and resultant trapping of low-molecular-weight radioactive compounds in the filter.

Yet another rapid method involves passage of the reaction mixture through a small (0.5 × 2 cm) column of Dowex-2 X8, 200–400 mesh, in the chloride form, equilibrated in water (Fleischer *et al.*, 1969). Sugar nucleotide and sugar phosphate are retained on the column while radioactive product and free sugar pass through the column and are counted. The "endogenous acceptor" control therefore will contain not only endogenous acceptor activity but also free sugar from sugar nucleotide breakdown. The amount of free sugar formed may not be constant from incubation to incubation, especially with crude enzyme preparations containing nucleotide pyrophosphatase, nonspecific phosphatases, glycosidases acting on radioactive product, and other hydrolytic enzymes. The assay is fast and convenient but should be used only when the glycosyltransferase preparation is relatively free of these interfering hydrolytic enzymes.

A preferable small column assay is the use of gel filtration such as columns (0.8 × 10 cm) of Sephadex G-50 (fine) equilibrated with buffer and 0.01% sodium azide (Schwyzer and Hill, 1977a). High-molecular-weight product elutes before low-molecular-weight radioactive compounds and can be readily counted. The assay is rapid and effective but cannot be used with low-molecular-weight acceptors.

Finally, reaction mixtures can be subjected to paper chromatography on ordinary filter paper (Freilich *et al.*, 1975a,b; 1977) or on DEAE ion-exchange paper (Bergeron *et al.*, 1973). The latter procedure is effective only with low-molecular-weight acceptors since sugar nucleotide and sugar phosphate remain at the origin and product must migrate away from the origin for the assay to work. The paper chromatographic approach is reliable but quite slow and tedious. It is useful in determining the nature of low-molecular-weight radioactive derivatives formed in an incubation but is not advisable as a routine procedure in enzyme purification or kinetic experiments.

It is obviously important to establish saturating conditions for both acceptor and sugar nucleotide and to maintain saturation throughout the course of the reaction. Acceptor concentrations must be individually optimized since excess acceptor may cause substrate-dependent inhibition (Treloar *et al.*, 1974; Freilich *et al.*, 1977; Hudgin and Schachter, 1971*a,b*). The high cost of radioactive sugar nucleotides and the commercial unavailability of certain nonradioactive sugar nucleotides (CMP-sialic acid, GDP-L-fucose and UDP-*N*-acetyl-D-galactosamine) have tempted some investigators into working with subsaturating amounts of sugar nucleotide; such assays result in semiquantitative data and unreliable kinetics. As mentioned above, steps must be taken to maintain saturating levels of sugar nucleotide and to minimize its hydrolysis.

Endogenous acceptors are often used in glycosyltransferase assays. While this approach is often unavoidable (see Chapter 2), it has certain pitfalls. Kinetics using the endogenous acceptor assay are useless since it is not known whether enzyme protein or acceptor is rate-limiting and since these two components of the assay cannot be varied independently. Further, there may be more than one endogenous acceptor in the enzyme preparation and therefore more than one enzyme may be assayed; the exact chemical nature of the reaction may be difficult and probably impossible to determine; and attempts at enzyme purification would be frustrated by separation of enzyme from endogenous substrate. For these reasons, the present review will be limited to studies involving well-defined exogenous acceptors.

Most glycosyltransferases (with the exception of sialyltransferases) require divalent cation for activity. Optimum cation concentrations must be determined, and it is advisable to do both exogenous and endogenous acceptor assays whenever conditions (cation concentration, type of cation, ionic strength, pH, etc.) are varied. The requirement for cation enables the convenient use of EDTA to terminate the reaction prior to assay. As will be discussed below, intracellular glycosyltransferases are membrane-bound and almost invariably require some sort of detergent and/or ultrasonic treatment for expression of optimum activity; such treatment is usually not required for extracellular glycosyltransferases such as are present in milk, colostrum, serum, amniotic fluid, etc. The tight association of glycosyltransferases to intracellular membranes has, until recently, frustrated attempts at purification of these enzymes to homogeneity. However, the application of sophisticated affinity chromatography techniques to detergent-solubilized transferases has led to successful purification of some transferases (Schwyzer and Hill, 1977a,b; Sadler et al., 1979).

Much of the work on glycosyltransferases has been carried out with large glycoprotein acceptors such as α_1-acid glycoprotein. These acceptors have many diverse oligosaccharide chains (Schmid et al., 1977), and it has always been recognized that more than one glycosyltransferase may be contributing to the assay. For example, the production of more than one positional isomer is a likely possibility when crude enzyme preparations are used with acceptors such as α_1-acid glycoprotein. This problem is best approached by using low-molecular-weight acceptors such as oligosaccharides or glycopeptides; for example, lactose can be used as an acceptor for sialyltransferases and the resulting (2,3) and (2,6) isomers of sialyllactose can be separated from one another (Hudgin and Schachter, 1972; Bartholomew et al., 1973; Carlson et al., 1973b; Paulson et al., 1977a,b), or glycopeptide acceptors can be used to distinguish between two very similar N-acetylglucosaminyltransferases (Narasimhan et al., 1977).

The glycosyltransferases show a high degree of specificity for the sugar being transferred, for the base component of the sugar nucleotide (in the few instances where this point has been investigated), and for the sugar residues at the nonreducing end of the acceptor. It is likely that a separate transferase is required for every known sugar–sugar linkage; this concept has been called the "one linkage–one enzyme" hypothesis (Schachter and Rodén, 1973). The hypothesis was originally proposed on the basis of indirect evidence for the existence of multiple transferases (Roseman, 1970) but has received strong support from recent work on highly purified enzymes (Paulson et al., 1977a,b; Sadler et al., 1979; Schwyzer and Hill, 1977a,b).

The following sections will deal with the properties of the various glycosyltransferases that have been detected and characterized to date. No attempt will be made to deal with such topics as the possible roles that glycosyltransferases might play in differentiation (Carlson et al., 1973a, Den et al., 1970), development, and oncogenic transformation.

2. GLYCOSYLTRANSFERASES INVOLVED IN ELONGATION OF N-GLYCOSIDICALLY LINKED OLIGOSACCHARIDES OF THE N-ACETYLLACTOSAMINE TYPE

As described in Chapter 1, N-glycosidically linked oligosaccharides occur in two major forms, the oligomannoside form and the N-acetyllactosamine form. These structures share a common core containing mannose and N-acetylglucosamine. The assembly of this core as a lipid-bound intermediate and its transfer to protein were discussed in Chapter

2. The following section deals with the glycosylation reactions that convert the oligosaccharide donated to protein by dolichol pyrophosphate oligosaccharide to a structure of the N-acetyllactosamine type. This subject was also discussed in Chapter 1 from a somewhat different perspective.

2.1. Processing of Protein-Bound Oligosaccharide Prior to Elongation

It appears that synthesis of many, and possibly all, N-glycosidically linked oligosaccharides may be initiated by the transfer of oligosaccharide from a dolichol pyrophosphate oligosaccharide to protein (see Chapter 2). The exact structure of the precursor dolichol pyrophosphate oligosaccharide may not be the same for all situations or tissues but appears to have the general structure $(Glc)_x(Man-\alpha-)_y-Man-\beta1,4-GlcNAc-\beta1,4-$ GlcNAc-pyrophosphate dolichol (see Chapters 1 and 2). Glucose-free dolichol pyrophosphate oligosaccharides have been isolated from various sources and have been shown to transfer oligosaccharide to protein. Glucose-containing dolichol pyrophosphate oligosaccharides have also been isolated from many sources (Parodi *et al.*, 1972; Behrens *et al.*, 1971; Herscovics *et al.*, 1977a,b; R. G. Spiro *et al.*, 1976; M. J. Spiro *et al.*, 1976a,b) and have recently been shown to be eight to nine times more effective in oligosaccharide transfer to endogenous protein acceptors than non-glucose-containing dolichol pyrophosphate oligosaccharides (Turco *et al.*, 1977).

Non-glucose-containing dolichol pyrophosphate oligosaccharides appear to be suitable precursors for the synthesis of the oligomannoside type of oligosaccharide. However, the N-acetyllactosamine type of oligosaccharide contains only three mannose residues and, if dolichol pyrophosphate oligosaccharides are also involved in the assembly of this type of structure, one must postulate an oligosaccharide processing mechanism for the removal of excess mannose residues. Further, N-glycosidically linked oligosaccharides do not contain glucose and if the precursor is a glucose-containing dolichol pyrophosphate oligosaccharide, as suggested by the work of Turco *et al.* (1977), then oligosaccharide processing would involve removal of both glucose and mannose residues. In fact, enzymes that remove these two sugars have been detected (see Chapters 1 and 2).

Recent work on the biosynthesis of the envelope glycoproteins of vesicular stomatitis virus (VSV) and Sindbis virus has provided strong support for the concept that the core structure of N-acetyllactosamine-type oligosaccharides is derived from a mannose-rich oligosaccharide by removal of mannose residues (Tabas *et al.*, 1978; Hunt *et al.*, 1978; Robbins *et al.*, 1977). The following sequence of reactions is suggested as a likely mechanism for the generation of the N-acetyllactosamine core structure:

$(Glc)_x(Man-\alpha)_y$-Man-β1,4-GlcNAc-β1,4-GlcNAc-pyrophosphate-dolichol

+ protein

$\rightarrow$ $(Glc)_x(Man-\alpha)_y$-Man-β1,4-GlcNAc-β1,4-GlcNAc-Asn-protein

+ dolichol pyrophosphate

$(Glc)_x(Man-\alpha)_y$-Man-β1,4-GlcNAc-β1,4-GlcNAc-Asn-protein

$$
\begin{array}{l}
\text{Man} \\
\quad\diagdown_{\alpha 1,3} \\
\rightarrow \qquad\qquad\diagdown\text{Man-}\beta 1,4\text{-GlcNAc-}\beta 1,4\text{-GlcNAc-Asn-protein} \\
\quad\diagup_{\alpha 1,6} \\
\text{Man}
\end{array}
$$

+ x(Glc) + (y − 2) (Man)

The $Man_3GlcNAc_2$-Asn-protein core structure subsequently appears in the Golgi apparatus for the elongation reactions, i.e., addition of further GlcNAc residues and of galactose, fucose, and sialic acid residues; the elongation process is discussed in Sections 2.2 to 2.6.

The assembly of enveloped viruses is an ideal system for the study of membrane glycoprotein synthesis. For example, VSV has only a single glycoprotein (protein G) which is inserted into the membrane enveloping this virus. Although the viral RNA genome codes for the peptide moiety of protein G, the glycosylation apparatus is derived from the host cell. Infection of the host cell is followed by a complete shutdown of host protein synthesis, thereby simplifying the task of studying viral protein synthesis. The G protein is glycosylated within the host cell's endomembrane system, inserted into the plasma membrane of the host cell, and eventually incorporated into the mature virus particle when the virus buds through the plasma membrane of the host cell. When cells infected with VSV were pulsed with radioactive mannose, a precursor form of the G protein was detected (Tabas $et\ al.$, 1978; Hunt $et\ al.$, 1978; Robbins $et\ al.$, 1977). Glycopeptides were prepared from pronase digests of precursor G protein, and endo-β-N-acetylglucosaminidase and exoglycosidase digests were subjected to molecular-weight analysis by gel filtration. It was concluded that precursor G protein contained oligosaccharides with more than three mannose residues per mole. Since mature G protein contains only N-acetyllactosamine-type oligosaccharides, it was concluded that the G protein oligosaccharides were derived from an oligomannoside structure by removal of mannose residues.

The generality of this processing mechanism has not been established, although it is interesting that Tabas *et al.* (1978) have reported a similar process in the assembly of IgG by a mouse myeloma cell line. There is evidence from work on α_1-antitrypsin storage in human liver (Hercz *et al.*, 1978) that the major intracellular location of oligosaccharide processing is within the rough endoplasmic reticulum; it is now also believed that a few mannose residues are removed in the Golgi apparatus (see Section 2.2). It is not known why some oligosaccharides remain in the oligomannoside form while others are processed down to the $Man_3GlcNAc_2Asn$ core and subsequently elongated into *N*-acetyllactosamine-type oligosaccharides.

2.2. N-Acetylglucosaminyltransferases and Control of Elongation

Johnston *et al.* (1966) first reported the presence in goat colostrum of an enzyme catalyzing the following reaction:

$$UDP\text{-}\alpha\text{-}GlcNAc + Man\text{-}\alpha\text{-}R \rightarrow GlcNAc\text{-}\beta\text{-}Man\text{-}\alpha\text{-}R + UDP$$

The product was shown to contain a β-linked GlcNAc residue by digestion with β-*N*-acetylglucosaminidase, but the linkage to mannose was not established (Johnston *et al.*, 1966, 1973). This linkage was considered likely because all effective acceptors contained a terminal α-linked mannose, e.g., α_1-acid glycoprotein or fetuin pretreated with sialidase, β-galactosidase, and β-*N*-acetylglucosaminidase. Among the many ineffective acceptors tested were native α_1-acid glycoprotein, sialidase-treated α_1-acid glycoprotein, sialidase-, β-galactosidase-treated α_1-acid glycoprotein, salivary mucins either before or after glycosidase treatment, and a large number of low-molecular-weight compounds; the latter contained a variety of free sugars and their glycosides, including methyl α- and β-mannopyranosides, L-amino acids, and $\alpha1,2$-linked α-mannotriose. It is interesting that native ovalbumin and ribonuclease B were also acceptors (Johnston *et al.*, 1973), but activity was 20% of the level obtained with glycosidase-treated α_1-acid glycoprotein. A similar enzyme activity has been found in many other mammalian tissues such as rat liver, intestine, kidney, lung, and mammary gland (Johnston *et al.*, 1973), pork liver and serum (Hudgin and Schachter, 1971c), guinea pig liver (Bosmann, 1970), rat serum (Mookerjea *et al.*, 1971), and human serum (Mookerjea *et al.*, 1972). The liver enzyme is strongly bound to membrane but can be solubilized with Triton X-100 or acetone treatment (Bosmann, 1970; Hudgin and Schachter, 1971c). The colostrum enzyme has been purified 200-fold (Johnston *et al.*, 1973). The enzyme is inhibited by EDTA and requires divalent cation for activity (Mn^{2+} > Mg^{2+} > Co^{2+}) (Hudgin and Schachter, 1971c).

Recent work with wild-type and lectin-resistant Chinese hamster ovary cells (Narasimhan et al., 1977; Stanley et al., 1975a,b; see also Chapter 4 for a detailed discussion of such mutants) has shown that the enzyme activity determined by the use of high-molecular-weight acceptors such as glycosidase-treated α_1-acid glycoprotein is due to at least two N-acetylglucosaminyltransferases designated as GlcNAc-transferases I and II. A line of Chinese hamster ovary cells (Pha^{R_I}) selected for resistance to a phytohemagglutinin from Phaseolus vulgaris has been shown to lack GlcNAc-transferase I but to contain normal levels of GlcNAc-transferase II; wild-type cells contain both transferases. The substrate specificities of these two enzymes were determined by using as acceptors low-molecular-weight glycopeptides prepared from human IgG (Narasimhan et al., 1977). It was shown that GlcNAc-transferase I transfers GlcNAc from UDP-GlcNAc to Man-α1,3-Man-β1,4-GlcNAc and to Man-α1,3-[Man-α1,6]Man-β1,4-GlcNAc-R where R is either —H or -GlcNAc-Asn (see Figure 1). GlcNAc-transferase II, however, had a specific requirement for the branched structure GlcNAc-β1,2-Man-α1,3-[Man-α1,6]Man-β1,4-GlcNAc-R (R has the same designation as above). It appears therefore that GlcNAc-transferase II cannot act until GlcNAc-transferase I has acted (Figure 1). In fact, as will be discussed further in Sections 2.3, 2.4, and 2.5, GlcNAc-transferase I controls the entire elongation process. If GlcNAc-transferase I is absent, as is the case in the Pha^{R_I} mutant, there is no addition of further GlcNAc residues nor of fucose, galactose, and sialic acid residues to the Man$_3$GlcNAc$_2$Asn core. A pathway encompassing these observations is shown in Figure 1.

The resolution of GlcNAc-transferases I and II depended on two factors: the use of somatic cell genetics and of well-defined low-molecular-weight acceptors. These approaches may prove useful in differentiating other glycosyltransferases. It should be stressed that the use of high-molecular-weight molecules (containing a variety of oligosaccharides) as acceptors may lead to the measurement of more than one glycosyltransferase; the differentiation of these transferases from one another requires the application of approaches such as have been described above, as well as others to be discussed in subsequent sections, e.g., separation of products, chemical analysis of products, and purification of glycosyltransferases to states of homogeneity.

Both GlcNAc-transferases I and II catalyze the synthesis of GlcNAc-β1,2-Man linkages, as determined by analysis on concanavalin A-Sepharose columns (Narasimhan et al., 1977). Preliminary data (J. Wilson, S. Narasimhan, and H. Schachter, unpublished data) indicate that GlcNAc-transferase I attaches GlcNAc preferentially to the Man-α1,3-

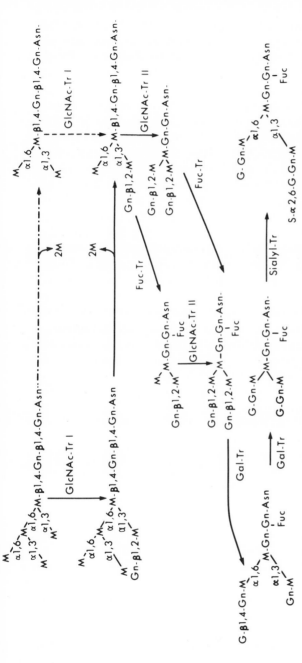

Figure 1. The elongation reactions required for the assembly of N-acetyllactosamine oligosaccharides. The scheme begins with the $Man_5GlcNAc_2Asn$-peptide structure which is believed to be an intermediate in the processing of protein-bound oligosaccharide. Dolichol pyrophosphate oligosaccharide is postulated to transfer a glucose-containing, mannose-rich oligosaccharide to protein within the rough endoplasmic reticulum. Oligosaccharide processing then occurs in the rough endoplasmic reticulum resulting in the removal of all the glucose residues and all but five mannose residues. In α_1-antitrypsin deficiency (Hercz et al., 1978), partially processed α_1-antitrypsin accumulates in the rough endoplasmic reticulum of the liver; analysis of this material indicates that removal of all the glucose and most of the mannose residues occurs within the rough endoplasmic reticulum. The $Man_5GlcNAc_2Asn$-peptide then moves to the Golgi apparatus where elongation takes place. There are two theoretically possible paths for the further processing of $Man_5GlcNAc_2Asn$-peptide. GlcNAc-transferase I has been highly purified from bovine colostrum (Harpaz and Schachter, 1979a) and has been shown to act on both $Man_5GlcNAc_2$-Asn and $Man_5GlcNAc_2Asn$. The enzyme has a lower K_m for $Man_5GlcNAc_2Asn$ indicating that the physiological path may be the one shown in continuous arrows. Direct evidence in support of this path has recently been obtained by Kornfeld's group [see Chapter 1 and Harpaz and Schachter (1979b)]. The glycosyltransferases responsible for the elongation process are discussed in detail in the text. Abbreviations: M, D-mannose; Gn, N-acetyl-D-glucosamine; G, D-galactose; S, sialic acid; Tr, transferase.

arm of Man-α1,3-[Man-α1,6]Man-β1,4-GlcNAc-R, as depicted in Figure 1. However, as reported by Narasimhan *et al.* (1977), there is some evidence that GlcNAc-transferase I may incorporate GlcNAc into Man-α1,6-terminal compounds if Man-α1,3-terminal compounds are not available. GlcNAc-transferase II attaches GlcNAc to the Man-α1,6 terminus of the product of GlcNAc-transferase I (Figure 1); it has not yet been established that the enzyme is specific for the Man-α1,6 terminus.

The biosynthesis of VSV envelope glycoprotein G has been studied in wild-type and Pha^{RI} Chinese hamster ovary cells (Hunt *et al.*, 1977; Tabas *et al.*, 1978). In wild-type cells, protein G is synthesized with *N*-acetyllactosamine-type oligosaccharides containing three mannose residues per oligosaccharide and sialyl-*N*-acetyllactosamine arms attached to the core structure. As might be expected in the absence of GlcNAc-transferase I, protein G produced in the Pha^{RI} cells lacks the sialyl-*N*-acetyllactosamine arms; surprisingly, however, protein G from the lectin-resistant cells also contains more mannose than expected (about five mannose residues per oligosaccharide). These findings indicate that oligosaccharide processing (see previous section) is incomplete in the absence of GlcNAc-transferase I. A likely explanation is that GlcNAc-transferase I may act on Man$_5$GlcNAc$_2$Asn rather than on Man$_3$GlcNAc$_2$Asn (Figure 1); the removal of the last two mannose residues is believed to occur in the Golgi apparatus after prior incorporation of GlcNAc by GlcNAc-transferase I. Thus, GlcNAc-transferase I has been shown to transfer GlcNAc to an ovalbumin glycopeptide with the following structure (Narasimhan *et al.*, 1977):

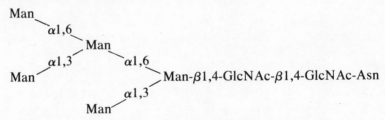

Also, Tai *et al.* (1977) have shown the presence in ovalbumin of the following structure:

```
Man
   \
    α1,6
        \
         Man
   α1,3 /    \
       /      α1,6
   Man            \
                   Man-β1,4-GlcNAc-β1,4-GlcNAc-Asn
             α1,3  |
                   |
   GlcNAc-β1,2-Man      GlcNAc
```

Direct evidence for this pathway has recently been obtained (Chapter 1; Harpaz and Schachter, 1979*b*).

2.3. Fucosyltransferases

When fucose occurs in N-glycosidically linked oligosaccharides, it is found only on N-acetyllactosamine structures, not on oligomannoside structures, and, almost invariably, linkage is to the most internal GlcNAc residue, i.e., the residue attached to asparagine (see Chapter 1). There have also been reports of fucose attached to galactose (Tsay *et al.*, 1976) and GlcNAc (Fournet *et al.*, 1979) residues more distally located on N-acetyllactosamine-type structures. The fucose–galactose linkages reported by Tsay *et al.* (1976) in cases of fucosidosis are probably anomalous products of the blood group H gene-dependent fucosyltransferase (see Section 3.3); this enzyme normally acts on O-glycosidically linked oligosaccharides but can also act on β-galactosyl-terminal N-glycosidically linked oligosaccharides (Munro and Schachter, 1973).

Jabbal and Schachter (1971) reported a fucosyltransferase in pork liver which transferred fucose from GDP-fucose to sialidase-, β-galactosidase-treated α_1-acid glycoprotein; this enzyme was termed a GDP-fucose: β-N-acetylglucosaminide fucosyltransferase since the acceptor molecule had a terminal β-GlcNAc residue. The linkage of fucose to acceptor was not established. The fucosyltransferase was subsequently shown to be highly enriched in the Golgi apparatus of rat and pork liver (Munro *et al.*, 1975) and was also shown to be present in human serum (Munro and Schachter, 1973; Chou *et al.*, 1977; Kessel *et al.*, 1977; Khilanani *et al.*, 1978) and human liver tumors (Kessel *et al.*, 1977).

Wilson *et al.* (1976) have shown that rat liver GDP-fucose: β-N-acetylglucosaminide fucosyltransferase catalyzes the following reaction:

GDP-β-L-fucose + GlcNAc-β1,2-Man-α1,3-

[Man-α1,6-]Man-β1,4-GlcNAc-β1,4-GlcNAc-Asn

→ GlcNAc-β1,2-Man-α1,3-[Man-α1,6-]Man-β1,4-

GlcNAc-β1,4-[Fuc-α-]GlcNAc-Asn + GDP

The attachment of fucose to the most internal GlcNAc residue was established by cleaving the product with an endo-β-N-acetylglucosaminidase. Attachment of fucose was by an α-linkage since fucose was released by α-fucosidase; however, it was recently shown that fucose is attached to C-6 of GlcNAc (G. Longmore and H. Schachter, unpublished data).

The fucosyltransferase can act on the products of either GlcNAc-transferase I or II (see Figure 1) but cannot act until at least one GlcNAc residue has been incorporated into the Man$_3$GlcNAc$_2$Asn core structure. Thus GlcNAc-transferase I controls the addition of fucose into the core.

This explains why the Pha[Rl] lectin-resistant mutant (which lacks GlcNAc-transferase I) cannot incorporate fucose into its N-glycosidically linked oligosaccharides. It is also evident why oligomannoside structures never contain fucose residues. Although the $Man_3GlcNAc_2Asn$ core is assembled in the rough endoplasmic reticulum at an early stage in the synthetic process, the addition of fucose to the core must await the action of GlcNAc-transferase I, a Golgi-localized enzyme, and is therefore also a late Golgi-localized process.

The specificity of the fucosyltransferase therefore explains some of the structural features of N-glycosidically linked oligosaccharides. It is, however, as yet unclear why some N-acetyllactosamine structures contain the fucose-GlcNAc-Asn structure (e.g., human IgG) while others do not (e.g., human α_1-acid glycoprotein); a possible explanation is the presence or absence of the appropriate fucosyltransferase.

It was reported several years ago (Shen et al., 1968) that human milk contained a fucosyltransferase catalyzing the reaction:

GDP-β-L-Fuc + Gal-β1,4-GlcNAc-R

$$\rightarrow \text{Gal-}\beta1,4\text{-(Fuc-}\alpha1,3)\text{GlcNAc-R} + \text{GDP}$$

The human milk α3-fucosyltransferase has recently been purified (Beyer et al., 1979) and the purified enzyme has been shown to incorporate fucose into sialidase-treated transferrin, presumably by formation of an α3 linkage to N-acetylglucosamine (Prieels et al., 1978; Paulson et al., 1978). The enzyme has an absolute requirement for the Gal-β1,4-GlcNAc terminus. Jabbal and Schachter (1971) reported a fucosyltransferase in pork liver which also showed a requirement for the Gal-β1,4-GlcNAc terminus, but the linkage synthesized by this enzyme was not established. The discovery of a Fuc-β1,3-GlcNAc linkage in α_1-acid glycoprotein (Fournet et al., 1979) suggests that liver probably has an α3-fucosyltransferase similar to the milk enzyme purified by Beyer et al. (1979) and that this enzyme is involved in the elongation of N-glycosyl oligosaccharides.

2.4. Galactosyltransferases

Galactose is almost invariably located at the same position in all N-glycosidically linked oligosaccharides, namely, as part of the sequence Gal-β-GlcNAc-β-Man-α-Man-β-GlcNAc (see Chapter 1). Galactose may be the terminal nonreducing sugar or may be penultimate to a sialic acid residue. Galactose is usually linked to N-acetylglucosamine by a β1,4 linkage but Gal-β1,6-GlcNAc sequences have also been described. Many tissues have been shown to contain an enzyme or enzymes capable of catalyzing the following reaction:

$$\text{UDP-}\alpha\text{-D-Gal} + \text{GlcNAc-}\beta\text{-R} \rightarrow \text{Gal-}\beta\text{-GlcNAc-}\beta\text{-R} + \text{UDP}$$

R can be —H or oligosaccharide, glycopeptide, or glycoprotein. In some studies, the product formed with N-acetylglucosamine has been characterized as Gal-β1,4-GlcNAc (N-acetyllactosamine); the synthesis of either Gal-β1,3-GlcNAc or Gal-β1,6-GlcNAc from UDP-Gal has not yet been reported. When large-molecular-weight acceptors such as sialidase-, β-galactosidase-treated α_1-acid glycoprotein have been used as acceptors, the product formed has not been characterized and, as mentioned previously, such assays may be measuring several glycosyltransferases, i.e., not only the 4'-galactosyltransferase but also 3'- and 6'-galactosyltransferases.

UDP-Gal: β-N-acetylglucosaminide galactosyltransferase activities are found inside most cells as membrane-bound enzymes (localized to the Golgi apparatus) and in a soluble form in milk, amniotic fluid, cerebrospinal fluid, saliva, urine, colostrum, and serum (references in Schachter and Rodén, 1973). The milk enzyme is a 4'-galactosyltransferase and is equivalent to the A protein of lactose synthetase (Brew *et al.*, 1968); the B protein is α-lactalbumin, which has no known enzymatic activity. Lactose synthetase A protein has a low affinity for glucose, although it can synthesize lactose at very high glucose concentrations; in the presence of α-lactalbumin, the affinity of the A protein for glucose is greatly enhanced (Ebner, 1973). The milk A protein in the absence of α-lactalbumin can transfer Gal from UDP-Gal to GlcNAc to make N-acetyllactosamine and can also transfer Gal to larger β-N-acetylglucosaminides. It is interesting that the UDP-Gal: β-N-acetylglucosaminide galactosyltransferases in liver, serum, and other tissues can make lactose in the presence of exogenous α-lactalbumin (Hudgin and Schachter, 1971b; Fitzgerald *et al.*, 1971), but the identity of these tissue enzymes with the milk A protein has not been definitively established; the physiological role of the galactosyltransferases in tissues other than lactating mammary gland is almost certainly the elongation of N-glycosidically linked oligosaccharides (Figure 1). It is evident that no galactose can be incorporated until at least one GlcNAc is incorporated into the core, i.e., GlcNAc-transferase I controls the addition of Gal residues.

There is no enzymological evidence for the attachment of galactose preferentially to the Man-α1,6-Man- branch (Figure 1); however, the structure Gal-β1,4-GlcNAc-β1,2-Man-α1,6-[GlcNAc-β1,2-Man-α1,3-] Man-β1,4-GlcNAc-β1,4-[Fuc-α1,6-]GlcNAc-Asn has been isolated from human and bovine IgG (Paulson *et al.*, 1977b; Tai *et al.*, 1955).

Bovine and human milk, bovine colostrum, and rat serum enzymes have been purified to homogeneity either by classical methods (Fitzgerald *et al.*, 1970*a*) or by affinity chromatography using either α-lactalbumin–Sepharose, UDP-hexanolamine–Sepharose or *N*-acetylglucosamine–Sepharose columns (Trayer and Hill, 1971; Andrews, 1970; Barker *et al.*, 1972; Khatra *et al.*, 1974; Geren *et al.*, 1976; Powell and Brew, 1974; Fraser and Mookerjea, 1976). The enzyme from swine mesentary lymph nodes has been purified on a column of *p*-aminophenyl-β-GlcNAc–Sepharose (Rao *et al.*, 1976). Purified milk galactosyltransferase has been shown to transfer Gal to GlcNAc and to oligosaccharides and glycoproteins with a terminal nonreducing β-GlcNAc residue, indicating that these activities are being catalyzed by a single enzyme.

The purified bovine milk galactosyltransferase has a specific activity of about 5.6 μmole/min per mg protein at 23°C with a purification factor of over 6000-fold (Geren *et al.*, 1976). The enzyme contains two major catalytically active forms with molecular weights estimated at 55,000–59,000 and 42,000–44,000 (Magee *et al.*, 1974). Both forms are glycoproteins and contain about 10–15% carbohydrate (Trayer and Hill, 1971; Lehman *et al.*, 1975); in fact, native galactosyltransferase catalyzed incorporation of galactose into sialidase-, β-galactosidase-treated galactosyltransferase (Geren *et al.*, 1977). The two forms of bovine milk galactosyltransferase resemble each other with regard to K_m for substrate, heat inactivation, and inhibition by sulfhydryl reagents (Magee *et al.*, 1974). There is evidence to suggest that the smaller form is derived from the larger form by proteolysis due to a trypsinlike milk protease which may be identical to plasmin (Magee *et al.*, 1976).

Powell and Brew (1974) found that bovine, ovine, or porcine colostrum was a richer source of galactosyltransferase than milk. They isolated from bovine colostrum a single glycosyltransferase component of molecular weight 51,000; since colostrum contains powerful protease inhibitors, they suggest that their enzyme corresponds to the largest milk component. Smith and Brew (1977) used 1% Triton X-100 to solubilize the membrane-bound galactosyltransferase of lactating sheep mammary gland Golgi membranes and purified the solubilized enzyme by combination of gel filtration and affinity chromatography. The pure preparation showed two components, a major band of molecular weight 65,000–69,000 and a minor component of molecular weight 53,000–55,000. The major component is thus appreciably larger than the largest component isolated from bovine milk, suggesting that the soluble component is derived from the membrane-bound component by proteolysis. The soluble milk enzyme and the Triton-solubilized Golgi enzyme have similar kinetic properties. An enzyme similar to the mammary gland enzyme has also been purified from fat globule membranes obtained from either bovine colostrum or milk (Powell *et al.*, 1977).

The interaction of bovine milk galactosyltransferase and α-lactalbumin is a unique example of enzyme modification and has been the subject of many studies (Ebner, 1973; Bell *et al.*, 1976; Trayer and Hill, 1971; Powell and Brew, 1975, 1976*a,b*; Khatra *et al.*, 1974; Ivatt and Rosemeyer, 1972). The interaction requires Mn^{2+} and either UDP-Gal or an acceptor molecule such as glucose or *N*-acetylglucosamine; for example, the galactosyltransferase will adhere to α-lactalbumin–Sepharose columns in buffers containing Mn^{2+} and either glucose or *N*-acetylglucosamine and can be eluted from such columns by the addition of EDTA and the omission of glucose or *N*-acetylglucosamine. Detailed kinetic studies have been carried out on the mechanism of action of galactosyltransferase in the absence of α-lactalbumin (Ebner, 1973; Khatra *et al.*, 1974; Geren *et al.*, 1975*a,b*; Tsopanakis and Herries, 1978). It appears that Mn^{2+} is always the first ligand to react with the enzyme followed by addition of either UDP-galactose or Mn^{2+}–UDP-galactose to the enzyme–Mn^{2+} complex. The acceptor (*N*-acetylglucosamine or β-*N*-acetylglucosaminide) then binds to form an enzyme–Mn^{2+}–UDP-Gal–acceptor complex which dissociates to release UDP and product. The addition of α-lactalbumin complicates the kinetics and serves to lower the K_m for glucose about 1000-fold (Ebner, 1973). Physiologically, lactose synthesis occurs only in the mammary gland at parturition and during lactation, when α-lactalbumin is formed presumably in response to a hormonal signal. Kinetic analysis (Bell *et al.*, 1976) has indicated that α-lactalbumin can bind either to an enzyme–Mn^{2+}–UDP-Gal complex or to an enzyme–Mn^{2+}–acceptor complex; these workers suggest that acceptor can add to enzyme–Mn^{2+} complex either before or after addition of UDP-Gal (a random equilibrium mechanism), in contrast to the ordered addition of UDP-Gal and acceptor proposed by other workers (see above). Kinetic studies on the mechanism of lactose synthetase have clarified the behavior of galactosyltransferase (the lactose synthetase A protein) on the various affinity adsorbents that have been used in its purification (Bell *et al.*, 1976; Barker *et al.*, 1972). Thus, for example, the transferase will bind weakly to α-lactalbumin–Sepharose in the presence of Mn^{2+} alone but, as mentioned above, binding is greatly enhanced by the presence of either glucose or *N*-acetylglucosamine. Also, the transferase requires Mn^{2+} for binding to UDP-hexanolamine–Sepharose but this binding is not affected by the presence or absence of glucose or *N*-acetylglucosamine. Finally, the transferase will bind to *N*-acetylglucosamine–Sepharose only in the presence of Mn^{2+} and UDP (or a suitable nucleotide analogue). The highly purified enzyme is unstable in the absence of substrates but can be stored for months at 4°C in the presence of chloroform and 5 mM *N*-acetylglucosamine.

2.5. Sialyltransferases

In N-glycosidically linked oligosaccharides, sialic acid is always found at the nonreducing terminus in an α-linkage to a subterminal galactosyl residue. The sialyl–galactosyl linkage is usually 2,3 or 2,6 (Montreuil, 1975) but reports of 2,4 (Sato *et al.*, 1967) and 2,2 (Isemura and Schmid, 1971) linkages have appeared.

Sialyltransferases catalyze the following general reaction:

$$\text{CMP-sialic acid} + \text{Gal-}\beta\text{-R} \rightarrow \text{Sialyl-}\alpha\text{-Gal-}\beta\text{-R} + \text{CMP}$$

Many tissues have been shown to contain membrane-bound enzymes capable of transferring sialic acid to various low-molecular-weight acceptors such as lactose and *N*-acetyllactosamine and to high-molecular-weight acceptors such as sialidase-treated α_1-acid glycoprotein or fetuin (Schachter and Rodén, 1973, for references); soluble forms of these enzymes have been described in goat, bovine, and human colostrum (Bartholomew *et al.*, 1973) and in pork and human serum (Hudgin and Schachter, 1971a; Kim *et al.*, 1971; Mookerjea *et al.*, 1972).

Crude goat, bovine, and human colostrum (Bartholomew *et al.*, 1973; Paulson *et al.*, 1977b) and rat, pork, bovine, and human liver (Hudgin and Schachter, 1972) have been shown to catalyze the synthesis of both sialyl-α2,3-lactose and sialyl-α2,6-lactose; further, the differential appearance of these two activities in the developing embryonic rat liver indicated that two separate enzymes were involved. It was therefore assumed that the action of crude sialyltransferases on high-molecular-weight acceptors such as sialidase-treated α_1-acid glycoprotein would result in both sialyl-α2,3-galactosyl and sialyl-α2,6-galactosyl linkages. A method for analyzing the product of sialyltransferase action on protein acceptors was recently developed (Stoffyn *et al.*, 1977; Van den Eijnden *et al.*, 1977). The terminal galactosyl residue of the acceptor was made radioactive by the use of galactose oxidase and tritiated sodium borohydride, and nonradioactive CMP-sialic acid was used in the sialyltransferase reaction; the linkage of sialic acid to galactose could then be determined on the small amount of product formed in the reaction. Rat liver microsomes were used as an enzyme source to transfer nonradioactive sialic acid to the tritiated sialic acid-free human α_1-acid glycoprotein, the product was isolated, and pronase glycopeptides were prepared; permethylation followed by hydrolysis yielded 2,3,4-trimethylgalactose as the only radioactive trimethylgalactose (Van den Eijnden *et al.*, 1977). Thus rat liver sialyltransferase transfers sialic acid only to the C-6 position of the terminal galactose of sialidase-treated human α_1-acid glycoprotein. Transferases capable of linking sialic acid to the terminal galactosyl residues of N-glycosidically linked oligosaccharides in linkages other than 2,6 have not yet been demonstrated. As will be discussed in Section 3.2, a porcine submaxillary gland α2,3-sialyltransferase has been purified that can transfer sialic acid in 2,3 linkage to the terminal β-

galactosyl of lactose, Gal-β1,3-GalNAc, and sialidase-treated porcine submaxillary mucin, but has little, if any, activity towards sialidase-treated human α_1-acid glycoprotein (Sadler *et al.*, 1979). It has also been reported that pork liver microsomes transfer sialic acid in α2,3 linkage to the terminal β-galactosyl residue of the Gal-β1,3-GalNAc disaccharides which are present in small amounts in sialidase-treated ovine submaxillary mucin (Van den Eijnden *et al.*, 1979). Thus, it is proposed that the enzyme in colostrum and liver which makes sialyl-α2,3-lactose may, in fact, be the enzyme which makes the sialyl-α2,3-galactosyl linkage in O-glycosidically linked oligosaccharides (Section 3.2) and is not involved in sialic acid incorporation into N-glycosidically linked oligosaccharides.

The only well-characterized sialyltransferase acting on N-glycosidically linked oligosaccharides is CMP-sialic acid:β-D-galactoside α2,6-sialyltransferase. This enzyme has recently been purified 440,000-fold to homogeneity from bovine colostrum (Paulson *et al.*, 1977*a*). Two homogeneous sialyltransferase preparations were obtained with apparent molecular weights of 56,000 and 43,000. Both forms had similar specific activities towards sialidase-treated α_1-acid glycoprotein of 26–28 μmole/min per mg of enzyme. The highly purified enzymes were readily denatured and were stored at a high protein concentration, pH 5.2–5.5, in plastic containers, at low temperature ($-20°C$) and in the presence of 0.2 mM CDP or 50% glycerol.

The dramatic purification achieved was due primarily to the use of CDP–agarose. This adsorbent was highly specific for the sialyltransferase and did not adsorb colostrum galactosyltransferase; conversely, UDP-hexanolamine–agarose, which specifically adsorbs the galactosyltransferase (see Section 2.4), did not adsorb the sialyltransferase. Affinity chromatography is proving to be a major advance in the purification of glycosyltransferases and will undoubtedly be used extensively in future work on these enzymes. Conditions for optimum effectiveness of an affinity column will have to be worked out individually for every enzyme. For example, the colostrum sialyltransferase is most stable below pH 6.0, but nonspecific protein adsorption to CDP-agarose is least above pH 6.5; the pH chosen for the initial adsorption (pH 6.5) was therefore a compromise. Adsorption to the second CDP-agarose column at pH 5.3 resulted in nonspecific adsorption to the column but purification was achieved by specific elution with CDP rather than with high concentrations of NaCl.

Bartholomew *et al.* (1973) purified the α2,6-sialyltransferase from goat and bovine colostrum 50- to 120-fold using gel filtration and ion-exchange chromatography. The enzyme had a pH optimum between 6.4 and 7.2, and showed no requirement for divalent cation. Both CMP-*N*-acetylneuraminic acid and CMP-*N*-glycolylneuraminic acid were effective sialic acid donors. A variety of glycoproteins with terminal β-gal-actosyl residues were effective acceptors, e.g., sialidase-treated α_1-acid glycoprotein, fetuin, transferrin, and ceruloplasmin. A large number of low-molecular-weight compounds were tested as acceptors; Gal-β1,4-GlcNAc was an excellent acceptor although not as effective as sialidase-treated α_1-acid glycoprotein, whereas Gal-β1,3-GlcNAc, Gal-β1,6-GlcNAc, Gal-β1,4-Glc, and other β-galactosides were poor substrates. It is interesting that the partially purified sialyltransferase produced pri-marily sialyl-α2,6-N-acetyllactosamine and little or no sialyl-α-2,3-N-ace-tyllactosamine, but produced both sialyl-α2,3-lactose and sialyl-α2,6-lac-tose (in a ratio of approximately 1:9). Bartholomew et al. (1973) could offer no explanation for this finding; however, it is likely that their preparation was contaminated with an α2,3-sialyltransferase which tran-sers sialic acid in α2,3 linkage to terminal β-galactosyl residues of lactose, Gal-β1,3-GalNAc, and O-glycosidically linked oligosaccharides but is sluggish towards N-acetyllactosamine and sialidase-treated α_1-acid gly-coprotein (see Section 3.2). Pork liver and serum sialyltransferase also showed a preference for Gal-β1,4-GlcNAc over lactose and the β1,3 and β1,6 isomers of Gal-GlcNAc (Hudgin and Schachter, 1971a).

Paulson et al. (1977b) have carried out detailed kinetic studies on their highly purified bovine colostrum α2,6-sialyltransferases; both forms of the enzyme have similar kinetic properties, suggesting that the low-molecular-weight form might be a proteolytic degradation product. The pure enzyme was more specific for Gal-β1,4-GlcNAc than the partially purified enzyme studied by Bartholomew et al. (1973), showing minimal activity towards Gal-β1,3-GlcNAc and Gal-β1,6-GlcNAc; product was, however, formed with the latter at high substrate concentrations (50–60 mM). Lactose had the same V_{max} as N-acetyllactosamine but a 30-fold higher K_m. The pure enzyme made only the 2,6 linkage with lactose and the three isomers of Gal-GlcNAc. This was in contrast to the partially purified colostrum enzyme which made a small amount of sialyl-α2,3-lactose in addition to sialyl-α2,6-lactose. The data on the purified enzyme therefore support the hypothesis that the crude enzyme preparation was contaminated with an α2,3-sialyltransferase.

Several glycoproteins known to contain Gal-β1,4-GlcNAc nonre-ducing termini were highly effective acceptors for the pure sialyltransfer-ase (Paulson et al., 1977b), e.g., sialidase-treated α_1-acid glycoprotein, fetuin, IgG, and IgM; native and sialidase-treated porcine and ovine submaxillary mucins as well as antifreeze glycoprotein were ineffective as acceptors, indicating that the purified colostrum enzyme was specific for N-glycosidically linked oligosaccharides and did not act on O-glycos-idically linked oligosaccharides.

As indicated in Figure 1, sialic acid incorporation depends on the prior action of galactosyltransferase which is in turn dependent on the action of GlcNAc-transferase I. There is no enzymological evidence to support the preferential incorporation of sialic acid into the Man-α1,3-Man- branch (Figure 1), but the asymmetrical structure shown as the final product in the scheme (Figure 1) has been isolated from various human immunoglobulins (Baenziger and Kornfeld, 1974; see also Chapter 1).

Kinetic analysis of the pure α2,6-sialyltransferase has indicated an equilibrium random order mechanism for the addition of CMP-sialic acid and acceptor to the enzyme (Paulson et al., 1977b). The K_m for CMP-sialic acid was 0.17 mM with sialidase-treated α_1-acid glycoprotein and 0.09 mM with N-acetyllactosamine; the K_m for glycoprotein acceptor was 1.64 mM and for N-acetyllactosamine was 13.8 mM.

Paulson et al. (1977b) have verified the observation of Hudgin and Schachter (1972) that rat liver makes both the 2,3 and 2,6 isomers of sialyllactose. Using rat liver Golgi preparations, they have shown, in addition, that sialyl-α2,3-lactose production was favored at low lactose concentrations, that Gal-β1,3-GlcNAc resulted in the production of sialyl-α2,3-Gal-β1,3-GlcNAc predominantly, while Gal-β1,4-GlcNAc and Gal-β1,6-GlcNAc resulted in the production of the 2,6 isomer and not the 2,3 isomer. These findings support the hypothesis suggested above that the rat liver enzyme making sialyl-α2,3-lactose is responsible for incorporation of sialic acid in α2,3 linkage to the galactosyl residue of O-glycosidically linked Gal-β1,3-GalNAc groups and does not act on N-acetyllactosamine or on N-glycosidically linked oligosaccharides; this enzyme apparently has a lower K_m for lactose than the α2,6-sialyltransferase. Rat liver does not appear to have an enzyme capable of making the sialyl-α2,3-galactosyl linkage in N-glycosidically linked oligosaccharides; perhaps rat liver and plasma proteins lack this grouping. Since human α_1-acid glycoprotein has been reported to contain α2,3-linked sialic acid, human liver may prove to be a better source for an enzyme capable of linking sialic acid in α2,3 linkage to the terminal galactosyl residue of N-acetyllactosamine and sialidase-treated α_1-acid glycoprotein.

The available data therefore suggest the existence of at least four sialyltransferases. Two enzymes are required for attaching sialic acid in 2,3 and 2,6 linkages to a galactosyl residue of N-glycosidically linked oligosaccharides and another two enzymes are required for making the sialyl-α2,6-GalNAc and sialyl-α2,3-Gal groupings present in O-glycosidically linked oligosaccharides (see Section 3.2). There may be additional enzymes for the 2,2 and 2,4 linkages. Proving the existence of these various enzymes will depend on further enzyme purifications and on the further development of methods for the separation and characterization of isomeric products of both low and high molecular weights.

The data presently available on highly purified glycosyltransferases strongly support the "one linkage–one glycosyltransferase" hypothesis; many additional enzymes will have to be purified and their substrate specificities will have to be determined. This task is a worthy one, not merely to test the hypothesis, but because purified glycosyltransferases can be powerful tools in probing the oligosaccharide structures of purified glycoconjugates and of glycoconjugates on cell surfaces. For example, Paulson *et al.* (1977c) have used the highly purified bovine colostrum α2,6-sialyltransferase to probe the structure of the rabbit hepatocyte plasma membrane glycoprotein responsible for binding plasma glycoproteins with exposed galactosyl residues and clearing them from the circulation (see Chapter 6). The liver binding protein has been called a mammalian lectin. It is a sialoglycoprotein which, on removal of sialic acid, is defective in its ability to bind galactosyl residues; the asialolectin is defective because it binds its own exposed galactosyl residues. It regained over 95% of its sialic acid residues and 80% of its binding activity on treatment with pure sialyltransferase and CMP-sialic acid. This study confirmed the role of sialic acid in binding and showed that most of the galactose in the lectin was linked β1,4 to *N*-acetylglucosamine.

2.6. The Golgi Apparatus as the Major Subcellular Site of Elongation

The transfer of oligosaccharide from dolichol pyrophosphate oligosaccharide to protein appears to be localized primarily within the rough endoplasmic reticulum (see Chapter 2), as predicted several years previously by kinetic experiments using radioactive glucosamine as a precursor in glycoprotein biosynthesis studies (see Schachter and Rodén, 1973, for a review of this literature). Further, several of these studies had indicated that some radioactive *N*-acetylglucosamine was incorporated into peptide which was still nascently attached to the ribosome; Kiely *et al.* (1976) have recently confirmed these older findings and have shown, in addition, that radioactive mannose was also incorporated into nascent peptide. These studies did not prove that all incorporation of core takes place on nascent peptide, but recent work on the biosynthesis of viral envelope glycoprotein strongly suggests that nascent peptide does not leave the polyribosome complex until the assembly of core sugars is complete (Rothman and Lodish, 1977; Sefton, 1977); however, studies on α_1-acid glycoprotein biosynthesis by rat liver have shown that only a little *N*-acetylglucosamine incorporation occurs on nascent peptide (Jamieson, 1977), indicating that the extent of nascent peptide glycosylation may vary from one system to another. It is, however, quite clear that the initial glycosylation of peptide is a membranous process; for example, the glycosylation of viral envelope glycoprotein depends absolutely on the presence of endoplasmic reticulum and on the vectorial transfer of peptide across the membrane (Katz *et al.*, 1977; Rothman and Lodish, 1977; Toneguzzo and Ghosh, 1977, 1978).

When the oligosaccharide has been incorporated into protein, it is subjected to oligosaccharide processing (see Section 2.1) and $Man_5GlcNAc_2Asn$-protein probably arrives at the Golgi apparatus for the elongation reactions shown in Figures 1 and 2. Several lines of evidence have implicated the Golgi apparatus as the major site for elongation. Only a brief review of this literature will be presented here (see Schachter and Rodén, 1973; Schachter, 1974*a,b*, for a more detailed discussion).

The Golgi apparatus was first implicated as a major glycosylation site by autoradiographic studies on tissues taken from animals that had been injected with radioactive sugars such as tritiated glucose or galactose (Whur *et al.*, 1969). Most sugars are, however, readily converted to other hexoses and hexosamines and do not give clear-cut incorporation patterns. In contrast, radioactive fucose is not converted to other sugars, although it can be degraded by an oxidative pathway in some species (Nwokoro and Schachter, 1975*a,b*). When tritiated fucose is used as a precursor in glycoprotein biosynthesis studies, a simple and precise incorporation pattern is usually obtained. For example, Haddad *et al.* (1971) used autoradiography to study the incorporation of tritiated fucose into rat thyroid and found that 85% of silver grains were localized over the Golgi apparatus within 3–5 min after administration of the label to the intact animal; radioactivity subsequently moved to the apical vesicles and finally to the colloid. It is important to note that autoradiographic studies with tritiated fucose show minimal incorporation into rough endoplasmic reticulum at either early or late times after injection. Incorporation of fucose into plasma membrane is also minimal at 2–5 min after injection but silver grains appear over the plasma membrane eventually; tritiated fucose is, in fact, an excellent label for the plasma membrane, and this is especially striking in nonsecretory cells (Bennett *et al.*, 1974; see also Chapter 5). These studies therefore indicate that fucosylation of glycoproteins occurs primarily within the Golgi apparatus and not in the rough endoplasmic reticulum or plasma membrane; fucosylated glycoproteins can either be secreted or stored in secretory granules or incorporated into the plasma membrane (Sturgess *et al.*, 1978).

Similar kinetic experiments can be carried out by subjecting tissues to subcellular fractionation and subsequent biochemical analysis at various times after administration of radioactive sugar. Incorporation studies have also been carried out in which the fate of radioactive glucosamine, sialic acid, and galactose has been followed with time. These various studies all show that sialic acid, galactose, and fucose are incorporated predominantly in the Golgi apparatus and that some N-acetylglucosamine also becomes protein-bound in this organelle.

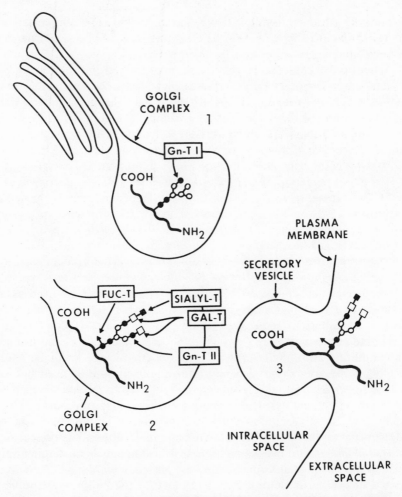

Figure 2. Terminal stages in the processing of a secretory glycoprotein in the Golgi apparatus. (1) The partially processed glycoprotein is transported to the Golgi apparatus for
elongation (see Figure 1). GlcNAc-transferase I (Gn-T I) adds a GlcNAc residue to the
oligosaccharide. (2) Two more mannose residues are removed and GlcNAc-transferase II
(Gn-T II), fucosyltransferase (FUC-T), galactosyltransferase (GAL-T), and sialyltransferase
(SIALYL-T) complete elongation. (3) A secretory vesicle transports the completed glycoprotein from the Golgi apparatus to the plasma membrane. Membrane fusion occurs,
followed by secretion of the glycoprotein. Abbreviations: Gn, N-acetyl-D-glucosamine; ●,
GlcNAc; ○, Man; ▲, Fuc; □, Gal; ■, sialic acid.

Studies on the subcellular localization of glycosyltransferases have
added strong confirmatory evidence for the role of the Golgi apparatus
in the elongation reactions shown in Figure 2. A typical subcellular
localization experiment on rat liver is shown in Table 1 (taken from
Munro *et al.*, 1975). It can be seen from Table 1 that five separate

glycoprotein glycosyltransferases were enriched 40- to 63-fold in rat liver Golgi preparations; the purity of the Golgi preparations can be assessed by the relatively low levels of other markers such as glucose-6-phosphatase (endoplasmic reticulum), glutamate dehydrogenase (mitochondria), 5'-nucleotidase (plasma membrane), acid phosphatase (lysosomes), RNA (rough endoplasmic reticulum), and DNA (nuclei). It is also evident that plasma membrane is not enriched in these transferases. Four of the five transferases in Table 1 are involved in the elongation reaction (Figure 2), namely, N-acetylglucosaminyltransferase acting on sialidase-, β-galactosidase-, β-N-acetylglucosaminidase-treated α_1-acid glycoprotein (this activity measures both GlcNAc-transferases I and II, see Section 2.2); fucosyltransferase acting on sialidase-,β-galactosidase-treated α_1-acid glycoprotein (Section 2.3); galactosyltransferase acting on GlcNAc (Section 2.4); and sialyltransferase acting on sialidase-treated α_1-acid glycoprotein (Section 2.5).

Analogous studies have been carried out in a variety of other tissues such as testis (Letts et al., 1974b) and thyroid (Chabaud et al., 1974), and similar results have been obtained. The Golgi apparatus is almost certainly the major site of the elongation reactions depicted in Figure 2. However, it cannot be ruled out that glycosylation reactions also occur to a limited extent in other organelles such as plasma membrane and mitochondria.

Glycosyltransferases are also found in various physiological fluids such as serum (Schachter, 1974a), cerebrospinal fluid (Den et al., 1970, 1975; Ko et al., 1973), and amniotic fluid (Nelson et al., 1974). The enzymes appear to be present in these fluids in a soluble form which does not require detergent for optimum activity. The source and function of these soluble transferases are not known, although there is evidence to suggest that damaged liver cells may release transferases into the serum (Mookerjea et al., 1972; Kim et al., 1972a,b). Although it has proved possible to purify transferases from fluids such as milk and serum, it has been suggested that these soluble enzymes may be proteolytic products of the native tissue enzymes (Section 2.4); further, in the case of one of the human blood group transferases (see Section 3.3), the enzyme purified from serum had a specific activity markedly lower than the enzyme isolated from detergent extracts of tissue (Schwyzer and Hill, 1977a).

3. GLYCOSYLTRANSFERASES INVOLVED IN SYNTHESIS OF O-GLYCOSIDICALLY LINKED OLIGOSACCHARIDES

The diverse structures contained within the O-glycosidically linked oligosaccharides have been described by Kornfeld and Kornfeld (Chapter

Table 1. Relative Specific Activities of Enzymes and Markers in Rat Liver Subcellular Fractions[a]

Subcellular fraction	Gal-T	Glc NAc-T	AGP (-SA, Gal) Fuc-T	AGP (-SA) Sialyl-T	ApoAla₁ (-SA) Sialyl-T	Glucose-6-phosphatase	Glutamate-dehydrogenase	5'-Nucleotidase	Acid phosphatase	RNA	DNA
Golgi	63	60	40	46	45	1.2	0.3	3.9	3.8	1.1	0.8
	(31)	(30)	(20)	(23)	(22)	(0.6)	(0.2)	(1.9)	(1.9)	(0.5)	(0.4)
Lysosomes–mitochondria	0.4	0.1	0.4	0.2	0.3	0.3	2.3	0.7	4.4	0.8	—
							(4.6)		(8.8)		
Nuclei	0.1	0.9	0.7	0.1	0.3	0.4	0.5	0.3	0.2	0.9	15
											(9.0)
Smooth microsomes	3.7	3.0	2.6	2.5	5.2	6.6	0.04	3.6	1.8	0.4	0.1
						(8.6)					
Rough microsomes	0.4	0.1	0.5	0.2	0.6	3.9	0.2	1.0	1.3	3.4	0.1
						(5.1)				(4.4)	
Plasma membrane	0.3	0.2	<0.3	0.3	0.2	0.9	—	7.5	1.1	—	—

[a] Rat liver homogenates were fractionated and the fractions were assayed. The data are expressed in relative specific activities, i.e., activity per milligram of protein in the fraction divided by activity per milligram of protein in the homogenate. Numbers in parentheses are the % yields relative to homogenate. The % yields of protein in the various fractions are: Golgi, 0.5; lysosomes–mitochondria, 2.0; nuclei, 0.6; smooth microsomes, 1.3; rough microsomes, 1.3. Abbreviations are: galactosyltransferase, Gal-T; N-acetylglucosaminyltransferase, GlcNAc-T; fucosyltransferase acting on sialidase-, α₁-acid glycoprotein, AGP(-SA, Gal)Fuc-T; sialyltransferase acting on sialidase-treated α₁-acid glycoprotein, AGP(-SA)Sialyl-T; sialyltransferase acting on sialidase-treated apolipoprotein-Ala₁, apoAla₁(-SA)Sialyl-T (after Munro et al., 1975).

1). These oligosaccharides can vary from the simple mono- and disaccharide structures present in ovine submaxillary mucin (OSM) to the complex structures found in human ovarian cyst mucins. The amino acid–carbohydrate linkages present in mammalian O-glycosidically linked oligosaccharides are usually serine(threonine)-N-acetyl-D-galactosamine, serine–xylose, and hydroxylysine–D-galactose. The former linkage is found in a variety of mucins derived from mucous glands and mucous membranes lining the gastrointestinal, respiratory, and genitourinary tracts. The serine–xylose linkage is found in proteoglycans (see Chapter 7). The hydroxylysine–galactose linkage is found in collagens and basement membranes. The glycosyltransferases involved in the biosynthesis of O-glycosidically linked oligosaccharides will be illustrated by considering two examples, i.e., the synthesis of ovine and porcine submaxillary gland mucins and of human blood group mucins. More detailed discussions on the synthesis of O-glycosidically linked oligosaccharides are available (Schachter and Rodén, 1973; Rodén and Horowitz, 1978).

3.1. Synthesis of Serine(Threonine)–N-Acetyl-D-Galactosamine Linkage

A large number of mucins have been isolated from mammalian mucous glands, mucous secretions, and mucous membranes. Invariably, these mucins have contained O-glycosidically linked oligosaccharides attached to a polypeptide core by a serine(threonine)–N-acetyl-D-galactosamine linkage. This linkage can be readily detected by its lability to mild alkaline hydrolysis. Oligosaccharides can be cleaved from the peptide core by mild alkaline hydrolysis in the presence of borohydride; the latter reagent converts the reducing end of the oligosaccharide to N-acetylgalactosaminitol and thereby prevents degradation of the oligosaccharide by alkali (the "peeling" reaction, Carlson et al., 1970).

Initiation of synthesis is catalyzed by a UDP-N-acetylgalactosamine:polypeptide α-N-acetylgalactosaminyltransferase which transfers N-acetylgalactosamine to polypeptide in O-glycosidic linkage:

UDP-α-GalNAc + Ser(Thr)-peptide
$$\rightarrow \text{GalNAc-}\alpha\text{-Ser(Thr)-peptide} + \text{UDP}$$

This mechanism of initiation is not as complex as that described for N-glycosidically linked oligosaccharides involving dolichol pyrophosphate oligosaccharide intermediates (Chapter 2). The N-acetylgalactosaminyltransferase was first described in ovine, porcine, and bovine submaxillary glands (McGuire and Roseman, 1967; Hagopian and Eylar, 1968a,b; 1969a,b) and has also been detected in HeLa cells, fibroblasts, and other tissues. The enzyme is membrane-bound but can be solubilized by non-ionic detergents.

The usual substrate used for assaying the enzyme is either ovine or bovine submaxillary mucin treated with sialidase and α-N-acetylgalactosaminidase to remove O-glycosidically linked sialyl-N-acetylgalactosamine disaccharides thereby exposing unsubstituted serine and threonine residues. The enzyme is very specific for these high-molecular-weight acceptors. Inactive as acceptors were 24 low-molecular-weight sugars, sugar derivatives, and amino acids, as well as over 30 glycoproteins and glycolipids including proteins and polypeptides rich in hydroxyamino acids such as albumin, dephosphorylated phosvitin, and poly-L-serine (McGuire and Roseman, 1967; Hagopian and Eylar, 1968b). An interesting exception is a basic protein isolated from bovine myelin which acts as an excellent acceptor for N-acetylgalactosamine (Hagopian et al., 1971); this protein is not a glycoprotein and presumably acts as an acceptor because a particular threonine residue is either in the correct amino acid sequence or three-dimensional environment. The nature of this environment is not as yet understood. Comparison of amino acid sequences near the serine(threonine)–N-acetylgalactosamine linkage regions of various O-glycosidically linked oligosaccharides has not identified a common sequence (Marshall, 1972, 1974; Goodwin and Watkins, 1974; Baenziger and Kornfeld, 1974). Hill et al. (1977a,b) have recently carried out extensive sequence studies on highly purified ovine submaxillary mucin and have also failed to detect any obvious homologies of amino acid sequence near the O-glycosidic linkage regions. Although the exact specificity requirements of this transferase are not known, it is of interest that the enzyme acts only on high-molecular-weight acceptors. Pronase digestion of glycosidase-treated mucin abolishes the acceptor activity of the preparation, although larger tryptic peptides have recently been shown to retain acceptor activity (Hill et al., 1977b).

The detergent-solubilized transferase has been partially purified from salivary gland extracts (McGuire and Roseman, 1967; Hagopian and Eylar, 1969b). A recent attempt to use affinity chromatography to purify the enzyme from porcine submaxillary gland led only to a 30-fold purification because the enzyme did not adhere to UDP-hexanolamine–Sepharose (Hill et al., 1977b). However, this partially purified enzyme preparation was relatively free of other mucin glycosyltransferases (these enzymes are discussed in Section 3.2).

Initiation of oligosaccharide is a critical step in the control of glycoprotein synthesis. Once the first sugar is incorporated, assembly of the oligosaccharide is determined by the substrate specificities of the remaining glycosyltransferases. It is therefore important to understand the factors controlling the specificity of the polypeptide α-N-acetylgalactosaminyltransferase. It will be of interest to study peptides smaller than the tryptic peptides of Hill *et al.* (1977*b*) and to determine the minimal requirements for acceptor activity. Since there do not appear to be any obvious sequence homologies, some sort of three-dimensional acceptor requirement must be considered; it has been suggested that the acceptor regions have little secondary structure and are thus accessible to the transferase (Hill *et al.*, 1977*b*).

3.2. Synthesis of Submaxillary Gland Mucins

Ovine and porcine submaxillary mucins have been thoroughly studied and the structures of the major oligosaccharides present in these mucins have been determined. Figure 3 summarizes the biosynthetic

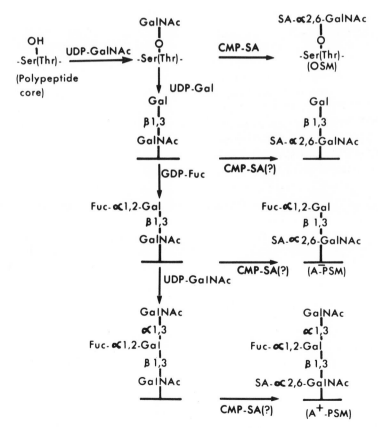

Figure 3. Biosynthesis of ovine submaxillary mucin (OSM) and porcine submaxillary mucin (PSM).

pathways for assembly of these oligosaccharides; at least five separate glycosyltransferases are involved.

The first enzyme is the α-N-acetylgalactosaminyltransferase discussed in the previous section. There is a branch point following the incorporation of N-acetylgalactosamine. If sialic acid is incorporated to form the sialyl-$\alpha2,6$-GalNAc disaccharide (Figure 3), further incorporation of carbohydrate ceases. Sialic acid incorporation is apparently a termination signal. Ovine submaxillary glands are relatively rich in the CMP-sialic acid:GalNAc-mucin $\alpha2,6$-sialyltransferase and therefore synthesize primarily the sialyl-$\alpha2,6$-GalNAc structure characteristic of ovine submaxillary mucin (McGuire, 1970; Schachter et al., 1971; Carlson et al., 1973c). Porcine submaxillary glands, however, have relatively more UDP-Gal:GalNAc-mucin $\beta1,3$-galactosyltransferase and therefore can incorporate a galactose residue before a sialic acid residue is incorporated. The galactosyltransferase cannot act on sialyl-$\alpha2,6$-GalNAc-mucin (Schachter et al., 1971). If a galactose residue is incorporated, further carbohydrate residues can be added, and the various structures present in porcine submaxillary mucin can be synthesized (Figure 3). It is probable (but not proved) that sialic acid incorporation at any stage of synthesis terminates further carbohydrate addition (Carlson et al., 1970). Thus, the relative contents of the $\alpha2,6$-sialyltransferase and $\beta1,3$-galactosyltransferase appear to control the type of oligosaccharide structure found in the final mucin product.

3.2.1. Mucin Sialyltransferases

CMP-sialic acid:GalNAc-mucin $\alpha2,6$-sialyltransferase (Figure 3) has been described in ovine, bovine, and porcine submaxillary glands (Carlson et al., 1973c). A suitable acceptor can be prepared by removal of sialic acid, either by mild acid hydrolysis or by sialidase, from ovine, bovine, or porcine submaxillary mucin. A partially purified (44-fold) enzyme from sheep glands was found to be free of endogenous mucin acceptors and showed no sialyltransferase activity towards either sialidase-treated α_1-acid glycoprotein or lactose (Carlson et al., 1973c). The mucin $\alpha2,6$-sialyltransferase is therefore a different enzyme from the CMP-sialic acid:β-D-galactoside $\alpha2,6$-sialyltransferase discussed in Section 2.5 and from the $\alpha2,3$-sialyltransferase acting on lactose (see Section 2.5 and discussion further on in the present section). The partially purified mucin $\alpha2,6$-sialyltransferase from sheep glands could use either CMP-N-acetylneuraminic acid or CMP-N-glycolylneuraminic acid as sialic acid donors (Carlson et al., 1973c), and extracts from bovine, porcine, and equine submaxillary glands were shown to transfer N-acetyl-, N-glycolyl-, N-acetyl-7(9)-O-acetyl, and N-acetyl-4-O-acetylneuraminic acids from their respective CMP-glycosides to endogenous acceptors with similar rates and kinetic constants (Schauer and Wember, 1973). Thus, the sialyltransferase does not appear to be specific for the acyl groups attached to neuraminic acid.

The mucin $\alpha 2,6$-sialyltransferase has been highly purified from detergent extracts of porcine submaxillary gland by affinity chromatography on CDP–agarose columns (Sadler et $al.$, 1979). The best substrate for this partially purified enzyme was found to be GalNAc-mucin prepared by removal of sialic acid from ovine submaxillary mucin or other salivary mucins. The enzyme does not act on sialidase-treated α_1-acid glycoprotein, lactose, or on various low-molecular-weight compounds with or without terminal GalNAc residues, as previously reported by Carlson et $al.$ (1973c). The disaccharide Gal-$\beta 1,3$-GalNAc had low acceptor activity whereas antifreeze glycoprotein, which contains many Gal-$\beta 1,3$-GalNAc units, was an excellent acceptor. Thus, the sialyltransferase probably acts both on GalNAc-mucin and on Gal-$\beta 1,3$-GalNAc-mucin, as depicted in Figure 3, although the linkage of sialic acid to Gal-$\beta 1,3$-GalNAc-mucin has not been established.

Although salivary mucins have not been reported to contain the structure sialyl-$\alpha 2,3$-Gal-$\beta 1,3$-(sialyl-$\alpha 2,6$-)GalNAc, the red cell membrane glycoprotein glycophorin contains many such tetrasaccharides (Marchesi et $al.$, 1976), and it is appropriate at this point to discuss the CMP-sialic acid:β-D-galactoside $\alpha 2,3$-sialyltransferase involved in the assembly of this O-glycosidically linked structure. Sadler et $al.$ (1979) purified two sialyltransferases from detergent extracts of porcine submaxillary glands by the use of CDP–agarose affinity chromatography. One enzyme was the mucin $\alpha 2,6$-sialyltransferase discussed above and the other was a CMP-sialic acid:β-D-galactoside $\alpha 2,3$-sialyltransferase. The latter enzyme transferred sialic acid in $\alpha 2,3$ linkage to the terminal β-D-galactosyl residues of lactose (Gal-$\beta 1,4$-Glc), Gal-$\beta 1,3$-GalNAc, antifreeze glycoprotein (containing many Gal-$\beta 1,3$-GalNAc disaccharides), and sialidase-treated porcine submaxillary mucin (which is also rich in Gal-$\beta 1,3$-GalNAc disaccharides) but had relatively little activity toward Gal-$\beta 1,4$-GlcNAc, Gal-$\beta 1,3$-GlcNAc, Gal-$\beta 1,6$-GlcNAc, and sialidase-treated α_1-acid glycoprotein (containing Gal-$\beta 1,4$-GlcNAc termini). The activity toward Gal-$\beta 1,3$-GalNAc and proteins containing these disaccharides strongly indicates the role of this enzyme in assembly of the sialyl-$\alpha 2,3$-galactose structure of O-glycosidically linked oligosaccharides. The absence of activity toward sialidase-treated α_1-acid glycoprotein shows that the enzyme is not responsible for synthesis of the sialyl-$\alpha 2,3$-galactose structure reported to be present in some N-glycosidically linked oligosaccharides (see Section 2.5). However, activity toward lactose, which contains a $\beta 1,4$ linkage, is a curious anomaly. The potent activity of the purified porcine submaxillary gland $\alpha 2,3$-sialyltransferase toward Gal-$\beta 1,3$-GalNAc and toward lactose (Gal-$\beta 1,4$-Glc) but not toward Gal-$\beta 1,4$-GlcNAc is an interesting example of substrate specificity which deserves further kinetic study.

The sialyl-$\alpha 2,3$ lactose synthesizing activity reported to be present in colostrum and liver (see Section 2.5) may be an enzyme which acts only on O-glycosidically linked oligosaccharides. Van den Eijnden *et al.* (1979) have indeed demonstrated a mucin $\alpha 2,3$-sialyltransferase making sialyl-$\alpha 2,3$-Gal-$\beta 1,3$-GalNAc in porcine liver microsomes. Paulson *et al.* (1977b) showed further that rat liver Golgi apparatus made predominantly sialyl-$\alpha 2,3$-lactose at low lactose concentrations (indicating that lactose had a relatively low K_m for the $\alpha 2,3$-sialyltransferase) and made predominantly sialyl-$\alpha 2,3$-Gal-$\beta 1,3$-GlcNAc with Gal-$\beta 1,3$-GlcNAc as acceptor and sialyl-$\alpha 2,6$-Gal-$\beta 1,4$-GlcNAc with *N*-acetyllactosamine as acceptor. As discussed in Section 2.5 the enzyme responsible for synthesis of the sialyl-$\alpha 2,3$-Gal-$\beta 1,4$-GlcNAc structure present in some N-glycosidically linked oligosaccharides remains to be demonstrated.

3.2.2. Galactosyl- and Fucosyltransferases

An important enzyme in the control of salivary mucin biosynthesis is the UDP-Gal:GalNAc-mucin $\beta 1,3$-galactosyltransferase which leads to the synthesis of porcine submaxillary mucin and similarly complex mucins (Figure 3). The galactosyltransferase cannot act on sialyl-GalNAc-mucin, thereby accounting for the role of sialic acid as a termination signal (Schachter *et al.*, 1971). The enzyme is strongly bound to membrane and has not as yet been purified. Competition studies have indicated that the mucin $\beta 1,3$-galactosyltransferase is a different enzyme from the $\beta 1,4$-galactosyltransferase discussed in Section 2.4.

The remaining two glycosyltransferases involved in porcine submaxillary mucin synthesis (Figure 3) are very similar to two enzymes which effect the synthesis of human blood group antigens H and A, respectively, and which are discussed in Section 3.3. The distinction between salivary mucin synthesis and human blood group antigen synthesis is, in fact, arbitrary. Mucins secreted in human saliva often carry human blood group antigenic activity and certain nonhuman species, such as the pig, can synthesize mucins with the human blood group immunological determinants.

Porcine submaxillary glands contain a GDP-Fuc:β-D-galactoside $\alpha 1,2$-fucosyltransferase analogous to the enzyme responsible for synthesis of human blood group antigen H. The enzyme can transfer fucose in $\alpha 1,2$ linkage to the terminal galactosyl residue of both low-molecular-weight acceptors such as lactose, Gal-$\beta 1,4$-GlcNAc, Gal-$\beta 1,3$-GlcNAc, and Gal-β1,6-GlcNAc and high-molecular-weight acceptors such as sial-idase-treated α_1-acid glycoprotein (McGuire, 1970). The fucosyltransfer-ase has recently been purified 180,000-fold from detergent extracts of porcine submaxillary gland by repeated affinity chromatography on GDP-hexanolamine–Sepharose (Beyer *et al.*, 1979). Although the human and porcine α1,2-fucosyltransferases resemble each other in their ability to transfer fucose to the terminal galactosyl residues of many different compounds, there is some evidence to suggest that the porcine submax-illary gland enzyme prefers Gal-β1,3-GlcNAc and Gal-β1,3-GalNAc as acceptors, whereas the human blood group enzyme shows no such pref-erence for a particular anomeric linkage between terminal galactose and penultimate sugar residue (McGuire, 1970; Beyer *et al.*, 1979). Thus the two fucosyltransferases may differ in structure.

Pork liver has also been shown to contain a fucosyltransferase acting on sialidase-treated α_1-acid glycoprotein; the enzyme can transfer fucose to Gal-β1,4-GlcNAc but has only weak activity towards lactose, Gal-β1,3-GlcNAc, and Gal-β1,6-GlcNAc (Jabbal and Schachter, 1971). This substrate specificity suggests that the enzyme is different from the GDP-Fuc:β-D-galactoside α1,2-fucosyltransferases discussed above. The lin-kage synthesized by the pork liver enzyme is not known; two possibilities are either the Fuc-α1,2-Gal linkage (Tsay *et al.*, 1976) or the Fuc-α1,3-GlcNAc linkage (Fournet *et al.*, 1979) in *N*-glycosyl oligosaccharides. The substrate specificity of the pork liver enzyme is similar to that of an α3-fucosyltransferase present in milk (Section 2.3; Beyer *et al.*, 1979; Prieels *et al.*, 1978; Paulson *et al.*, 1978), suggesting that the pork liver enzyme is responsible for the Fuc-α1,3-GlcNAc linkage.

Thus, human blood group *H* gene-dependent α1,2-fucosyltransfer-ase, human milk α1,3-fucosyltransferase, porcine submaxillary gland α1,2-fucosyltransferase, and a porcine liver fucosyltransferase all transfer fucose to sialidase-treated α_1-acid glycoprotein. The four enzymes differ, however, in their activity towards lactose and the (β1,3), (β1,4), and (β1,6) isomers of Gal-GlcNAc.

3.2.3. Blood Group A-Dependent α1,3-*N*-Acetylgalactosaminyltransferase

The final enzyme in the synthesis of porcine submaxillary mucin (Figure 3) is UDP-*N*-acetylgalactosamine:mucin α1,3-*N*-acetylgalacto-saminyltransferase which catalyzes the following reaction:

$$\text{UDP-}\alpha\text{GalNAc} + \text{Fuc-}\alpha\text{1,2-Gal-}\beta\text{-R}$$

$$\rightarrow \text{GalNAc-}\alpha\text{1,3-(Fuc-}\alpha\text{1,2-)Gal-}\beta\text{-R} + \text{UDP}$$

This enzyme is similar to the human blood group A-dependent $\alpha 1,3$-N-acetylgalactosaminyltransferase discussed in Section 3.3. Pig submaxillary glands can be classified into blood group A-positive and blood group A-negative glands according to the ability of aqueous extracts to inhibit the human blood group A anti-A hemagglutination reaction. An $\alpha 1,3$-N-acetylgalactosaminyltransferase is present in A-positive glands (Carlson *et al.*, 1970; McGuire, 1970) which converts A-negative mucin to A-positive mucin (Figure 3).

The substrate specificity of the porcine submaxillary gland enzyme is identical to that of the human enzyme, i.e., N-acetylgalactosamine is transferred in $\alpha 1,3$ linkage to the terminal galactosyl residue of both low- and high-molecular-weight acceptors with the structure Fuc-$\alpha 1,2$-Gal-β-R; a variety of other β-galactosides were ineffective acceptors, and structures with two neighboring fucosyl residues, such as Fuc-$\alpha 1,2$-Gal-$\beta 1,4$-(Fuc-$\alpha 1,3$-)Glc, were also found to be poor acceptors.

The porcine submaxillary gland blood group A-dependent $\alpha 1,3$-N-acetylgalactosaminyltransferase has recently been purified 38,000-fold by repeated affinity chromatography on UDP-hexanolamine–agarose (Schwyzer and Hill, 1977a). Glands were extracted with Triton X-100, and the detergent was present in all column chromatography buffers. Removal of Triton X-100 led to irreversible inactivation of the enzyme. The yield of protein from 1 kg of glands was 0.15–0.3 mg, and the yield of enzyme ranged from 9 to 16%. The specific activity of the pure enzyme ranged from 28 to 35 μmole/min per mg protein, which is 55,000 times that reported for a similar enzyme isolated from human serum (Whitehead *et al.*, 1974a,b). The purified enzyme is a glycoprotein with an apparent molecular weight of about 100,000. It requires divalent cation (Mn^{2+}, Cd^{2+}, or Zn^{2+}) for activity.

The purified porcine submaxillary gland enzyme can transfer GalNAc to oligosaccharides, glycoproteins, and glycolipids with the Fuc-$\alpha 1,2$-Gal-β-R determinant (Schwyzer and Hill, 1977b); however, as mentioned above, it will not act on compounds with two neighboring fucose residues such as Fuc-$\alpha 1,2$-Gal-$\beta 1,3$-(Fuc-$\alpha 1,4$-)GlcNAc-$\beta 1,3$-Gal-$\beta 1,4$-Glc, a milk oligosaccharide carrying blood group Le^b specificity.

Schwyzer and Hill (1977b) have reported that the enzyme acts on both native and sialic acid-free blood group A-negative porcine submaxillary mucin and have claimed that GalNAc can be incorporated into sialic acid-containing oligosaccharides; this contrasts with the reports of Carlson *et al.* (1970) and Takasaki *et al.* (1978) that sialic acid-containing oligosaccharides cannot accept GalNAc. The role of sialic acid as a termination signal at all stages of assembly (Figure 3) therefore remains to be established.

The pure blood group A transferase can be used to convert type O erythrocytes to type A erythrocytes (Schwyzer and Hill, 1977*b*). The material on the red cell membrane into which the transferase incorporates radioactive GalNAc migrates in sodium dodecyl sulfate–urea polyacrylamide gel electrophoresis as several bands ranging in molecular weight from below 20,000 to about 100,000. The lower-molecular-weight material was thought to be glycolipid but the larger material, comprising the majority of the radioactive product, was identified as glycoprotein on the basis of mobility in gel electrophoresis. This may be an erroneous conclusion since a recent report (Dejter-Juszynski *et al.*, 1978) has suggested that the majority of the ABH antigenic activity of human erythrocytes resides in large glycolipids that migrate like glycoproteins in gel electrophoresis. However, reports by Takasaki and Kobata (1976) and Takasaki *et al.* (1978) have shown that blood group A *N*-acetylgalactosaminyltransferase from human milk can transfer radioactive GalNAc to acceptors on human blood type O erythrocyte membranes with the electrophoretic mobility of glycoproteins and that the oligosaccharides of these products are mild-alkali labile; it appears therefore that some ABH antigenic activity on the erythrocyte may be due to O-glycosidically linked protein-bound oligosaccharides. Glycosphingolipids are stable to mild alkali.

3.3. Synthesis of Human Blood Group Oligosaccharides

The ABO blood group system was discovered when Landsteiner observed in 1900 that the serum of some individuals agglutinated the red cells of others. Many other blood group systems have been demonstrated since that discovery and a recent edition of *Blood Groups in Man* (Race and Sanger, 1975) lists over 160 antigens distributed among 20 or more systems.

The A, B, H(O) and Lewis (Le) antigens occur as O-glycosidically linked oligosaccharides attached to water-soluble mucins found in various exocrine secretions such as saliva, gastric juice, meconium, and ovarian cyst fluids. Ovarian cyst fluids provided relatively large amounts of mucin for structural studies and led to the elucidation of the antigenic determinants responsible for A, B, H, Le[a], and Le[b] activities shown in Table 2 (Watkins, 1974*a,b*). The immunological determinants of the ABH(O)-Lewis, MN, P, Ii, and possibly the En antigens are believed to be oligosaccharides; the following discussion will, however, be limited to the ABO-Lewis system.

The term "blood group" was coined because the ABH(O)-Lewis antigens were originally detected on the red cell membrane. The term is almost certainly a misnomer since these antigens are widely distributed throughout the body and occur in three chemical forms: (1) on the surfaces of red cells, and other cells, as glycosphingolipids and also possibly

Table 2. Determinants for Blood Group Antigens H, A, B, Le^a, and Le^b

	Structure	
Specificity	Type 1	Type 2
H	Gal-1,3-β-GlcNAc- | 1,2-α Fuc	Gal-1,4-β-GlcNAc- | 1,2-α Fuc
A	GalNAc-1,3-α-Gal-1,3-β-GlcNAc- | 1,2α Fuc	GalNAc-1,3-α-Gal-1,4-β-GlcNAc- | 1,2-α Fuc
B	Gal-1,3-α-Gal-1,3-β-GlcNAc- | 1,2α Fuc	Gal-1,3-α-Gal-1,4-β-GlcNAc- | 1,2-α Fuc
Le^a	Gal-1,3-β-GlcNAc- | 1,4-α Fuc	
Le^b	Gal-1,3-β-GlcNAc- | 1,2-α | 1,4-α Fuc Fuc	

as glycoproteins (see discussion in the previous section); (2) as oligosaccharides in milk and urine; and (3) as O-glycosidically linked oligosaccharides attached to mucins secreted in the gastrointestinal, genitourinary, and respiratory tracts.

The oligosaccharides of blood group mucins are large and complex (Kabat, 1970). The synthesis of the serine(threonine)-N-acetylgalactosamine linkage of mucins has been discussed in Section 3.1, and little is known about the assembly of the core of the blood group mucins. This discussion will deal with the glycosyltransferases determined by the H, A, B, and Le genes; these transferases catalyze the assembly of the blood group antigenic determinants.

3.3.1. Genes Regulating Synthesis of ABH and Lewis Determinants

Four independent gene systems— ABO, Hh, $Lele$, and $Sese$—control the synthesis of the ABH(O)-Lewis antigenic determinants. The H gene controls the appearance of GDP-fucose:β-galactoside α1,2-fucosyltransferase:

$$\text{GDP-}\beta\text{-fucose} + \text{Gal-}\beta\text{-R} \rightarrow \text{Fuc-}\alpha1,2\text{-Gal-}\beta\text{-R} + \text{GDP}$$

The h gene is inactive. Individuals with the rare O_h (Bombay) phenotype (genotype hh) cannot make H antigen in any of their cells, and, as will be discussed below, therefore cannot synthesize either the A or B antigens even though they possess the A- or B-dependent transferases.

The *Sese* locus does not appear to code for a functional transferase. It somehow controls the appearance of the *H* transferase in some organs but not in others. The *Se* gene allows the appearance of *H* transferase in exocrine cells such as the salivary glands; thus, people possessing the *Se* gene have ABH blood group antigens in their saliva and are called "secretors." "Nonsecretors" have the genotype *sese*, but the designation is a misnomer since these people do secrete mucins in their saliva although these mucins do not possess ABH antigenic activity. Nonsecretors with the *Le* gene secrete mucins with Le^a determinant (Table 2). The *Se* gene is not required for the expression of the *H* gene in hemopoietic tissues and therefore does not affect the appearance of ABH antigens on red cells.

The *ABO* locus is responsible for two transferases, the *A*-dependent $\alpha1,3$-*N*-acetylgalactosaminyltransferase and the *B*-dependent $\alpha1,3$-galactosyltransferase:

$$\text{UDP-}\alpha\text{-GalNAc} + \text{Fuc-}\alpha1,2\text{-Gal-}\beta\text{-R}$$

$$\rightarrow \text{GalNAc-}\alpha1,3\text{-(Fuc-}\alpha1,2\text{-)Gal-}\beta\text{-R} + \text{UDP}$$

$$\text{UDP-}\alpha\text{-Gal} + \text{Fuc-}\alpha1,2\text{-Gal-}\beta\text{-R}$$

$$\rightarrow \text{Gal-}\alpha1,3\text{-(Fuc-}\alpha1,2\text{-)Gal-}\beta\text{-R} + \text{UDP}$$

H antigen is the precursor of both A and B antigens; thus, a person with the *hh* genotype cannot make the A or B antigens even though he possesses the appropriate transferases. The *O* gene is inactive.

The *Le* gene codes for an $\alpha1,4$-fucosyltransferase catalyzing the following reaction:

$$\text{GDP-}\beta\text{-Fucose} + \text{Gal-}\beta1,3\text{-GlcNAc-R}$$

$$\rightarrow \text{Gal-}\beta1,3\text{-(Fuc-}\alpha1,4\text{-)GlcNAc-R} + \text{GDP}$$

The *le* gene is inactive. It should be noted that the Le determinant can only appear on the Gal-$\beta1,3$-GlcNAc structure (Type 1, Table 2).

The four gene systems discussed above control the appearance of blood group determinants in all three types of complex carbohydrates: glycoproteins (mucins), glycosphingolipids, and oligosaccharides. Thus, the various glycosyltransferases controlled by these genes act on both glycoproteins and glycolipids.

Figure 4 summarizes the biosynthetic path for secreted ABH-Lewis glycoproteins. The scheme considers the fate of both Type 1 and Type 2 precursor substances.

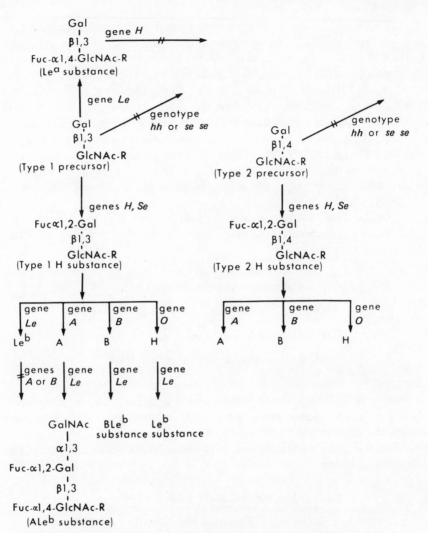

Figure 4. Biosynthetic scheme for secreted human blood group antigens. For these materials, the *Se* gene controls the appearance of *H*-dependent α1,2-fucosyltransferase.

3.3.2. *H*-Dependent α1,2-Fucosyltransferase

The *H* transferase has been detected in the milk, submaxillary glands, and gastric mucosa of secretors but is absent from these tissues in nonsecretors. However, the enzyme is present in equivalent amounts in the serum and bone marrow of both secretors and nonsecretors, indicating that the *Se* gene is not the structural gene for the fucosyltransferase but merely controls its appearance in certain exocrine tissues (Munro and Schachter, 1973; Schenkel-Brunner *et al.*, 1972; Pacuszka and Kościelak, 1974). Proof that the *H* gene is the structural gene for the fucosyltransferase was obtained by showing that the serum of O_h (genotype *hh*) donors lacked the enzyme (Munro and Schachter, 1973; Schenkel-Brunner *et al.*, 1972).

The human $\alpha 1,2$-fucosyltransferase has not as yet been purified. However, as discussed in Section 3.2.2, a similar enzyme has been purified 180,000-fold from porcine submaxillary glands (Beyer *et al.*, 1979).

Almost any β-galactoside of either low- or high-molecular-weight serves as an acceptor for the human *H* transferase. However, the use of acceptors such as lactose or Gal-$\beta 1,3$-GlcNAc-R can lead to complications if the enzyme preparation is crude and contains other fucosyltransferases such as the $\alpha 1,3$-fucosyltransferase, which converts lactose to Gal-$\beta 1,4$-(Fuc-$\alpha 1,3$-)Glc, or the *Le*-dependent $\alpha 1,4$-fucosyltransferase, which converts Gal-$\beta 1,3$-GlcNAc-R to Gal-$\beta 1,3$-(Fuc-$\alpha 1,4$-)GlcNAc-R. Chester *et al.* (1976) have suggested the use of phenyl-β-D-galactopyranoside as a specific acceptor for the *H* transferase.

Some β-galactosides are, however, poor acceptors. For example, lactosylceramide is not a substrate although other glycosphingolipids are effective (Basu *et al.*, 1975; Pacuszka and Kościelak, 1976). Also, a fucose residue on the sugar penultimate to the terminal β-galactosyl residue, e.g., as in the Le[a] antigen Gal-$\beta 1,3$-(Fuc-$\alpha 1,4$)-GlcNAc-R, partially or completely blocks the action of the *H* transferase (Watkins, 1974*a,b*; Schenkel-Brunner *et al.*, 1972); thus, the *H* transferase cannot act on Le[a] substance (Figure 4).

3.3.3. *A*-Dependent $\alpha 1,3$-*N*-Acetylgalactosaminyltransferase

An enzyme catalyzing the synthesis of the A determinant GalNAc-$\alpha 1,3$-(Fuc-$\alpha 1,2$-)Gal-R has been found in a variety of tissues and species, e.g., porcine submaxillary gland (Section 3.2.3), rat intestinal mucosa, canine tracheobronchial mucosa, human milk, human submaxillary gland, human and porcine gastric mucosa, human ovarian cysts, and human serum (Watkins, 1974*a,b*). The enzyme is present in humans with the blood group *A* gene whether or not they possess the *Se* gene.

The *A* transferases all share a common requirement for substrates with the H determinant Fuc-$\alpha 1,2$-Gal-β-R. Ineffective acceptors lack either a terminal β-galactosyl residue or an $\alpha 1,2$-linked fucosyl residue or contain a fucosyl residue adjacent to the $\alpha 1,2$-linked fucosyl residue, e.g., the Le[b] determinant Fuc-$\alpha 1,2$-Gal-$\beta 1,3$-(Fuc-$\alpha 1,4$-)GlcNAc-R. Thus, the *A* transferase cannot act on Le[b] substance (Figure 4).

The *A* transferase has been purified 1000-fold from human plasma (Whitehead *et al.*, 1974*a,b*) by a single step involving adsorption to Sepharose 4B and elution with UDP. As discussed in Section 3.2.3, a more potent preparation has recently been obtained from porcine submaxillary glands (Schwyzer and Hill, 1977*a*). Both preparations had similar molecular weights of about 100,000 and showed identical substrate specificities and dependence on divalent cation.

The human serum enzymes catalyzing synthesis of the A_1 and A_2 antigens have similar substrate specificities but are qualitatively different enzymes (Schachter *et al.*, 1973) and have in fact been separated (Topping and Watkins, 1975).

3.3.4. *B*-Dependent α1,3-Galactosyltransferase

The *B* transferase has also been found in a variety of tissues and species, e.g., human and baboon gastric mucosa and submaxillary glands, human milk, human ovarian cyst fluids, and human serum (Watkins, 1974*a,b*). This enzyme, like the *A* transferase, is independent of the *Se* gene and requires the H determinant Fuc-α1,2-Gal-R as substrate. In fact, the *A* and *B* transferases have identical substrate requirements and neither enzyme can act on Le^b substance (Figure 4).

The *B* transferase has been purified to homogeneity from human plasma and has been shown to cross-react immunologically with the human plasma *A* transferase (Nagai *et al.*, 1976). Both enzymes cross-react immunologically with an enzymatically inactive protein isolated from type O plasma. The two enzymes have also been shown to have similar physical properties and almost identical amino acid compositions.

3.3.5. *Le*-Dependent α1,4-Fucosyltransferase

The *Le* transferase has been found in human milk, submaxillary glands, and gastric mucosa (Watkins, 1974*a,b*), and its presence has been correlated with the *Le* gene. The enzyme is independent of the *Se* gene. The enzyme attaches Fuc in α1,4 linkage to the GlcNAc residue of Gal-β1,3-GlcNAc-R and Fuc-α1,2-Gal-β1,3-GlcNAc-R to form the Le^a and Le^b determinants, respectively (Figure 4).

Human milk, submaxillary glands, gastric mucosa, and serum contain an α1,3-fucosyltransferase (Watkins, 1974*a,b*) that catalyzes the reaction:

$$\text{GDP-}\beta\text{-Fuc} + \text{Gal-}\beta\text{1,4-GlcNAc}$$

$$\rightarrow \text{Gal-}\beta\text{1,4-(Fuc-}\alpha\text{1,3-)GlcNAc-R} + \text{GDP}$$

Structures analogous to the product of this reaction have been detected in ovarian cyst glycoproteins and milk oligosaccharides, although there is no blood group antigen or blood group gene known to be associated with these structures. The human milk α1,3-fucosyltransferase has been separated from the *H*-dependent α1,2-fucosyltransferase by ion-exchange chromatography on carboxymethyl–Sephadex and has been purified to over 50,000-fold by affinity chromatography on GDP-hexanolamine–Sepharose (Beyer *et al.*, 1979). Although the α1,2-fucosyltransferase does not have a strict requirement for cation, the α1,3-fucosyltransferase is inactive in the absence of divalent cation. The purified α1,3-fucosyltransferase can transfer fucose in α1,3 linkage to either the GlcNAc residue of Gal-β1,4-GlcNAc-R or the Glc residue of Gal-β1,4-Glc-R. Analogous to the *Le* transferase, the α1,3-fucosyltransferase can act on acceptors either with or without a Fuc-α1,2-Gal sequence.

As discussed in Sections 2.3 and 3.2.2, the milk α1,3-fucosyltransferase can incorporate fucose into sialidase-treated transferrin. Although the linkage synthesized in this reaction has not been rigorously established, there is evidence that a fucose residue was incorporated in α1,3 linkage to a peripheral GlcNAc residue of an *N*-acetyllactosamine oligosaccharide (Prieels *et al.*, 1978; Paulson *et al.*, 1978). Such a structure has recently been described in α_1-acid glycoprotein (Fournet *et al.*, 1979). Although *N*-acetyllactosamine oligosaccharides are more likely to have a fucose residue attached to the GlcNAc residue nearest the asparagine (Montreuil, 1975; Section 2.3), it appears that peripheral fucose residues can also occur in these structures. An enzyme similar to the milk α1,3-fucosyltransferase is probably responsible for the synthesis of this linkage. It is not known why some glycoproteins (e.g., IgG) have a fucose attached to a core GlcNAc residue while others (e.g., α_1-acid glycoprotein) lack the core fucose but have a peripheral Fuc-GlcNAc linkage. Many other factors that control the structural differences between glycoproteins are not understood and require further study.

PART B

1. INTRODUCTION

Glycosylation reactions of the type discussed in Part A of this chapter also play a key role in assembly of the oligosaccharide chains found in a very different class of molecules, namely, the glycosphingolipids. Because of limitations of space this review will be restricted to consideration of one subclass of glycosphingolipids, the gangliosides. The reader is referred to several excellent reviews of glycosphinoglipids in general (Porcellati *et al.*, 1975; Volk and Schneck, 1975; Svennerholm, 1964; Sturgeon, 1977; Witting, 1975; Ledeen and Yu, 1973; Wherrett, 1976; Fishman and Brady, 1976; Murray *et al.*, 1973).

In the late 1930s, Klenk (reviewed by Klenk, 1969) identified a new class of glycosphingolipids that accumulated in high concentrations in ganglion cells of patients with certain lipid storage diseases. Klenk termed these compounds "gangliosides," and they were subsequently isolated from normal brain. The distinctive feature of these glycosphingolipids was the presence of N-acylneuraminic acid (sialic acid) in the oligosaccharide moiety. Since their discovery, these compounds have been extensively examined with respect to their structures, properties, and metabolism. Many new gangliosides have been identified, and much has been learned about their accumulation in lipid storage diseases. Their distribution, originally thought to be limited to nervous tissue, has been shown to extend to a wide variety of tissues, and they are probably present in all cell types in eukaryotic organisms.

This brief review will focus on some of the novel gangliosides that have been isolated in recent years, the biosynthesis of the gangliosides, their distribution in normal and pathological tissues, and in normal and tumorigenic tissue cultured cell lines; potential functions of the gangliosides will also be considered.

2. NEW GANGLIOSIDES

In the earlier work of Klenk, Svennerholm, Wiegandt, Kuhn, and many others (reviewed by Svennerholm, 1964), attention was focused on the major gangliosides, particularly of brain. The difficulties encountered in quantitative extraction of these compounds from tissues and their subsequent purification to homogeneity were emphasized, as well as the fact that each so-called "homogeneous" ganglioside actually consisted of a class of compounds containing a variety of sphingosine and fatty acid residues. However, each of the isolated substances contained only one oligosaccharide chain, and once these were available it was possible to determine the detailed structures of the gangliosides.

As new techniques have been developed, the older isolation procedures have improved (Witting, 1975; Ledeen and Yu, 1973; Tjaden *et al.*, 1977; Zanetta *et al.*, 1977; Kawamura and Taketomi, 1977). Thus ion-exchange resins (usually modified cellulose or dextrans) are particularly useful in the first stages of purification (Ledeen and Yu, 1973), and final purification almost invariably involves the use of thin-layer chromatography. Mild alkaline hydrolysis is also frequently used before the final steps in purification to remove phospholipids and other contaminants from the gangliosides.

The structure of a major human brain ganglioside is shown in Figure 5, and the structures of the other members of this class of compounds are given in Figure 6, using of the nomenclature of Svennerholm (1964). These gangliosides consist of glycosylceramide derivatives and contain at least one, and perhaps as many as four, sialic acid residues attached to di-, tri-, or tetrasaccharides linked in turn to ceramide at the C-1 position. The tetraglycosylceramide is: Gal-β1,3-GalNAc-β1,4-Gal-β1,4-Glc-β1,1'-ceramide. In these and all remaining structures, all glycosides are D-pyranosides with the exception of L-fucopyranoside.

For many years, the N-acetylgalactosamine gangliosides listed in Figure 6 were regarded not only as the major gangliosides of a variety of tissues and cell types, but perhaps the only gangliosides present in these tissues, even though there were isolated reports that other gangliosides had been detected.

In a series of reports from the laboratories of Yamakawa, Wiegandt, Svennerholm, and others (reviewed by Wherrett, 1976; and Hakomori and Young, 1978), it has been shown that human erythrocytes, plasma, brain, muscle, spleen, liver, kidney, and peripheral nerve contain gangliosides with glucosamine replacing galactosamine. For example, the glucosamine ganglioside is the major ganglioside of human femoral nerve (Li et al., 1973; Månsson et al., 1974). In these compounds (reviewed by

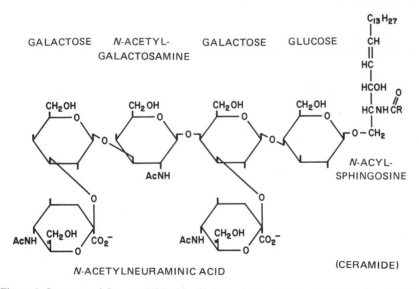

Figure 5. Structure of G_{D1a} established by Kuhn and Wiegandt (see Svennerholm, 1964).

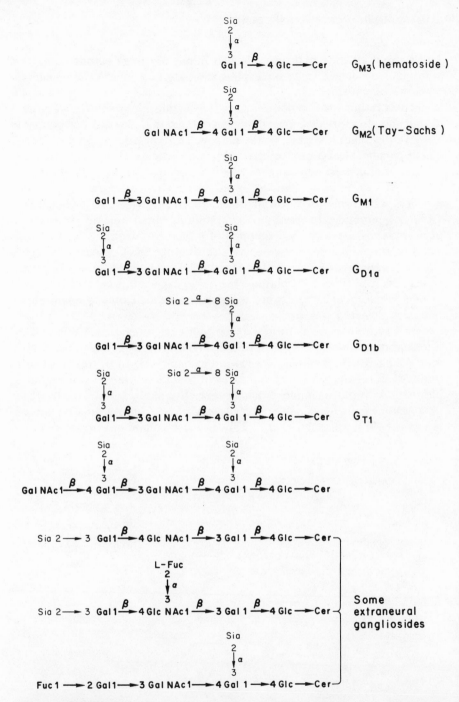

Figure 6. Structures of major gangliosides. Cer, ceramide; Sia, sialic acid (*N*-acetyl- or *N*-glycolylneuraminic acid; or *N*-acyl-*O*-acylneuraminic acid). The nomenclature of Svennerholm (G_{M3}, etc.) is used.

Wherrett, 1976), the oligosaccharide chains of the glycosylceramides all contain the sequence Gal-GlcNAc-Gal-Glc-ceramide. Although the disaccharide linked to ceramide (Gal-Glc) is lactose, similar to the GalNAc species, other linkages differ: Gal-β1,3-GalNAc-β1,4-Gal- vs. Gal-β1,4-(sometimes 1,3-)GlcNAc-β1,3-Gal. Moreover, in the GlcNAc gangliosides, one or two sialic acid residues are linked to the terminal galactose unit, whereas the GalNAc gangliosides always contain a sialic acid residue linked to the internal galactose residue. The terminal galactose residue is sialylated in the di-, tri-, and tetrasialyl GalNAc gangliosides. Other differences between the GlcNAc and GalNAc gangliosides have been reported, such as the occurrence of branch points in the neutral oligosaccharide backbone of the GlcNAc gangliosides, and that the neutral oligosaccharide portion of the molecule contains six glycose units.

Details of the fine structural differences between the GalNAc and GlcNAc classes have been emphasized because these undoubtedly result in major differences in the three-dimensional structures, which may serve to delineate the physiological functions of these two sets of molecules.

The GlcNAc gangliosides should also be compared with the serum-type glycoproteins discussed in Part A of this chapter, because these molecules contain similar or identical terminal trisaccharide units: sialyl-(α2,3 or 2,6)-Gal-(β1,4 or 1,3)-GlcNAc-. It is difficult to believe that these identities are fortuitous and without physiological significance.

With the exception of the sialic acid residues, the neutral oligosaccharides in the GlcNAc gangliosides contain the tetrasaccharide proposed to be present in many or all of the blood group active glycosphingolipids. The latter (reviewed by Hakamori and Young, 1978) are responsible for the blood types of human erythrocyte and other cell types, and consist of very long, sometimes branched, oligosaccharide chains linked to ceramide. However, thus far, there have been no reports on the occurrence of sialic acid in these lipids. The presence of the same tetrasaccharide unit in the GlcNAc gangliosides and blood group active lipids does suggest, however, either a common biosynthetic pathway or a related physiological function, or both.

L-Fucose occurs rarely in the gangliosides, although it is commonly found in the blood group active glycosphingolipids. Its presence has been reported in both a GalNAc ganglioside from boar testis (Suzuki et al., 1975) and the GlcNAc ganglioside from human kidney (Rauvala, 1976a).

The simplest ganglioside thus far reported is designated "G_7," and consists of sialic acid linked to galactosylceramide (galactocerebroside). This compound was isolated with other brain gangliosides from central nervous system myelin; G_7 comprised 15% of the total ganglioside fraction from human myelin. In a recent report (Fong et al., 1976), it was shown that G_7 was not present in the ganglioside fractions of peripheral nerve myelin obtained from three sources; the mixed gangliosides contained both the GalNAc and GlcNAc type. The conclusion that myelin does contain gangliosides is an important one, since a number of earlier reports indicated that it did not. The authors of the papers (reviewed by Fong *et al.*, 1976) who have detected the gangliosides in these membranes ascribe their results to improved methodology.

3. OCCURRENCE AND SUBCELLULAR DISTRIBUTION

In considering possible physiological functions of the gangliosides, as well as their biosynthetic pathways, one can ask certain obvious questions. Can a single cell generate more than one ganglioside and more than a single class of gangliosides? How do ganglioside patterns change during development and differentiation?

Although a positive answer to the first question is suggested by studies on intact tissue, it should be emphasized that these tissues may contain more than one cell type and each may give rise to different ganglioside patterns. Studies with tissue cultured cells (described in Sections 6 and 7) provide unequivocal evidence that cells can generate more than one ganglioside (and ganglioside class). The change in ganglioside patterns during development is described in Section 7.

In the earliest studies on the gangliosides, the gray matter from human brain was shown to contain a mixture of gangliosides (e.g., see Figure 7). Some of the more recent studies show that the pattern from other tissues can be equally or even more complex. An extensive study on the glycosphingolipids from pig adipose tissue (Ohashi and Yamakawa, 1977) showed that 11 species of gangliosides were present in a suitably fractionated extract. G_{D1a} predominated, together with G_{M3}, three species of fucose-containing gangliosides, two containing GlcNAc, as well as others related or identical to the brain gangliosides. Two species of trisialoganglioside were detected. Human kidney gangliosides are also complex. Five were identified (Rauvala, 1976*b*), three being identical with the common brain type gangliosides G_{M3}, G_{D3}, G_{D1}; two GlcNAc gangliosides, one of which contained fucose, were also detected. As a final example, six gangliosides have been characterized from bovine thyroid tissue (Van Dessel *et al.*, 1976).

The subcellular localization of the gangliosides is of special interest. Studies of this type are, of course, always hampered by the difficulties in isolating homogeneous subcellular fractions, but some of the recent findings in this area are given below.

In earlier studies from several laboratories (see reviews cited in the Introduction to Part B) evidence was presented that gangliosides were present in the plasma membrane. This conclusion is substantiated by more recent work on subcellular fractionation, the interactions between whole cells and toxins, such as cholera toxin (discussed below), and the interactions between whole cells and antibodies directed at gangliosides. The oligosaccharide chains of the gangliosides are not always available to macromolecules such as antibodies; in some instances at least, these sugars are masked.

Rat brain was homogenized and fractionated (Avrova et al., 1973), and gangliosides were detected both in synaptosomal membranes, in confirmation of earlier work, and in other membrane systems as well, i.e., microsomes, mitochondria, and myelin. The synaptosomal membranes contained the highest content of gangliosides, and these were different in composition from those in the other membranes in that the synaptosomes contained much higher ratios of disialogangliosides relative to the total ganglioside content.

An immunohistochemical study (De Baecque et al., 1976) of rat cerebellar cortex at the light-microscopic level has been reported that offers considerable promise for the future. Purified rabbit antibodies to G_{M1} were incubated with brain sections, and the antibodies fixed to the section were then detected with horseradish peroxidase coupled to anti-rabbit IgG antibodies, followed by incubation with diaminobenzidine, which yields precipitates on interacting with the peroxidase. Discrete areas of the sections were stained, particularly in the granular layer. If this technique can be extended so that it can be used at the level of the electron microscope, and if other ganglioside antibodies can be used as well, it is likely that a precise mapping of specific ganglioside distribution will be achieved. The composition and location of gangliosides at or near the synaptic junction may then be revealed.

4. GANGLIOSIDES AS MEMBRANE COMPONENTS

As noted above, many early reports indicated that the oligosaccharide chains of gangliosides are not always available on the cell surface for interaction with macromolecules such as antibodies and enzymes. For example, the ganglioside sialyl residues on milk fat globules and erythrocyte membranes are resistant to sialidase (Tomich et al., 1976). As with other cells, however, when the membranes were first incubated with trypsin or extracted with a mixture of EDTA and mercaptoethanol, the ganglioside sialic acid moieties became sensitive to the sialidase treatment. The conclusion that ganglioside oligosaccharide chains are shielded by cell surface proteins is not consistent with the ability of these cells to interact with toxins (see Section 8).

Interactions between gangliosides in the plasma membranes of whole cells and extracellular macromolecules may result from the distribution, orientation, and packing of the ganglioside molecules in the membrane, rather than removal of a shielding effect by membrane proteins. In other words, the ability of cell surface gangliosides to act as receptors for enzymes, toxins, hormones, antibodies, etc., may be a function of interactions between gangliosides *per se,* or between gangliosides and phospholipids and cholesterol, rather than between gangliosides and membrane proteins.

This point is illustrated by the action of sialidase on gangliosides. Despite earlier reports (from this laboratory), the accepted concept was that the sialic acid glycosidic linkage to the internal galactose residue in G_{M1} and G_{M2} was resistant to *Clostridium perfringens* sialidase, whereas this residue could be cleaved from hematoside (G_{M3}) by the enzyme. In other words, the GalNAc residue in the longer chain gangliosides protected this sialyl–galactose bond from the action of the hydrolase. However, a recent study (Rauvala, 1976c) shows that this sialidase is indeed active on the sialic acid glycosidic linkage in G_{M1} and G_{M2}, provided that the substrates are below their critical micelle concentrations, which lie somewhere between 10^{-4} and 10^{-5} M.

An intriguing and important question concerns the packing of gangliosides in the plasma membrane. There have been relatively few physical studies with pure preparations of gangliosides or of ganglioside–lipid mixtures. Molecular models show that gangliosides such as G_{D1a} are quite different from the phospholipids, in that the polar head groups of the gangliosides occupy about the same volume as do the hydrophobic chains in the molecule (Curatolo *et al.*, 1977). Whether the large polar groups interfere significantly with the packing of gangliosides in the lipid bilayer is not known, but it appears likely that aggregation of gangliosides within the bilayer by multivalent reagents such as antibodies or cholera toxin could have a profound effect on the permeability of the lipid layer. Indeed, observations of this type have been made by studying the permeability of mixed ganglioside–phospholipid liposomes after interaction with cholera toxin (discussed below).

Physical studies with mixed bovine brain gangliosides (Curatolo *et al.*, 1977) show that the physical organization of these compounds depends on the water content. At high concentrations of water they form prolate ellipsoid micelles with estimated masses of about 2.5–4 $\times$ 10^5 daltons. At low water-to-ganglioside ratios (i.e., 18–50% water), which reflect the more likely situation in natural membranes, the gangliosides form long cylinders with the oligosaccharide chains on the surfaces of the cylinders exposed to the aqueous environment. The behavior of the gangliosides as a function of water content and temperature was very different from the behavior of the phospholipids. For example, with the gangliosides lamellar structures were not observed under any conditions. The physical studies also suggest that the hydrocarbon chains were packed differently from the packing observed with phospholipids. Furthermore, in the gangliosides, rigid packing of the hydrocarbon chains under any conditions appears highly unlikely. The potential ability of gangliosides to disrupt the ordered structure in phospholipid bilayers is also discussed in this study.

As a final point, the physical studies reviewed above suggest that experiments where exogenous gangliosides are incorporated into whole cells could result in important, perhaps major changes in the structure of the plasma membrane of such cells. Resulting changes in cell physiology may be due to physical changes in the bilayer rather than to changes in putative "receptors" in the cell membrane, or because of interactions between the gangliosides and other cell components such as adenylate cyclase.

5. BIOSYNTHESIS AND DEGRADATION

Our original proposals (Kaufman et al., 1966; Roseman, 1970; Steigerwald et al., 1975) concerning the biosynthesis of the gangliosides now appear to be generally accepted (Caputto et al., 1974; Fishman, 1974; Fishman and Brady, 1976; Coleman et al., 1975). The initial suggestions were based on the in vitro properties, particularly specificities, of glycosyltransferases found in membrane preparations, usually isolated from chicken embryonic brain homogenates. The following general conclusions were drawn: (1) Gangliosides are synthesized in a stepwise manner, by the addition of a glycose unit at each step to the nonreducing terminus of the ceramide oligosaccharide, or at a branch point, as illustrated in Figure 7. (2) Each glycosyltransferase is specific not only for a sugar nucleotide, but also for the acceptor molecule. Thus, two different sialyltransferases are responsible for incorporating the two sialic acid residues of G_{D1}, and likewise, each galactose residue is incorporated by a different galactosyltransferase. In this manner, the product of each step is the substrate for the next step in the sequence. (3) Since all of the enzymes were found in the same particulate fraction, it was proposed that the gangliosides were synthesized by a multienzyme complex, termed a multiglycosyltransferase complex. Furthermore, based again on substrate specificities and location of each glycosyltransferase, different multiglycosyltransferase complexes would be responsible for the synthesis of the gangliosides, the terminal trisaccharide unit in serum-type glycoproteins, and the oligosaccharide chains of the mucins.

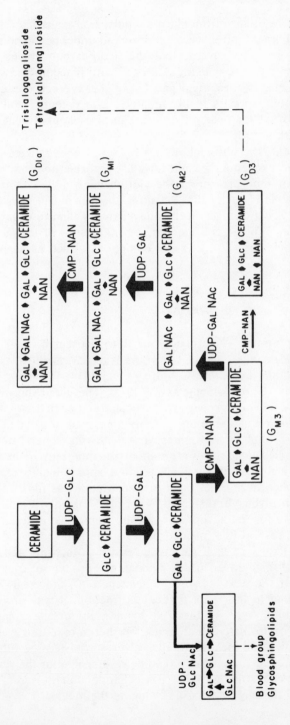

Figure 7. Proposed pathway for the synthesis of gangliosides from ceramide. The broad solid arrows indicate the main pathway from ceramide to disialoganglioside G_{D1a}. The thin solid arrows show branches from the main pathway that lead to the formation of G_{D3} and probably also to tri- and tetragangliosides, and to the formation of a triglycosylceramide and possibly to the formation of blood group glycosphingolipids.

Several attempts have been made to detect lipid intermediates such as dolichol derivatives (see Chapter 2) in the biosynthesis of the gangliosides, but thus far have been without success. It should be emphasized, however, that all of the preparations that synthesize gangliosides are particulate. Since no soluble, homogeneous glycosyltransferases involved in the stepwise synthesis have been obtained, it is possible that each of the reactions shown in Figure 7 is the net result of more than one step; for example, a glycose unit could be transferred first to an endogenous membrane-bound lipid, followed by a second transfer to the exogenous glycosylceramide acceptor, in a manner analogous to the dolichol system involved in glycoprotein synthesis. However, a substantial portion of the oligosaccharide moiety of glycoproteins is first synthesized *de novo* during the time that it is linked to dolichol and is then transferred to the protein acceptor (see Chapter 2). If dolichol or a similar isoprenoid was involved in ganglioside synthesis, direct transfer of single glycose units to exogenous ceramide derivatives, as has actually been observed experimentally, would be highly unlikely. Thus, while still possible, the dolichol pathway does not appear to be involved in ganglioside synthesis.

As noted earlier in this review, the simplest ganglioside, G_7, or sialylgalactocerebroside, is one of the major gangliosides in the white matter of human brain. The composition of the fatty acyl and sphingosine residues, unusual for gangliosides, closely resembled the composition of galactocerebroside from the same source, suggesting that cerebroside was the precursor of G_7. Yu and Lee (1976) showed that mouse brain microsomes contain a sialyltransferase capable of converting galactocerebroside to G_7 in the presence of CMP-sialic acid. Liver preparations also contained this activity.

The subcellular location of glycosyltransferases involved in ganglioside synthesis is of interest. In the original studies on this subject (Den *et al.*, 1970), a synaptosome-enriched fraction from chicken embryonic brain homogenates was found to contain the bulk of the four glycosyltransferases specific for ganglioside synthesis (as well as three involved in the synthesis of glycoproteins). More recent studies with homogenates from rat liver (Wilkinson *et al.*, 1976; Richardson *et al.*, 1977; Keenan *et al.*, 1974) have demonstrated a 50-fold enrichment in the Golgi fraction of one of the galactosyl- and one of the sialyltransferases.

Are glycosyltransferases present on the cell surface? This issue remains controversial. It was first proposed (Roseman, 1970) that such enzymes might occur on cell surfaces as part of a mechanism for regulating complex intercellular reactions, such as cell–cell recognition and adhesion. Evidence (Roth *et al.*, 1971) was then presented for the existence of two UDP-galactose:glycoprotein glycosyltransferases on the surface of chick embryonic neural retina cells, and this has been followed by numerous papers on this subject (reviewed by Shur and Roth, 1975). It is generally concluded that glycosyltransferases do occur on cell surfaces, although there is some disagreement on this point. Few of the endogenous acceptors in these experiments have been characterized, and it is therefore not clear whether transferases of this type, even if they do occur on cell surfaces, can utilize gangliosides as acceptors.

The earlier reviews on the gangliosides summarize a large number of *in vivo* experiments on ganglioside biosynthesis and degradation. These results are in general agreement with the pathway shown in Figure 7, and also with the concept of the multiglycosyltransferase complex. *In vivo* experiments can provide additional information. For example, although G_{D3} represents 50% of the mammalian retinal gangliosides, it had a sphingosine and fatty acid composition different from the other retinal gangliosides G_{M1}, G_{D1a}, G_{D1b}, and G_{T1}. Moreover, G_{D3} was not found in bovine optic nerve, which arises from the neural retina, whereas the other gangliosides were present in the optic nerve. The metabolic interrelationships of the gangliosides were studied (Holm and Månnson, 1974) by injecting labeled N-acetylmannosamine into the vitreous humor of rabbits and isolating the retinal gangliosides at various times after the injection. The labeling pattern of the sialic acid moieties in all of the gangliosides were similar, except for G_{D3}. The latter was labeled more rapidly and much more extensively. The authors conclude that these results present further evidence for the localization of G_{D3} in neural retinal cells other than those involved in the formation of the axons of the optic nerve, i.e., the ganglion cells. In a second report from the same laboratory (Holm, 1977), the half-lives of the sialic acid moieties in the retinal gangliosides were studied. G_{D3} showed a biphasic rate of loss of the isotope, the first phase between 2 and 4 weeks after the injection with a half-life of 2 weeks, and the second after 4 weeks with a half-life of 8–10 weeks. The other retinal gangliosides showed only the second phase, i.e., a half-life of degradation of 8–10 weeks.

While several theoretical pathways can be proposed for the degradation of gangliosides in mammalian tissues, all of the available results are consistent with one pathway, which is essentially the reverse of the biosynthetic pathway. Terminal glycosidic linkages in the complex gangliosides are cleaved one at a time by specific exoglycosidases. A defect in one of these glycosidases results in an accumulation of the corresponding ganglioside substrate, or a gangliosidosis. At least five clinical forms of gangliosidoses have been identified and, with the possible exception of one which may result from a defect in synthesis, have been partially or fully characterized as being due to defects in specific hydrolases. Thus, these and other results argue in favor of a stepwise degradative sequence for the gangliosides, at least in nervous tissue. If there are endoglycosidases which can cleave internal glycosidic bonds in the gangliosides, they are apparently not important physiologically in *these tissues* since their action would result in degradation of the stored compounds.

There are several recent reports concerned with protein "activators" for lysosomal hydrolases. These activators have no catalytic activity *per se*, but significantly stimulate the activity of a hydrolase. Such activators have been reported for glucocerebrosidase (Ho and O'Brien, 1971), a sulfatase (Fischer and Jatzkewitz, 1975), and for gangliosidases (Li and Li, 1976; Hechtman, 1977). In recent studies (Sandhoff *et al.*, 1977; Conzelmann and Sandhoff, 1978) a new activator was described which stimulated the action of hexosaminidase A on G_{M2} 20-fold, but did not affect the activity of the isoenzyme, hexosaminidase B, on this substrate. Furthermore, one of the variants of Tay–Sachs disease was shown to have normal levels of the hexosaminidases, but was defective in the protein activator. It seems unlikely that the activator protein serves simply as a surfactant in this case, since both hexosaminidases were stimulated by detergents such as sodium taurodeoxycholate. From these results, the authors conclude that the new activator interacts specifically with hexosaminidase A. The mechanism of action of these novel enzyme activators remains to be elucidated, but the new findings indicate their important physiological functions.

6. EFFECT OF CELL TRANSFORMATION ON GANGLIOSIDE SYNTHESIS

Early reports indicated that there was a striking correlation between cell transformation (to oncogenicity) and changes in the gangliosides in such cells. Transformed cells contained simpler gangliosides than those found in the normal parental cell lines (reviewed by Fishman, 1974; and Fishman and Brady, 1976). Transformed cells exhibit certain phenotypic properties, including the following: a marked increase in their ability to give rise to tumors when injected into appropriate host animals, striking changes in morphology on tissue culture plates, lack of growth control under specified conditions, and ability to grow in soft agar. Because of the putative correlation between ganglioside patterns and transformation, it was concluded that gangliosides in cell surface membranes probably participated in these physiological processes.

In the early work, it was also shown that the change in the patterns to simpler gangliosides resulting from cell transformation occurred because the cells lost one of the glycosyltransferases in the biosynthetic pathway (Figure 7). Thus, parental cell lines, which had a full complement of these enzymes were capable of synthesizing the complete spectrum of gangliosides, whereas many of the transformed cell lines either lost the N-acetylgalactosaminyltransferase completely or contained much less of this enzyme than did the parental cells and therefore could not synthesize molecules more complex than G_{M3}. As a result G_{M3} was the sole ganglioside, or if a little of the enzyme was present, the major component.

As might be expected, these results aroused great interest and quickly became the focus of numerous studies. Ultimately, the putative correlation between transformation and ganglioside patterns became a matter of controversy, which is still not unequivocally resolved. For example, it was shown that revertants of transformed cells, which regained normal morphology and growth patterns, neither regained the defective enzyme nor the ability to synthesize the complex gangliosides (Den *et al.*, 1974). Other recent papers in this area can be summarized as follows:

The original studies included a number of papers on 3T3 mouse fibroblasts and their SV40 viral transformants. In a reinvestigation of the problem (Itaya and Hakomori, 1976) gangliosides were measured in a temperature-sensitive transformed mutant line and its parental cell line after growth of both cell types at 32°C (permissive temperature) and 37°C (nonpermissive). Qualitatively, there were no differences in gangliosides in the two cell lines. When grown at either temperature, both cell lines contained G_{M3}, G_{M2}, G_{M1}, G_{D1a}, G_{D1b}, and G_T. Quantitatively, the ganglioside patterns were also the same, with the possible exception of G_{M2}. In this case, the ratios of G_{M2} found in cells grown at 32°C compared to 37°C were: parental cell line, 70%; temperature-sensitive mutant cells, 33%. There was also a difference in the amount of G_{M2} exposed on the cell surfaces as determined by treating the intact cells with galactose oxidase and sodium borotritide, followed by isolation of the gangliosides. Although the authors suggest that the observed quantitative changes of G_{M2} correlated with expression of transformation, this reviewer was much more impressed with the lack of major changes in ganglioside patterns, both qualitatively and quantitatively, which would lead to the opposite conclusion.

Ganglioside patterns in rat embryo cells, normal and transformed (by viral, chemical, and a combination of these means), have been examined (Langenbach, 1975). All four lines showed a complex pattern of gangliosides, although there were quantitative differences between the normal and transformed cells, and between the different transformants.

Five human glioma cell lines were analyzed (Manuelidis *et al.*, 1977). Three of these maintained their characteristic morphology during early subculture, and their ganglioside patterns were similar to the pattern obtained from normal human brain tissue. Two other lines, which had been maintained for some time in culture and which showed morphology entirely different from the original primary explants (they become epitheliallike on repeated passage), had much less total ganglioside and a much simpler ganglioside pattern. Other studies from the same laboratory had previously shown that gangliosides were lost from glioblastoma cells as they were repeatedly passed in culture. From these results, the authors suggest the possibility that ganglioside composition correlates with morphology, although the same results suggest that ganglioside content does not correlate with tumorigenicity.

A careful and extensive study (Murray et al., 1973) with both normal and transformed 3T3 cells may provide the answer to the apparent controversy. Murray et al. (1973) reported that there is a considerable variation in ganglioside content and pattern among clonal isolates from a single cell line. One of the SV40-transformed clonal isolates showed the expected simpler ganglioside pattern, while two others showed the complex pattern found in the parental 3T3 cells. These and other results in this study indicate that changes in ganglioside content are not required for cell transformation. Where such changes are found, it is not clear whether they are directly related to transformation or represent secondary effects unrelated to the primary event.

Analyses of gangliosides in tumor tissues have also been performed and some of the results follow:

Twenty-five human gastric and 11 human colonic adenocarcinomas were analyzed (Keränen et al., 1976). The ganglioside patterns of the carcinomas closely resembled those of normal tissue, and in fact, most of the tumors contained significantly more total ganglioside than the normal tissue (the ganglioside content was lower in only six cases).

Increased quantities of gangliosides were found in two rat hepatomas (Dyatlovitskaya et al., 1976; Dnistrian et al., 1976). The ratios of individual gangliosides (relative to total gangliosides) were different in the hepatomas from those in normal rat liver. In addition, there was no detectable G_T in the hepatomas, whereas G_{D1b} was found in the tumor tissue and was not detected in the normal liver. It was further shown (Dyatlovitskaya et al., 1975, 1976) that the change in ganglioside levels in the tumor did not result from rapid growth of the tissue since in "normal" rapidly growing tissue, i.e., regenerating rat liver, the ganglioside content was decreased. The increased levels of gangliosides in tumors were also reflected in an increased ganglioside content of blood sera of animals bearing hepatomas (Skipski et al., 1975) and mammary carcinomas (Kloppel et al., 1977). In the latter case, surgical removal of the mammary carcinomas from mice and humans resulted in a decline in the levels of serum gangliosides.

7. GANGLIOSIDES IN DEVELOPMENT AND DIFFERENTIATION

Since nervous tissue is a particularly rich source of gangliosides, and since this tissue undergoes major changes in structure and function during growth and development, several laboratories have analyzed brain and other nervous tissues for gangliosides as a function of age.

An extensive, well-documented series of studies (Svennerholm, 1964) of human brain gave the following results. The increases in gangliosides and sphingomyelin content parallel the increase in weight of human brain from 0 to 21 days. During the period characterized by a very rapid outgrowth of dendrites and axons, and establishment of neuronal connections, i.e., between the 15th fetal week and 6 months postpartum, the ganglioside concentration increased by 300% while phospholipids increased by about 100%. There was no proportionate increase in individual gangliosides, but instead some, such as G_{D1a}, increased dramatically (as percentage of the total gangliosides), while others, such as G_{T1} and G_{D1b}, decreased. Other results of these studies showed that minor gangliosides were present in higher concentrations than previously supposed and that white matter contains lower relative amounts of di- and trisialogangliosides.

In another study on human brain (Yusuf et al., 1977) the ganglioside content of the forebrain, cerebellum, and brain stem was analyzed as a function of age from 13 weeks gestation to 26 months postpartum. Plateau values were reached in each of these tissues at different times during growth. Furthermore, there were significant differences in the pattern of gangliosides, depending on the tissue. Thus, the brain stem showed little change in ganglioside composition during development, while in the forebrain there was a gradual increase in the proportion of G_{D1a}, and G_{T1} became the major ganglioside in the cerebellum.

Similar studies have been conducted on rat and pig brain (Merat and Dickerson, 1973), rat cerebellum (Vincendon et al., 1975), chick retina and brain (Dreyfus et al., 1975), and in myelin in the brain of the developing mouse (Yu and Yen, 1975). Intestinal mucosa also represents a differentiating tissue, as crypt cells migrate and change to villus cells. The major ganglioside of rat intestinal mucosa was found to be G_{M3}, and its content in crypt cells was found to be significantly lower than that in villus cells (Glickman and Bouhours, 1976). In a potentially interesting communication (Whatley et al., 1976), it was reported that cloned rat myoblast cells contained G_{M3}, G_{M2}, G_{M1}, and G_{D1a}. During differentiation to myotubes, there was essentially no change in the content of the gangliosides, with the exception of G_{D1a}, which showed a transient threefold increase. Furthermore, mutant cell lines unable to fuse did not contain

G_{DIa}. The authors suggest that the latter ganglioside may be involved in the fusion process.

Gangliosides have also been studied as possible mediators of retinal–tectal specificity. In this classical example of neuronal specificity, ganglion cells in the retina establish specific topological synaptic connections with areas of the optic tectum. An analysis of the chick embryonic tecta (6 through 12 days) dissected along the topological gradient showed that gangliosides increased in complexity with age; this effect of aging was particularly noticeable with polysialogangliosides (Irwin et al., 1976).

In this study, complex carbohydrates (not identified) showed strikingly asymmetrical topological distributions as early as 8 days of embryonic life, and even more interesting, the amounts of these compounds tended to fluctuate or oscillate within 24 hours or less. The authors suggest that the complex carbohydrates fluctuate according to topological position and developmental state in a complex, relatively rapid manner. In this connection, it has been suggested (Marchase, 1977) that a ganglioside, G_{M2}, is one of a pair of molecules (the other being a binding protein for N-acetylgalactosamine residues) responsible for retinotectal adhesive specificity. The specificity is presumably based on concentration gradients of these complementary molecules in the neuroretinal cells and in the tecta, although no such gradient could be detected experimentally by measuring ganglioside levels in the corresponding halves of these tissues.

8. GANGLIOSIDES AS MEMBRANE RECEPTORS FOR TOXINS AND HORMONES

Perhaps the most exciting and provocative area of research on gangliosides today is concerned with the functions of these molecules as cell plasma membrane receptors for certain toxins and hormones.

While this field of research is only about 5 years old, a substantial literature has already been accumulated (see, for example, reviews by Fishman and Brady, 1976; and Kohn, 1978). In studies which were conducted in the 1960s, van Heyningen and his collaborators showed that gangliosides could be used to neutralize cholera and tetanus toxins. In 1973, studies from three laboratories (Holmgren et al., 1973; King and van Heyningen, 1973; Cuatrecasas, 1973a,b,c,d) showed that G_{M1} was by far the most active ganglioside insofar as binding to the toxin was concerned (500-fold more active than any other ganglioside). These studies showed that the toxin was specific for the oligosaccharide chain of G_{M1}, but a subsequent study (Holmgren et al., 1974) showed that the sphingosine portion of the molecule was also essential; G_{M1} and its N-acetyl analogue (i.e., where the fatty acyl chain in the ceramide moiety was replaced by acetyl) were 20,000-fold more effective in interacting with the toxin than the oligosaccharide chain of G_{M1}.

These early reports were followed by papers from many laboratories on the effect of cholera toxin on a wide variety of cell types (references to many of these publications are given in the reviews, while additional references are given below). This work may be summarized as follows: (1) With few exceptions (discussed below), there is apparently general agreement that the cell membrane receptor for cholera toxin is G_{M1} in all cell types studied. (2) The intact toxin molecule consists of two protomers, a B protomer containing 4–6 identical subunits (56,000 daltons) and an A protomer, molecular weight 29,000, composed of two nonidentical subunits. (3) One key consequence of the interaction between cholera toxin and cells is a rapid increase in cyclic AMP within the cytoplasm, which is the result of activation of adenylate cyclase by the toxin. In cell-free extracts, the membrane-bound adenylate cyclase is activated either by the intact toxin or by the A protomer; the B protomer is inactive. It appears likely (Moss et al., 1977a,b) that the A protomer activates either adenylate cyclase or a protein which regulates this enzyme by transferring an ADP-ribose moiety from NAD to the acceptor protein molecule. It had earlier been shown that diphtheria toxin (Kandel et al., 1974) is also an ADP-ribosylating enzyme, and it is now suggested, although not completely established, that protomer A of cholera toxin acts in the same way (Moss et al., 1977a,b). (4) The function of the B protomer is to bind to G_{M1} in the cell membrane. After formation of this complex, it is presumed that the A protomer dissociates from the complex, penetrates into the membrane, and activates adenylate cyclase as described above. (5) In recent experiments (Gill and Meren, 1978; Cassel and Pfeuffer, 1978) it has been shown that the toxin incorporates ADP-ribose into a number of pigeon erythrocyte membrane proteins concomitant with activation of the adenylate cyclase. The major protein that incorporated the ADP-ribose is apparently the GTP binding protein which regulates adenylate cyclase. Arginine may be the amino acid residue which accepts the ADP-ribosyl moiety (Moss and Vaughn, 1977).

These conclusions are not accepted by all workers in the field. Thus, evidence has been presented (Kanfer et al., 1976) against the idea that G_{M1} is the cholera toxin receptor in adipocytes. These cells responded to the toxin (increased rate and extent of lipolysis) but contained no detectable G_{M1}; adipocytes contain G_{M3} as the major ganglioside as well as substantial quantities of G_{M2}. Furthermore, although exogenously added G_{M1} enhanced the effect of cholera toxin on whole cells (as has been previously observed with a number of other cell types), the kinetics of the stimulation (lipolysis) did not conform to the activity observed with untreated cells. The results obtained with the adipocytes have, however, been questioned (Pacuszka *et al.*, 1978*a*). By using a more sensitive analytical method, the latter workers have detected G_{M1} in adipocytes, albeit at low concentrations.

The mechanism of action of cholera toxin has also been questioned by Tait and van Heyningen (1978) since these workers could detect no NADase in the toxin. [In the absence of protein acceptors, the A promoter of cholera toxin acts as an NAD hydrolase and "transfers" the ADP-ribose moiety to water (Moss *et al.*, 1977*a,b*).]

Thus, there appears to be some doubt concerning the binding of cholera toxin to G_{M1} in whole cells and on the mechanism by which the toxin stimulates adenylate cyclase. However, at this time the bulk of the evidence strongly favors the model described above.

While other bacterial toxins have not been studied as extensively as cholera toxin, the available data suggest that these may act similarly. Tetanus toxin contains a heavy and a light polypeptide chain. This toxin binds to gangliosides containing two internal sialic acid residues such as G_{D1b} and G_{T1} (W. E. van Heyningen, 1974; S. van Heyningen, 1976). An elegant analysis of the phenomenon (Helting *et al.*, 1977) showed that the binding was extremely tight (half-saturation of the ganglioside at 5×10^{-8} M), that the oligosaccharide portion of the molecule by itself did not bind to the toxin, and that the heavy chain of the toxin was responsible for the binding. Earlier work had suggested that the light chain was responsible for the physiological effect after binding to cells had occurred.

Other reports on toxin binding to gangliosides suggested that the corresponding gangliosides were the receptors for various toxins in the cell membranes. For example, the hemolytic activity of staphylococcal alpha-toxin was specifically inhibited (Kato and Masahara, 1976) by a GlcNAc ganglioside of the following structure: sialyl-α2,3-Galβ1,4-GlcNAc-β1,3-Gal-β1,4-Glc-ceramide. Removal of the sialyl residue (which gives paragloboside) led to a major loss in activity (about 1700-fold). The references and reviews cited above list other toxins with similar abilities to bind to gangliosides.

These studies therefore strongly implicate the gangliosides as the cell membrane receptors for various toxins. Although these experiments are intrinsically of major interest, they do not in themselves reveal the biological roles of the gangliosides, since cells are not normally exposed to toxins. One possibility is that the toxins are chemical analogues of natural substances such as hormones and interferon and that the gangliosides act as membrane receptors for interferon and at least certain hormones. The review (Kohn, 1978) on this subject is extensive, and no attempt will be made to summarize it, except for a few brief comments.

While it was originally demonstrated that thyrotropin binds to gangliosides (Mullin *et al.*, 1976) and that gangliosides may be the receptors in thyroid tissue responsible for the binding of this hormone (Meldolesi *et al.*, 1976), these conclusions are apparently in doubt according to more recent work from at least one of these laboratories (Pacuszka *et al.*, 1978*b*). Furthermore, as discussed in the review (Kohn, 1978) and elsewhere (Meldolesi *et al.*, 1977), binding of the hormone to thyroid plasma membranes involves at least one component in addition to gangliosides, namely, a glycoprotein fraction. From this, and other results, hormone binding to plasma membranes and intact cells may or may not involve gangliosides, but even if these glycolipids are involved, the binding site is more complex than is required for toxins, particularly for cholera toxin. At this time, it appears reasonable to conclude that hormone binding to cells via gangliosides is an immensely important but unresolved issue.

One final point must be considered with respect to toxin (and possibly hormone) binding to cells. Is the total physiological effect expressed only by activation of adenylate cyclase? For hormones, at least, there is no question that a concomitant major effect is on cell permeability (Kohn, 1978). This may also be true for the effect of the toxins. Liposomes containing gangliosides and preloaded with glucose were treated with cholera toxin or the A or B protomers (Ohsawa *et al.*, 1977; Moss *et al.*, 1976, 1977*b*). A portion of the trapped glucose was released from the liposomes by this treatment, but only by the cholera toxin or the B protomer and only when the liposomes contained the appropriate receptor, G_{M1}. These effects on the permeability of model membrane systems are in accord with the physical studies discussed in an earlier section of this review. That is, gangliosides apparently do not pack in the lipid bilayer in the same way as phospholipids. One can therefore speculate as follows. In the untreated cell membrane, gangliosides are normally present at low concentrations (excluding synapses, etc.), and thus are dispersed as monomolecular species. When cells are treated with bi- or polyvalent reagents which bind to gangliosides, such as toxins or antibodies, the gangliosides are clustered. At these points, there may be sufficient perturbation of the lipid bilayer to result in an important, perhaps even a drastic, change in cell permeability.

If toxins and possibly hormones can induce such changes in permeability, the latter may be at least as important and take place much more quickly than do changes in cyclic AMP levels within the cell.

In conclusion, the physiological functions of gangliosides and other glycosphingolipids (as well as glycoproteins for that matter) remain largely unknown. Recent work in this area, however, suggests that these substances are probably involved in a wide variety of important processes, and that at least some of these functions will become clarified over the course of the next decade or two.

ACKNOWLEDGMENTS

Part B of this chapter would not have been written without the expert help of Dr. Pamela Talalay. One of us (S.R.) also wishes to express his gratitude to Mr. Mark Zimmerman and Mrs. Dorothy Regula for their assistance. Some of the studies described in Part B were supported by a grant from the NIH, AM09851.

9. REFERENCES

Andrews, P., 1970, Purification of lactose synthetase A protein from human milk and demonstration of its interaction with alpha-lactalbumin, *FEBS Lett.* **9**:297.

Avrova, N. F., Chenykaeva, E. Yu, and Obukhova, E. L., 1973, Ganglioside composition and content of rat-brain subcellular fractions, *J. Neurochem.* **20**:997.

Baenziger, J., and Kornfeld, S., 1974, Structure of the carbohydrate units of IgA₁ immunoglobulin. II. Structure of the O-glycosidically linked oligosaccharide units, *J. Biol. Chem.* **249**:7270.

Barker, R., Olsen, K. W., Shaper, J. H., and Hill, R. L., 1972, Agarose derivatives of uridine diphosphate and *N*-acetylglucosamine for the purification of a galactosyltransferase, *J. Biol. Chem.* **247**:7135.

Bartholomew, B. A., Jourdian, G. W., and Roseman, S., 1973, The sialic acids. XV. Transfer of sialic acid to glycoproteins by a sialyltransferase from colostrum, *J. Biol. Chem.* **248**:5751.

Basu, S., Basu, M., and Chien, J. L., 1975, Enzymatic synthesis of a blood group H-related glycosphingolipid by an α-fucosyltransferase from bovine spleen, *J. Biol. Chem.* **250**:2956.

Behrens, N. H., Parodi, A. J., and Leloir, L. F., 1971, Glucose transfer from dolichol monophosphate glucose: The product formed with endogenous microsomal acceptor, *Proc. Natl. Acad. Sci. U.S.A.* **68**:2857.

Bell, J. E., Beyer, T. A., and Hill, R. L., 1976, The kinetic mechanism of bovine milk galactosyltransferase, the role of α-lactalbumin, *J. Biol. Chem.* **251**:3003.

Bennett, G., Leblond, C. P., and Haddad, A., 1974, Migration of glycoprotein from the Golgi apparatus to the surface of various cell types as shown by radioautography after labeled fucose injection into rats, *J. Cell Biol.* **60**:258.

Bergeron, J. J. M., Ehrenreich, J. H., Siekevitz, P., and Palade, G. E., 1973, Golgi fractions prepared from rat liver homogenates. II. Biochemical characterization, *J. Cell Biol.* **59**:73.

Beyer, T. A., Prieels, J. P., and Hill, R. L., 1979, Characterization of two highly purified fucosyltransferases, in: *Glycoconjugate Research: Proceedings of the Fourth International Symposium on Glycoconjugates,* Vol. II (J. D. Gregory and R. W. Jealoz, eds.), pp. 641–643, Academic Press, New York.

Bischoff, E., Tranh-Thi, T. A., and Decker, K. F. A., 1975, Nucleotide pyrophosphatase of rat liver, *Eur. J. Biochem.* **51**:353.

Bischoff, E., Wilkening, J., Tranh-Thi, T. A., and Decker, K., 1976, Differentiation of the nucleotide pyrophosphatases of rat liver plasma membrane and endoplasmic reticulum by enzymic iodination, *Eur. J. Biochem.* **62**:279.

Bosmann, H. B., 1970, Glycoprotein biosynthesis: Purification and properties of glycoprotein *N*-acetylglucosaminyltransferases from guinea pig liver utilizing endogenous and exogenous acceptors, *Eur. J. Biochem.* **14**:33.

Brew, K., Vanaman, T. C., and Hill, R. L., 1968, The role of α-lactalbumin and the A protein in lactose synthetase: A unique mechanism for the control of a biological reaction, *Proc. Natl. Acad. Sci. U.S.A.* **59**:491.

Buscher, H. P., Casals-Stenzel, J., Schauer, R., and Mestres-Ventura, P., 1977, Biosynthesis of *N*-glycolylneuraminic acid in porcine submandibular glands, *Eur. J. Biochem.* **77**:297.

Caputto, R., Maccioni, H. J., and Arce, A., 1974, Biosynthesis of brain gangliosides, *Mol. Cell Biochem.* **4**:97.

Carlson, D. M., McGuire, E. J., and Jourdian, G. W., 1966, Sheep submaxillary gland sialyltransferase, in: *Methods in Enzymology,* Vol. VIII, *Complex Carbohydrates* (E. F. Neufeld and V. Ginsburg, eds.), pp. 361–365, Academic Press, New York.

Carlson, D. M., Iyer, R. N., and Mayo, J., 1970, Carbohydrate compositions of epithelial mucins, in: *Blood and Tissue Antigens* (D. Aminoff, ed.), pp. 229–247, Academic Press, New York.

Carlson, D. M., David, J., and Rutter, W. J., 1973*a*, Galactosyltransferase activities in pancreas, liver and gut of the developing rat, *Arch. Biochem. Biophys.* **157**:605.

Carlson, D. M., Jourdian, G. W., and Roseman, S., 1973*b*, The sialic acids. XIV. Synthesis of sialyl-lactose by a sialyltransferase from rat mammary gland, *J. Biol. Chem.* **248**:5742.

Carlson, D. M., McGuire, E. J., Jourdian, G. W., and Roseman, S., 1973*c*, The sialic acids. XVI. Isolation of a mucin sialyltransferase from sheep submaxillary gland, *J. Biol. Chem.* **248**:5763.

Cassel, D., and Pfeuffer, T., 1978, Mechanism of cholera toxin action: Covalent modification of the guanyl nucleotide-binding protein of the adenylate cyclase system, *Proc. Natl. Acad. Sci. U.S.A.* **75**:2669.

Chabaud, O., Bouchilloux, S., Ronin, C., and Ferrand, M., 1974, Localization in a Golgirich thyroid fraction of sialyl-, galactosyl-, and *N*-acetylglucosaminyltransferases, *Biochimie* **56**:119.

Chester, M. A., Yates, A. D., and Watkins, W. M., 1976, Phenyl β-D-galactopyranoside as an acceptor substrate for the blood group *H* gene-associated GDP-L-fucose: β-D-galactosyl α-2-L-fucosyltransferase, *Eur. J. Biochem.* **69**:583.

Chou, T. H., Murphy, C., and Kessel, D., 1977, Selective inhibition of a plasma fucosyltransferase by *N*-ethylmaleimide, *Biochem. Biophys. Res. Commun.* **74**:1001.

Coleman, P. L., Fishman, P. H., Brady, R. O., and Todaro, G. J., 1975, Altered ganglioside biosynthesis in mouse cell cultures following transformation with chemical carcinogens and X-irradiation, *J. Biol. Chem.* **250**:55.

Conzelmann, E., and Sandhoff, K., 1978, AB variant of infantile G_{M2} gangliosidosis: Deficiency of a factor necessary for stimulation of hexosaminidase A-catalyzed degradation of ganglioside G_{M2} and glycolipid G_{A2}, *Proc. Natl. Acad. Sci. U.S.A.* **75**:3979.

Corfield, A. P., Ferreira do Amaral, C., Wember, M., and Schauer, R., 1976, The metabolism of *O*-acyl-*N*-acylneuraminic acids, biosynthesis of *O*-acylated sialic acids in bovine and equine submandibular glands, *Eur. J. Biochem.* **68**:597.

Cuatrecasas, P., 1973*a*, Interaction of *Vibrio cholerae* enterotoxin with cell membranes, *Biochemistry* **12**:3547.

Cuatrecasas, P., 1973*b*, Gangliosides and membrane receptors for cholera toxin, *Biochemistry* **12**:3558.

Cuatrecasas, P., 1973c, Cholera toxin-fat cell interaction and the mechanism of activation of the lipolytic response, *Biochemistry* **12**:3567.

Cuatrecasas, P., 1973d, *Vibrio cholerae* choleragenoid. Mechanism of inhibition of cholera toxin action, *Biochemistry* **12**:3577.

Curatolo, W., Small, D. M., and Shipley, G., 1977, Phase behavior and structural characteristics of hydrated bovine brain gangliosides, *Biochim. Biophys. Acta* **468**:11.

De Baecque, C., Johnson, A. B., Naiki, M., Schwarting, G., and Marcus, D. M., 1976, Ganglioside localization in cerebellar cortex: An immunoperoxidase study with antibody to G_{M1} ganglioside, *Brain Res.* **114**:117.

Decker, K., and Bischoff, E., 1972, Purification and properties of nucleotide pyrophosphatase from rat liver plasma membranes, *FEBS Lett.* **21**:9598.

Dejter-Juszynski, M., Harpaz, N., Flowers, H. M., and Sharon, N., 1978, Blood group ABH-specific macroglycolipids of human erythrocytes. Isolation in high yield from a crude membrane glycoprotein fraction, *Eur. J. Biochem.* **83**:363.

Den, H., Kaufman, B., and Roseman, S., 1970, Properties of some glycosyltransferases in embryonic chicken brain, *J. Biol. Chem.* **245**:6607.

Den, H., Sela, B.-A., Roseman, S., and Sachs, L., 1974, Blocks in ganglioside synthesis in transformed hamster cells and their revertants, *J. Biol. Chem.* **249**:659.

Den, H., Kaufman, B., McGuire, E. J., and Roseman, S., 1975, The sialic acids. XVIII. Subcellular distribution of seven glycosyltransferases in embryonic chicken brain, *J. Biol. Chem.* **250**:739.

Dnistrian, A. M., Skipski, V. P., Barclay, M., Essner, E. S., and Stock, C. C., 1975, Gangliosides of plasma membranes from normal rat liver and Morris hepatoma, *Biochem. Biophys. Res. Commun.* **64**:367.

Dreyfus, H., Urban, P. F., Edel-Harth, S., and Mandel, P., 1975, Developmental patterns of gangliosides and of phospholipids in chick retina and brain, *J. Neurochem.* **25**:245.

Dyatlovitskaya, E. V., Morgenrot, U., Novikov, A. M., Mal'kova, V. P., and Bergel'son, L. D., 1975, Gangliosides of the regenerating liver, *Dokl. Akad. Nauk. SSSR.* **223**:1481.

Dyatlovitskaya, E. V., Novikov, A. M., Gorkova, N. P., and Bergelson, L. D., 1976, Gangliosides of hepatoma 27, normal and regenerating rat liver, *Eur. J. Biochem.* **63**:357.

Ebner, K. E., 1973, Lactose synthetase, in: *The Enzymes* (P. D. Boyer, ed.), Vol. 9, 3rd ed., pp. 363–377, Academic Press, New York.

Evans, W. H., 1974, Nucleotide pyrophosphatase, a sialoglycoprotein located on the hepatocyte surface, *Nature* **250**:391.

Evans, W. H., Hood, D. O., and Gurd, J. W., 1973, Purification and properties of a mouse liver plasma-membrane glycoprotein hydrolysing nucleotide pyrophosphate and phosphodiester bonds, *Biochem. J.* **135**:819.

Fischer, G., and Jatzkewitz, H., 1975, The activator of cerebroside sulphatase. Purification from human liver and identification as a protein, *Hoppe-Seylers Z. Physiol. Chem.* **356**:605.

Fishman, P. H., 1974, Normal and abnormal biosynthesis of gangliosides, *Chem. Phys. Lipids* **13**:305.

Fishman, P. H., and Brady, R. O., 1976, Biosynthesis and function of gangliosides, *Science* **194**:906.

Fitzgerald, D. K., Brodbeck, U., Kiyosawa, I., Mawal, R., Colvin, B., and Ebner, K. E., 1970a, Alpha-lactalbumin and the lactose synthetase reaction, *J. Biol. Chem.* **245**:2103.

Fitzgerald, D. K., Colvin, B., Mawal, R., and Ebner, K. E., 1970b, Enzymic assay for galactosyl transferase activity of lactose synthetase and α-lactalbumin in purified and crude systems, *Anal. Biochem.* **36**:43.

Fitzgerald, D. K., McKenzie, L., and Ebner, K. E., 1971, Galactosyltransferase activity in a variety of sources, *Biochim. Biophys. Acta* **235**:425.

Fleischer, B., Fleischer, S., and Ozawa, H., 1969, Isolation and characterization of Golgi membranes from bovine liver, *J. Cell Biol.* **43**:59.

Fong, J. W., Ledeen, R. W., Kundu, S. K., and Brostoff, S. W., 1976, Gangliosides of peripheral nerve myelin, *J. Neurochem.* **26**:157.

Fournet, B., Strecker, G., Spik, G., Montreuil, J., Schmid, K., Binette, J. P., Dorland, L., Haverkamp, J., Schut, B. L., and Vliegenthart, J. F. G., 1979, Structure of ten glycopeptides from orosomucoid, in: *Glycoconjugate Research: Proceedings of the Fourth International Symposium on Glycoconjugates*, Vol. I (J. D. Gregory and R. W. Jealoz, eds.), pp. 149–156, Academic Press, New York.

Fraser, I. H., and Mookerjea, S., 1976, Studies on the purification and properties of UDP-galactose-glycoprotein galactosyltransferase from rat liver and serum, *Biochem. J.* **156**:347.

Freilich, L. S., Lewis, R. G., Reppucci, A. C., Jr., and Silbert, J. E., 1975a, Glycosaminoglycan-synthesizing activity of an isolated Golgi preparation from cultured mast cells, *Biochem. Biophys. Res. Commun.* **63**:663.

Freilich, L. S., Richmond, M. E., Reppucci, A. C., Jr., and Silbert, J. E., 1975b, A micro method for simultaneous determination of galactosyl-transferase and 5'-nucleotidase activities in cell fractions, *Biochem. J.* **146**:741.

Freilich, L. S., Lewis, R. G., Reppucci, A. C., Jr., and Silbert, J. E., 1977, Galactosyltransferase of a Golgi fraction from cultured neoplastic mast cells, *J. Cell Biol.* **72**:655.

Geren, C. R., Geren, L. M., and Ebner, K. E., 1975a, Galactosyltransferase: Confirmation of equilibrium-ordered mechanism, *Biochem. Biophys. Res. Commun.* **66**:139.

Geren, C. R., Magee, S. C., and Ebner, K. E., 1975b, Circular dichroism changes in galactosyltransferase upon substrate binding, *Biochemistry* **14**:1461.

Geren, C. R., Magee, S. C., and Ebner, K. E., 1976, Hydrophobic chromatography of galactosyltransferase, *Arch. Biochem. Biophys.* **172**:149.

Geren, C. R., Geren, L. M., Lee, D., and Ebner, K. E., 1977, Incorporation of galactose into galactosyltransferase, *Biochim. Biophys. Acta* **497**:128.

Gill, D. M., and Meren, R., 1978, ADP-Ribosylation of membrane proteins catalyzed by cholera toxin: Basis of the activation of adenylate cyclase, *Proc. Natl. Acad. Sci. U.S.A.* **75**:3050.

Glickman, R. M., and Bouhours, J. F., 1976, Characterization, distribution, and biosynthesis of the major ganglioside of rat intestinal mucosa, *Biochim. Biophys. Acta* **424**:17.

Goodwin, S. D., and Watkins, W. M., 1974, The peptide moiety of blood group specific glycoproteins. Some amino acid sequences in the regions carrying the carbohydrate chains, *Eur. J. Biochem.* **47**:371.

Haddad, A., Smith, M. D., Herscovics, A., Nadler, N. J., and Leblond, C. P., 1971, Radioautographic study of *in vivo* and *in vitro* incorporation of fucose-^{3}H into thyroglobulin by rat thyroid follicular cells, *J. Cell Biol.* **49**:856.

Hagopian, A., and Eylar, E. H., 1968a, Glycoprotein biosynthesis: The basic protein encephalitogen from bovine spinal cord, a receptor for the polypeptidyl:N-acetylgalactosaminyltransferase from bovine submaxillary glands, *Arch. Biochem. Biophys.* **126**:785.

Hagopian, A., and Eylar, E. H., 1968b, Glycoprotein biosynthesis: Studies on the receptor specificity of the polypeptidyl:N-acetylgalactosaminyltransferase from bovine submaxillary glands, *Arch. Biochem. Biophys.* **128**:422.

Hagopian, A., and Eylar, E. H., 1969a, Glycoprotein biosynthesis: The solubilization of glycosyltransferases from membranes of HeLa cells and bovine submaxillary glycoproteins, *Arch. Biochem. Biophys.* **129**:447.

Hagopian, A., and Eylar, E. H., 1969b, Glycoprotein biosynthesis: The purification and characterization of a polypeptide:N-acetylgalactosaminyltransferase from bovine submaxillary glands, *Arch. Biochem. Biophys.* **129**:515.

Hagopian, A., Westall, F. C., Whitehead, J. S., and Eylar, E. H., 1971, Glycosylation of the A1 protein from myelin by a polypeptide N-acetylgalactosaminyltransferase. Identification of the receptor sequence, *J. Biol. Chem.* **246**:2519.

Hakomori, S., and Young, W. W., Jr., 1978, Tumor-associated glycolipid antigens and modified blood group antigens, *Scand. J. Immunol.* **7**:97.

Harpaz, N., and Schachter, H., 1979a, Purification of UDP-GlcNAc:α-D-mannoside GlcNAc-transferase I from bovine colostrum and separation from UDP-GlcNAc:α-D-mannoside GlcNAc-transferase II, *Fed. Proc.* **38**:292.

Harpaz, N., and Schachter, H., 1979b, Processing of "hybrid" Asn-linked oligosaccharides by an N-acetylglucosaminyltransferase I-dependent α-D-mannosidase in rat liver Golgi-enriched membranes, in: *Proceedings of the Fifth International Symposium on Glycoconjugates,* (R. Schauer, P. Boer, E. Buddecke, M. F. Kramer, J. F. G. Vliegenthart, and H. Wiegandt, eds.), pp. 307–308, Georg Thieme Publishers, Stuttgart, Germany.

Hassid, W. Z., 1967, Biosynthesis of complex saccharides, in: *Metabolic Pathways* (D. M. Greenberg, ed.), Vol. 1, 3rd ed., pp. 307–392, Academic Press, New York.

Hechtman, P., 1977, Characterization of an activating factor required for hydrolysis of G_{M2} ganglioside catalyzed by hexosaminidase A, *Can. J. Biochem.* **55**:315.

Helting, T. B., Zwisler, O., and Weigandt, H., 1977, Structure of tetanus toxin. II. Toxin binding to ganglioside, *J. Biol. Chem.* **252**:194.

Hercz, A., Katona, E., Cutz, E., Wilson, J. R., and Barton, M., 1978, α_1-Anti trypsin. The presence of excess mannose in the z variant isolated from liver, *Science* **4362**:1229.

Herscovics, A., Bugge, B., and Jeanloz, R. W., 1977a, Glucosyltransferase activity in calf pancreas microsomes, *J. Biol. Chem.* **252**:2271.

Herscovics, A., Golovtchenko, A. M., Warren, C. D., Bugge, B., and Jeanloz, R. W., 1977b, Mannosyltransferase activity in calf pancreas microsomes, *J. Biol. Chem.* **252**:224.

Hill, H. D., Jr., Reynolds, J., and Hill, R. L., 1977a, Purification, composition, molecular weight, and subunit structure of ovine submaxillary mucin, *J. Biol. Chem.* **252**:3791.

Hill, H. D., Jr., Schwyzer, M., Steinman, H. M., and Hill, R. L., 1977b, Ovine submaxillary mucin. Primary structure and peptide substrates of UDP-N-acetylgalactosamine:mucin transferase, *J. Biol. Chem.* **252**:3799.

Ho, M. W., and O'Brien, J. S., 1971, Gaucher's disease: Deficiency of "acid" β-glucosidase and reconstitution of enzyme activity *in vitro, Proc. Natl. Acad. Sci. U.S.A.* **68**:2810.

Holm, M., 1977, Biodegradation of the major rabbit retinal gangliosides, studied *in vivo, FEBS Lett.* **77**:225.

Holm, M., and Månsson, J.-E., 1974, Differences in the incorporation of N-[acetyl-³H]mannosamine into the sialic acid of the major retinal gangliosides, studied *in vivo, FEBS Lett.* **46**:200.

Holmgren, J., Lönnroth, J.-E., and Svennerholm, L., 1973, Tissue receptor for cholera exotoxin: Postulated structure from studies with G_{M1} ganglioside and related glycolipids, *Infect. Immun.* **8**:208.

Holmgren, J., Månsson, J.-E., and Svennerholm, L., 1974, Tissue receptor for cholera exotoxin: Structural requirements for G_{M1} ganglioside in toxin binding and inactivation, *Med. Biol.* **52**:229.

Hudgin, R. L., and Schachter, H., 1971a, Porcine sugar nucleotide:glycoprotein glycosyltransferases. I. Blood serum and liver sialyltransferase, *Can. J. Biochem.* **49**:829.

Hudgin, R. L., and Schachter, H., 1971b, Porcine sugar nucleotide:glycoprotein glycosyltransferases. II. Blood serum and liver galactosyltransferase, *Can. J. Biochem.* **49**:838.

Hudgin, R. L., and Schachter, H., 1971c, Porcine sugar nucleotide:glycoprotein glycosyltransferase. III. Blood serum and liver N-acetylglucosaminyltransferase, *Can. J. Biochem.* **49**:847.

Hudgin, R. L., and Schachter, H., 1972, Evidence for two CMP-N-acetylneuraminic acid:lactose sialyltransferases in rat, porcine, bovine and human liver, *Can. J. Biochem.* **50**:1024.

Hunt, L. A., Etchison, J. R., Robertson, J., Robertson, M., and Summers, D. F., 1977, Host-cell processing of envelope glycoprotein of vesicular stomatitis virus, Abstracts, 174th National Meeting of the American Chemical Society, Chicago, Illinois.

Hunt, L. A., Etchison, J. R., and Summers, D. F., 1978, Oligosaccharide chains are trimmed during synthesis of the envelope glycoprotein of vesicular stomatitis virus, *Proc. Natl. Acad. Sci. U.S.A.* **75**:754.

Irwin, L. N., Chen, H., and Barraco, R. A., 1976, Ganglioside, protein, hexose, and sialic acid changes in the trisected optic tectum of the chick embryo, *Dev. Biol.* **49**:29.

Isemura, M., and Schmid, K., 1971, Studies on the carbohydrate moiety of α_1-acid glycoprotein (orosomucoid) by using alkaline hydrolysis and deamination by nitrous acid, *Biochem. J.* **124**:591.

Itaya, K., and Hakomori, S., 1976, Gangliosides and "galactoprotein A" ("LETS" protein) of temperature-sensitive mutant of transformed 3T3 cells, *FEBS Lett.* **66**:65.

Ivatt, R. J., and Rosemeyer, M. A., 1972, The complex formed between the A and B proteins of lactose synthetase, *FEBS Lett.* **28**:195.

Jabbal, I., and Schachter, H., 1971, Pork liver GDP-L-fucose glycoprotein fucosyltransferases, *J. Biol. Chem.* **246**:5154.

Jamieson, J. C., 1977, Studies on the site of addition of sialic acid and glucosamine to rat α_1-acid glycoprotein, *Can. J. Biochem.* **55**:408.

Johnston, I. R., McGuire, E. J., Jourdian, G. W., and Roseman, S., 1966, Incorporation of N-acetyl-D-glucosamine into glycoproteins, *J. Biol. Chem.* **241**:5735.

Johnston, I. R., McGuire, E. J., and Roseman, S., 1973, Sialic acids. XVII. A uridine diphosphate N-acetylglucosamine:glycoprotein N-acetylglucosaminyltransferase from goat colostrum, *J. Biol. Chem.* **248**:7281.

Kabat, E. A., 1970, The carbohydrate moiety of the water-soluble human A, B, H, Le[a] and Le[b] substances, in: *Blood and Tissue Antigens* (D. Aminoff, ed.), pp. 187–203, Academic Press, New York.

Kandel, J., Collier, R. J., and Chung, D. W., 1974, Interaction of fragment A from diphtheria toxin with nicotinamide adenine dinucleotide, *J. Biol. Chem.* **249**:2088.

Kanfer, J. N., Carter, T. P., and Katzen, H. M., 1976, Lipolytic action of cholera toxin on fat cells, Re-examination of the concept implicating G_{M1} ganglioside as the native membrane receptor, *J. Biol. Chem.* **251**:7610.

Kato, I., and Masaharu, N., 1976, Ganglioside and rabbit erythrocyte membrane receptor for staphylococcal alpha-toxin, *Infect. Immun.* **13**:289.

Katz, F. N., Rothman, J. E., Lingappa, V. R., Blobel, G., and Lodish, H. F., 1977, Membrane assembly *in vitro*: Synthesis, glycosylation and asymmetric insertion of a transmembrane protein, *Proc. Natl. Acad. Sci. U.S.A.* **74**:3278.

Kaufman, B., Basu, S., and Roseman, S., 1966, Studies on the biosynthesis of gangliosides, in: *Inborn Disorders of Sphingolipid Metabolism*, Proceedings 3rd International Symposium on the Cerebral Sphingolipidoses, pp. 193–213, Pergamon Press, New York.

Kawamura, N., and Taketomi, T., 1977, A new procedure for the isolation of brain gangliosides, and determination of their long chain base compositions, *J. Biochem. Tokyo* **81**:1217.

Kean, E. L., 1970, Nuclear cytidine 5'-monophosphosialic acid synthetase, *J. Biol. Chem.* **245**:2301.

Kean, E. L., and Bighouse, K. J., 1974, Cytidine 5'-monophosphosialic acid hydrolase, subcelluar location and properties, *J. Biol. Chem.* **249**:7813.

Kean, E. L., and Bruner, W. E., 1971, Cytidine 5'-monophosphosialic acid synthetase, activity and localization in the nucleate fragments of the unfertilized sea urchin egg, *Exp. Cell Res.* **69**:384.

Keenan, T. W., Morré, D. J., and Basu, S., 1974, Ganglioside biosynthesis. Concentration of glycosphingolipid glycosyltransferases in Golgi apparatus from rat liver, *J. Biol. Chem.* **249**:310.

Keränen, A., Lempiner, M., and Puro, K., 1976, Ganglioside pattern and neuraminic acid content of human gastric and colonic carcinoma, *Clin. Chim. Acta* **70**:103.

Kessel, D., Sykes, E., and Henderson, M., 1977, Glycosyltransferases levels in tumors metastatic to liver and in uninvolved liver tissue, *J. Natl. Cancer Inst.* **59**:29.

Khatra, B. S., Herries, D. G., and Brew, K., 1974, Some kinetic properties of human milk galactosyltransferase, *Eur. J. Biochem.* **44**:537.

Khilanani, P., Chou, T. H., and Kessel, D., 1978, Guanosine diphosphate-L-fucose plasma: *N*-Acetylglucosaminide fucosyltransferase as an index of bone marrow hyperplasia after chemotherapy, *Cancer Res.* **38**:181.

Kiely, M. L., McKnight, G. S., and Schimke, R. T., 1976, Studies on the attachment of carbohydrate to ovalbumin nascent chains in hen oviduct, *J. Biol. Chem.* **251**:5490.

Kim, Y. S., Perdomo, J., Bella, A., Jr., and Nordberg, J., 1971, Properties of a CMP-*N*-acetylneuraminic acid:glycoprotein sialytransferase in human serum and erythrocyte membranes, *Biochim. Biophys. Acta* **244**:505.

Kim, Y. S., Perdomo, J., and Whitehead, J. S., 1972a, Glycosyltransferases in human blood. I. Galactosyltransferase in human serum and erythrocyte membranes, *J. Clin. Invest.* **51**;2024.

Kim, Y. S., Perdomo, J., Whitehead, J. S., and Curtis, K. J., 1972b, Glycosyltransferases in human blood. II. Study of serum galactosyltransferase and *N*-acetylgalactosaminyl-transferase in patients with liver diseases, *J. Clin. Invest.* **51**:2033.

King, C. A., and van Heyningen, W. E., 1973, Deactivation of cholera toxin by a sialidase-resistant monosialosylganglioside, *J. Infect. Dis.* **127**:639.

Klenk, E., 1969, On cerebrosides and gangliosides, *Progr. Chem. Fats Other Lipids* **10**:411.

Kloppel, T. M., Keenan, T. W., Freeman, M. J., and Morré, J., 1977, Glycolipid-bound sialic acid in serum: Increased levels in mice and humans bearing mammary carcinomas, *Proc. Natl. Acad. Sci. U.S.A.* **74**:3011.

Ko, G. K. W., Raghupathy, E., and McKean, C. M., 1973, UDP-galactose:glycoprotein galactosyltransferase and UDP-*N*-acetylgalactosamine:protein *N*-acetylgalactosaminyltransferase activities of human cerebrospinal fluid, *Can. J. Biochem.* **51**:1460.

Kohn, L. D., 1978, Relationships in the structure and function of receptors for glycoprotein hormones, bacterial toxins and interferon, in: *Receptors and Recognition* (P. Cuatrecasas and M. F. Greaves, eds.), Series A, Vol. 5, pp. 133–212, Chapman and Hall, London.

Langenbach, R., 1975, Gangliosides of chemically and virally transformed rat embryo cells, *Biochim. Biophys. Acta* **388**:231.

Ledeen, R., and Yu, R. K., 1973, Isolation and purification of gangliosides, in: *Biological Diagnosis of Brain Disorders. The Future of the Brain Sciences* (S. Bogoch, ed.), pp. 372–376, Spectrum, New York.

Lehman, E. D., Hudson, B. G., and Ebner, K. E., 1975, Studies on the carbohydrate structure of bovine milk galactosyltransferase, *FEBS Lett.* **54**:65.

Leloir, L. F., and Cardini, C. E., 1957, Biosynthesis of glycogen from uridine diphosphate glucose, *J. Am. Chem. Soc.* **79**:6340.

Letts, P. J., Meistrich, M. L., Bruce, W. R., and Schachter, H., 1974a, Glycoprotein glycosyltransferase levels during spermatogenesis in mice, *Biochim. Biophys. Acta* **343**:192.

Letts, P. J., Pinteric, L., and Schachter, H., 1974b, Localization of glycoprotein glycosyltransferases in the Golgi apparatus of rat and mouse testis, *Biochim. Biophys. Acta* **372**:304.

Levy, J. A., Carminatti, H., Cantarella, A. I., Behrens, N. H., Leloir, L. F., and Tabora, E., 1974, Mannose transfer to lipid linked di-N-acetylchitobiose, *Biochem. Biophys. Res. Commun.* **60:**118.

Li, S. C., and Li, Y.-T., 1976, An activator stimulating the enzymic hydrolysis of spingoglycolipids, *J. Biol. Chem.* **251:**1159.

Li, Y.-T., Månsson, J.-E., Vanier, M.-T., and Svennerholm, L., 1973, Structure of the major glucosamine-containing ganglioside of human tissues, *J. Biol. Chem.* **248:**2634.

Magee, S. C., Mawal, R., and Ebner, K. E., 1974, Multiple forms of galactosyltransferase from bovine milk, *Biochemistry* **13:**99.

Magee, S. C., Geren, C. R., and Ebner, K. E., 1976, Plasmin and the conversion of the molecular forms of bovine milk galactosyltransferase, *Biochim. Biophys. Acta* **420:**187.

Månsson, J.-E., Holmgren, J., Li, Y.-T., Vanier, M.-T., and Svennerholm, L., 1974, Chemical and immunological characterization of the major glucosamine-containing ganglioside of human tissues, *Med. Biol.* **52:**240.

Manuelidis, L., Yu, R. K., and Manuelidis, E. E., 1977, Ganglioside content and pattern in human gliomas in culture. Correlation of morphological changes with altered gangliosides, *Acta Neuropathol.* **38:**129.

Marchase, R. B., 1977, Biochemical investigations of retinotectal adhesive specificity, *J. Cell Biol.* **75:**237.

Marchesi, V. T., Furthmayr, H., and Tomita, M., 1976, The red cell membrane, *Annu. Rev. Biochem.* **45:**667.

Marshall, R. D., 1972, Glycoproteins, *Annu. Rev. Biochem.* **41:**673.

Marshall, R. D., 1974, The nature and metabolism of the carbohydrate–peptide linkages of glycoproteins, *Biochem. Soc. Symp.* **40:**17.

McGuire, E. J., 1970, Biosynthesis of submaxillary mucins, in: *Blood and Tissue Antigens* (D. Aminoff, ed.), pp. 461–478, Academic Press, New York.

McGuire, E. J., and Roseman, S., 1967, Enzymatic synthesis of the protein–hexosamine linkage in sheep submaxillary mucin, *J. Biol. Chem.* **242:**3745.

Meldolesi, M. F., Fishman, P. H., Aloj, S. M., Kohn, L. D., and Brady, R. O., 1976, Relationship of gangliosides to the structure and function of thyrotropin receptors: Their absence on plasma membranes of a thyroid tumor defective in thyrotropin receptor activity, *Proc. Natl. Acad. Sci. U.S.A.* **73:**4060.

Meldolesi, M. F., Fishman, P. H., Aloj, S. M., Ledley, F. D., Lee, G., Bradley, R. M., Brady, R. O., and Kohn, L. D., 1977, Separation of the glycoprotein and ganglioside components of thyrotropin receptor activity in plasma membranes, *Biochem. Biophys. Res. Commun.* **75:**581.

Mendicino, J., and Rao, A. K., 1975, Regulation of the synthesis of nucleoside diphosphate sugars in reticulo-endothelial tissues, *Eur. J. Biochem.* **51:**547.

Merat, A., and Dickerson, J. W. T., 1973, The effect of development on the gangliosides of rat and pig brain, *J. Neurochem.* **20:**873.

Montreuil, J., 1975, Recent data on the structure of the carbohydrate moiety of glycoproteins. Metabolic and biological implications, *Pure Appl. Chem.* **42:**431.

Mookerjea, S., and Yung, J. W. M., 1975, Studies on uridine diphosphate-galactose pyrophosphatase and uridine diphosphate-galactose:glycoprotein galactosyltransferase activities in microsomal membranes, *Arch. Biochem. Biophys.* **166:**223.

Mookerjea, S., Chow, A., and Hudgin, R. L., 1971, Occurrence of UDP-N-acetylglucosamine:glycoprotein N-acetylglucosaminyltransferase activity in human and rat sera, *Can. J. Biochem.* **49:**297.

Mookerjea, S., Michaels, M. A., Hudgin, R. L., Moscarello, M. A., Chow, A., and Schachter, H., 1972, The levels of nucleotide-sugar:glycoprotein sialyl- and N-acetylglucosaminyltransferases in normal and pathological human sera, *Can. J. Biochem.* **50:**738.

Moss, J., and Vaughan, M., 1977, Mechanism of action of choleragen. Evidence for ADP-ribosyltransferase activity with arginine as an acceptor, *J. Biol. Chem.* **252:**2455.

Moss, J., Fishman, P. H., Richards, R. L., Alving, C. R., Vaughan, M., and Brady, R. O., 1976, Choleragen-mediated release of trapped glucose from liposomes containing ganglioside G_{M1}, *Proc. Natl. Acad. Sci. U.S.A.* **73:**3480.

Moss, J., Osborne, J. C., Jr., Fishman, P. H., Brewer, H. B., Jr., Vaughan, M., and Brady, R. O., 1977a, Effect of gangliosides and substrate analogues on the hydrolysis of nicotinamide adenine dinucleotide by choleragen, *Proc. Natl. Acad. Sci. U.S.A.* **77:**74.

Moss, J., Richards, R. L., Alving, C. R., and Fishman, P. H., 1977b, Effect of the A and B protomers of choleragen on release of trapped glucose from liposomes containing or lacking ganglioside G_{M1}, *J. Biol. Chem.* **252:**797.

Mullin, B. R., Fishman, P. H., Lee, G., Aloj, S. M., Ledley, F. D., Winand, R. J., Kohn, L. D., and Brady, R. O., 1976, Thyrotropin-ganglioside interactions and their relationship to the structure and function of thyrotropin receptors, *Proc. Natl. Acad. Sci. U.S.A.* **73:**842.

Munro, J. R., and Schachter, H., 1973, The presence of two GDP-L-fucose:glycoprotein fucosyltransferases in human serum, *Arch. Biochem. Biophys.* **156:**534.

Munro, J. R., Narasimhan, S., Wetmore, S., Riordan, J. R., and Schachter, H., 1975, Intracellular localization of GDP-L-fucose:glycoprotein and CMP-sialic acid apolipoprotein glycosyltransferases in rat and pork liver, *Arch. Biochem. Biophys.* **169:**269.

Murray, R. K., Yogeeswaran, G., Sheinin, R., and Schimmer, B. P., 1973, Glycosphingolipids of clonal lines of transformed mouse fibroblasts and mouse adrenocortical cells, in: *Tumor Lipids, Biochemistry and Metabolism* (R. Wood, ed.), pp. 285–302, American Oil Chemist's Society Press, Champaign, Illinois.

Nagai, M., Muensch, H., and Yoshida, A., 1976, Isolation and characterization of blood group glycosyltransferases and genetic mechanism of blood group determination, *Fed. Proc.* **35:**1441.

Narasimhan, S., Stanley, P., and Schachter, H., 1977, Control of glycoprotein synthesis, lectin-resistant mutant containing only one of two distinct N-acetylglucosaminyltransferase activities present in wild type Chinese hamster ovary cells, *J. Biol. Chem.* **252:**3926.

Nelson, J. D., Jato-Rodriguez, J. J., and Mookerjea, S., 1974, The occurrence and properties of soluble UDP-galactose:glycoprotein galactosyltransferase in human amniotic fluid, *Can. J. Biochem.* **52:**42.

Nwokoro, N. A., and Schachter, H., 1975a, L-Fucose metabolism in mammals. Purification of pork liver 2-keto-3-deoxy-L-fuconate:NAD oxidoreductase by NAD-agarose affinity chromatography, *J. Biol. Chem.* **250:**6185.

Nwokoro, N. A., and Schachter, H., 1975b, L-Fucose metabolism in mammals. Kinetic studies on pork liver 2-keto-3-deoxy-L-fuconate:NAD oxidoreductase, *J. Biol. Chem.* **250:**6191.

Ohashi, M., and Yamakawa, T., 1977, Isolation and characterization of glycosphingolipids in pig adipose tissue, *J. Biochem. Tokyo* **81:**1675.

Ohsawa, T., Yoshitaka, N., and Weigandt, H., 1977, Functional incorporation of gangliosides into liposomes, *Jpn. J. Exp. Med.* **47:**221.

Pacuszka, T., and Kóscielak, J., 1974, α1,2-Fucosyltransferase of human bone marrow, *FEBS Lett.* **41:**348.

Pacuszka, T., and Kóscielak, J., 1976, Enzymatic synthesis of two fucose-containing glycolipids with fucosyltransferases of human serum, *Eur. J. Biochem.* **64:**499.

Pacuszka, T., Moss, J., and Fishman, P. H., 1978a, A sensitive method for the detection of G_{M1}-ganglioside in rat adipocyte preparations based on its interaction with choleragen, *J. Biol. Chem.* **253:**5103.

Pacuszka, T., Osborne, J. C., Jr., Brady, R. O., and Fishman, P. H., 1978b, Interaction of human chorionic gonadotropin with membrane components of rat testes, Proc. Natl. Acad. Sci. U.S.A. 75:764.

Parodi, A. J., Behrens, N. H., Leloir, L. F., and Carminatti, H., 1972, The role of polyprenol-bound saccharides as intermediates in glycoprotein synthesis in liver, Proc. Natl. Acad. Sci. U.S.A. 69:3268.

Paulson, J. C., Beranek, W. E., and Hill, R. L., 1977a, Purification of a sialyltransferase from bovine colustrum by affinity chromatography on CDP-agarose, J. Biol. Chem. 252:2356.

Paulson, J. C., Rearick, J. I., and Hill, R. L., 1977b, Enzymatic properties of β-D-galactoside α2-6 sialyltransferase from bovine colostrum, J. Biol. Chem. 252:2363.

Paulson, J. C., Hill, R. L., Tanabe, T., and Ashwell, G., 1977c, Reactivation of asialorabbit liver binding protein by resialylation with β-D-galactoside α2-6 sialyltransferase, J. Biol. Chem. 252:8624.

Paulson, J. C., Prieels, J. P., Glasgow, L. R., and Hill, R. L., 1978, Sialyl- and fucosyltransferases in the biosynthesis of asparaginyl-linked oligosaccharides in glycoproteins, J. Biol. Chem. 252:5617.

Porcellati, G., Ceccarelli, G., and Tettamanti, G. (eds.), 1975, Ganglioside Function. Biochemical and Pharmocological Implications, Plenum Press, New York.

Powell, J. T., and Brew, K., 1974, The preparation and characterization of two forms of bovine galactosyltransferase, Eur. J. Biochem. 48:217.

Powell, J. T., and Brew, K., 1975, On the interaction of α-lactalbumin and galactosyltransferase during lactose synthesis, J. Biol. Chem. 250:6337.

Powell, J. T., and Brew, K., 1976a, Metal ion activation of galactosyltransferase, J. Biol. Chem. 251:3645.

Powell, J. T., and Brew, K., 1976b, A comparison of the interactions of galactosyltransferase with a glycoprotein substrate (ovalbumin) and with alpha-lactalbumin, J. Biol. Chem. 251:3653.

Powell, J. T., Jarlfors, U., and Brew, K., 1977, Enzymic characteristics of fat globule membranes from bovine colostrum and bovine milk, J. Cell Biol. 72:617.

Prieels, J. P., Pizzo, S. V., Glasgow, L. R., Paulson, J. C., and Hill, R. L., 1978, Hepatic receptor that specifically binds oligosaccharides containing fucosyl α1-3 N-acetylglucosamine linkages, Proc. Natl. Acad. Sci. U.S.A. 75:2215.

Race, R. R., and Sanger, R., 1975, Blood Groups in Man, 6th ed., Blackwell Scientific, Oxford.

Rao, A. K., Garver, F., and Mendicino, J., 1976, Biosynthesis of the carbohydrate units of immunoglobulins. 1. Purification and properties of galactosyltransferases from swine mesentary lymph nodes, Biochemistry 15:5001.

Rauvala, H., 1976a, The fucoganglioside of human kidney, FEBS Lett. 62:161.

Rauvala, H., 1976b, Gangliosides of human kidney, J. Biol. Chem. 251:7517.

Rauvala, H., 1976c, Action of Clostridium perfringens neuraminidase on gangliosides G_{M1} and G_{M2} above and below the critical micelle concentration of substrate, FEBS Lett. 65:229.

Richardson, C. L., Keenan, T. W., and Morré, D. J., 1977, Ganglioside biosynthesis. Characterization of CMP-N-acetylneuraminic acid:lactosylceramide sialyltransferase in Golgi apparatus from rat liver, Biochim. Biophys. Acta 488:88.

Robbins, P. W., Hubbard, S. C., Turco, S. J., and Wirth, D. F., 1977, Proposal for a common oligosaccharide intermediate in the synthesis of membrane glycoproteins, Cell 12:893.

Rodén, L., and Horowitz, M. I., 1978, Proteoglycans and structural glycoproteins, in: The Glycoconjugates (M. Horowitz and W. Pigman, eds.), Vol. 2, pp. 3–71, Academic Press, New York.

Roseman, S., 1970, The synthesis of complex carbohydrates by multglycosyltransferase systems and their potential function in intercellular adhesion, *Chem. Phys. Lipids* **5**:270.

Roseman, S., Carlson, D. M., Jourdian, G. W., McGuire, E. J., Kaufman, B., Basu, S., and Bartholomew, B., 1966, Animal sialic acid transferases (sialyltransferases), in: *Methods in Enzymology,* Vol. VIII, *Complex Carbohydrates* (E. F. Neufeld and V. Ginsberg, eds.), pp. 354–372, Academic Press, New York.

Roth, S., McGuire, E. J., and Roseman, S., 1971, Evidence for cell-surface glycosyltransferases: Their potential role in cellular recognition, *J. Cell Biol.* **51**:536.

Rothman, J. E., and Lodish, H. F., 1977, Synchronized transmembrane insertion and glycosylation of a nascent membrane protein, *Nature* **269**:775.

Sadler, J. E., Rearick, J. I., Paulson, J. C., and Hill, R. L., 1979, Purification and characterization of two sialyltransferase activities from porcine submaxillary glands, in: *Glycoconjugate Research: Proceedings of the Fourth International Symposium on Glycoconjugates,* Vol. II (J. D. Gregory and R. W. Jealoz, eds.), pp. 763–766, Academic Press, New York.

Sandhoff, K., Conzelmann, E., and Nehrkorn, H., 1977, Specificity of human liver hexosaminidases A and B against glycosphingolipids G_{M2} and G_{A2}. Purification of the enzymes by affinity chromatography employing specific elution, *Hoppe-Seyler's Z. Physiol. Chem.* **358**:779.

Sato, T., Yosizawa, Z., Masubuchi, M., and Yamauchi, F., 1967, Structure of the carbohydrate moiety of the α_1-acid glycoprotein of human plasma, *Carbohydr. Res.* **5**:387.

Schachter, H., 1974a, The subcellular sites of glycosylation, *Biochem. Soc. Symp.* **40**:57.

Schachter, H., 1974b, Glycosylation of glycoproteins during intracellular transport of secretory products, *Adv. Cytopharmacol.* **2**:207.

Schachter, H., and Rodén, L., 1973, The biosynthesis of animal glycoproteins, in: *Metabolic Conjugation and Metabolic Hydrolysis* (W. H. Fishman, ed.), pp. 1–149, Academic Press, New York.

Schachter, H., McGuire, E. J., and Roseman, S., 1971, Sialic acids. XIII. A uridine diphosphate D-galactose: mucin galactosyltransferase from porcine submaxillary gland, *J. Biol. Chem.* **246**:5321.

Schachter, H., Michaels, M. A., Tilley, C. A., Crookston, M. C., and Crookston, J. H., 1973, Qualitative differences in the N-acetyl-D-galactosaminyltransferases produced by human A^1 and A^2 genes, *Proc. Natl. Acad. Sci. U.S.A.* **70**:220.

Schauer, R., and Wember, M., 1973, Studies on the substrate specificity of acylneuraminate cytidyltransferase and sialyltransferase of submandibular glands from cow, pig and horse, *Hoppe-Seyler's Z. Physiol. Chem.* **354**:1405.

Schauer, R., Wember, M., and Ferreira do Amaral, C., 1972, Synthesis of CMP-glycosides of radioactive N-acetyl-, N-glycoloyl-, N-acetyl-7-O-acetyl-, and N-acetyl-8-O-acetylneuraminic acids by CMP-sialate synthase from bovine submaxillary glands, *Hoppe-Seyler's Z. Physiol. Chem.* **353**:883.

Schauer, R., Buscher, H. P., and Casals-Stenzel, J., 1974, Sialic acids: Their analysis and enzymic modification in relation to the synthesis of submandibular gland glycoproteins, *Biochem. Soc. Symp.* **40**:87.

Schenkel-Brunner, H., Chester, M. A., and Watkins, W. M., 1972, α-L-Fucosyltransferases in human serum from donors of different ABO, secretor and Lewis blood group phenotypes, *Eur. J. Biochem.* **30**:269.

Schmid, K., Nimberg, R. B., Kimura, A., Yamaguchi, H., and Binette, J. P., 1977, The carbohydrate units of human plasma α_1-acid glycoprotein, *Biochim. Biophys. Acta* **492**:291.

Schwyzer, M., and Hill, R. L., 1977a, Porcine A blood group-specific N-acetylgalactosaminyltransferase. I. Purification from porcine submaxillary glands, *J. Biol. Chem.* **252**:2338.

Schwyzer, M., and Hill, R. L., 1977*b*, Porcine A blood group-specific *N*-acetylgalactosa-minyltransferase. II. Enzymatic properties, *J. Biol. Chem.* **252:**2346.

Sefton, B., 1977, Immediate glycosylation of Sindbis virus membrane proteins, *Cell* **10:**659.

Shen, L., Grollman, E. F., and Ginsburg, V., 1968, An enzymatic basis for secretor status and blood group substance specificity in humans, *Proc. Natl. Acad. Sci. U.S.A.* **59:**224.

Shoyab, M., and Bachhawat, B. K., 1967, Age dependent changes in the level of cytidine 5'-monophospho-*N*-acetyl neuraminic acid synthesizing and degrading enzymes and bound sialic acid in rat liver, *Indian J. Biochem.* **4:**142.

Shoyab, M., and Bachhawat, B. K., 1969, Partial purification and properties of cytidine 5'-monophospho-*N*-acetyl neuraminic acid degrading enzyme from sheep liver, *Indian J. Biochem.* **6:**56.

Shur, B. D., and Roth, S., 1975, Cell surface glycosyltransferases, *Biochim. Biophys. Acta* **455:**473.

Skipski, V. P., Katopodis, N., Prendergast, J. S., and Stock, C. C., 1975, Gangliosides in blood serum of normal rats and Morris hepatoma 5123ts-bearing rats, *Biochem. Biophys. Res. Commun.* **67:**1122.

Smith, C. A., and Brew, K., 1977, Isolation and characteristics of galactosyltransferase from Golgi membranes of lactating sheep mammary glands, *J. Biol. Chem.* **252:**7294.

Spiro, M. J., Spiro, R. G., and Bhoyroo, V. D., 1976*a*, Lipid–saccharide intermediates in glycoprotein biosynthesis. I. Formation of an oligosaccharide–lipid by thyroid slices and evaluation of its role in protein glycosylation, *J. Biol. Chem.* **251:**6400.

Spiro, M. J., Spiro, R. G., and Bhoyroo, V. D., 1976*b*, Lipid–saccharide intermediates in glycoprotein biosynthesis. III. Comparison of oligosaccharide–lipids formed by slices from several tissues, *J. Biol. Chem.* **251:**6420.

Spiro, R. G., Spiro, M. J., and Bhoyroo, V. D., 1976, Lipid–saccharide intermediates in glycoprotein biosynthesis. II. Studies on the structure of an oligosaccharide–lipid from thyroid, *J. Biol. Chem.* **251:**6409.

Stanley, P., Caillibot, V., and Siminovitch, L., 1975*a*, Selection and characterization of eight phenotypically distinct lines of lectin-resistant Chinese hamster ovary cells, *Cell* **6:**121.

Stanley, P., Narasimhan, S., Siminovitch, L., and Schachter, H., 1975*b*, Chinese hamster ovary cells selected for resistance to the cytotoxicity of phytohemagglutinin are deficient in a UDP-*N*-acetylglucosamine-glycoprotein *N*-acetylglucosaminyltransferase activity, *Proc. Natl. Acad. Sci. U.S.A.* **72:**3323.

Steigerwald, J. C., Basu, S., Kaufman, B., and Roseman, S., 1975, Sialic acids. XIX. Enzymatic synthesis of Tay–Sachs ganglioside, *J. Biol. Chem.* **250:**6727.

Stoffyn, P., van den Eijnden, D., and Stoffyn, A., 1977, Specificity of sialyltransferase in glycoprotein biosynthesis, *Fed Proc.* **36:**744.

Sturgess, J., Moscarello, M. A., and Schachter, H., 1978, The structure and biosynthesis of membrane glycoproteins, *Curr. Top. Membr. Transp.* **11:**15.

Sturgeon, R. J., 1977, Glycolipids and gangliosides, *Carbohydr. Chem.* **9:**397.

Suzuki, A., Ishizuka, I., and Yamakawa, T., 1975, Isolation and characterization of a ganglioside containing fucose from boar testis, *J. Biochem. Tokyo* **78:**947.

Svennerholm, L., 1964, The gangliosides, *J. Lipid Res.* **5:**145.

Svennerholm, L., 1974, Sphingolipid changes during development, *Mod. Probl. Paediatr.* **13:**104.

Tabas, I., Schlesinger, S., and Kornfeld, S., 1978, Processing of high mannose oligosaccharides to form complex type oligosaccharides on the newly synthesized polypeptides of the vesicular stomatitis virus G protein and the IgG heavy chain, *J. Biol. Chem.* **253:**716.

Tai, T., Ito, S., Yamashita, K., Muramatsu, T., and Kobata, A., 1975, Asparagine-linked oligosaccharide chains of IgG: A revised structure, *Biochem. Biophys. Res. Commun.* **65:**968.

Tai, T., Yamashita, K., Ito, S., and Kobata, A., 1977, Structures of the carbohydrate moiety of ovalbumin glycopeptide III and the difference in specificity of endo-β-N-acetylglucosaminidases C_{II} and H, *J. Biol. Chem.* **252**:6687.

Tait, R. M., and van Heyningen, S., 1978, The adenylate cyclase-activating activity of cholera toxin is not associated with a nicotinamide-adenine nucleotide glycohydrolase activity, *Biochem. J.* **174**:1059.

Takasaki, S., and Kobata, A., 1976, Chemical characterization and distribution of ABO blood group active glycoprotein in human erythrocyte membrane, *J. Biol. Chem.* **251**:3610.

Takasaki, S., Yamashita, K., and Kobata, A., 1978, The sugar chain structures of ABO blood group active glycoproteins obtained from human erythrocyte membrane, *J. Biol. Chem.* **253**:6086.

Tjaden, U. R., Krol, J. H., Van Hoeven, R. P., Oomen-Meulemans, E. P. M., and Emmelot, P., 1977, High-pressure liquid chromatography of glycosphingolipids (with special reference to gangliosides), *J. Chromatogr.* **136**:233.

Tomich, J. M., Mather, I. H., and Keenan, T. W., 1976, Proteins mask gangliosides in milk fat globule and erythrocyte membranes, *Biochim. Biophys. Acta* **433**:357.

Toneguzzo, F., and Ghosh, H. P., 1977, Synthesis and glycosylation *in vitro* of glycoprotein of vesicular stomatitis virus, *Proc. Natl. Acad. Sci. U.S.A.* **74**:1516.

Toneguzzo, F., and Ghosh, H. P., 1978, *In vitro* synthesis of vesicular stomatitis virus membrane glycoprotein and insertion into membranes, *Proc. Natl. Acad. Sci. U.S.A.* **75**:715.

Topping, M. D., and Watkins, W. M., 1975, Isoelectric points of the human blood group A^1, A^2 and B gene-associated glycosyltransferases in ovarian cyst fluids and serum, *Biochem. Biophys. Res. Commun.* **64**:89.

Trayer, I. P., and Hill, R. L., 1971, The purification and properties of the A protein of lactose synthetase, *J. Biol. Chem.* **246**:6666.

Treloar, M., Sturgess, J. M., and Moscarello, M. A., 1974, An effect of puromycin on galactosyltransferase of Golgi-rich fractions from rat liver, *J. Biol. Chem.* **249**:6628.

Tsay, G. C., Dawson, G., and Sung, S. S. J., 1976, Structure of the accumulating oligosaccharide in fucosidosis, *J. Biol. Chem.* **251**:5852.

Tsopanakis, A. D., and Herries, D. G., 1978, Bovine galactosyltransferase, substrate–manganese complexes and the role of manganese ions in the mechanism, *Eur. J. Biochem.* **83**:179.

Turco, S. J., Stetson, B., and Robbins, P. W., 1977, Comparative rates of transfer of lipid-linked oligosaccharides to endogenous glycoprotein acceptors *in vitro*, *Proc. Natl. Acad. Sci. U.S.A.* **74**:4411.

Van den Eijnden, D. H., 1973, Subcellular localization of cytidine-5'-monophospho-N-acetylneuraminic acid synthetase in calf brain, *J. Neurochem.* **21**:949.

Van den Eijnden, D. H., Meems, L., and Roukema, P. A., 1972, Regional distribution of cytidine-5'-monophospho-N-acetylneuraminic acid synthetase in calf brain, *J. Neurochem.* **19**:1649.

Van den Eijnden, D. H., Stoffyn, P., Stoffyn, A., and Schiphorst, W. E. C. M., 1977, Specificity of sialyltransferase: Structure of α_1-acid glycoprotein sialylated *in vitro*, *Eur. J. Biochem.* **81**:1.

Van den Eijnden, D. H., Dieleman, B., and Schiphorst, W. C. M., 1979, Sialylation of desialylated ovine submaxillary mucin by porcine liver sialyltransferase *in vitro*, in: *Glycoconjugate Research: Proceedings of the Fourth International Symposium on Glycoconjugates,* Vol. II (J. D. Gregory and R. W. Jealoz, eds.), pp. 829–834, Academic Press, New York.

Van Dessel, G., Lagrou, A., Hilderson, H. J., and Dierick, W., 1976, Chemical characterization of six ganglioside components from bovine thyroid tissue, *Arch. Int. Physiol. Biochim.* **84**:674.

Van Dijk, W., Ferwerda, W., and Van den Eijnden, D. H., 1973, Subcellular and regional distribution of CMP-N-acetylneuraminic acid synthetase in the calf kidney, *Biochim. Biophys. Acta* **315:**162.

Van Dijk, W., Maier, H., and Van den Eijnden, D. H., 1976, Properties and subcellular localization of CMP-N-acetylneuraminic acid hydrolase of calf kidney, *Biochim. Biophys. Acta* **444:**816.

van Heyningen, S., 1976, Binding of ganglioside by the chains of tetanus toxin, *FEBS Lett.* **68**(1):5.

van Heyningen, W. E., 1974, Gangliosides as membrane receptors for tetanus toxin, cholera toxin and serotonin, *Nature* **249:**415.

Vincendon, G., Ghandour, M. S., Robert, J., Gombos, G., and Rebel, G., 1975, Changes of ganglioside pattern during postnatal development of rat cerebellum, *Biochem. Soc. Trans.* **3:**696.

Volk, B. W., and Schneck, L. (eds.), 1975, *The Gangliosides,* Plenum Press, New York.

Watkins, W. M., 1974*a*, Genetic regulation of the structure of blood group-specific glycoproteins, *Biochem. Soc. Symp.* **40:**125.

Watkins, W. M., 1974*b*, Blood group substances: Their nature and genetics, in: *The Red Blood Cell* (D. M. Surgenor, ed.), 2nd ed., pp. 293–360, Academic Press, New York.

Wedgwood, J. F., Warren, C. D., Jeanloz, R. W., and Strominger, J. L., 1974, Enzymatic utilization of P^1-di-N-acetylchitobiosyl P^2-dolichol pyrophosphate and its chemical synthesis, *Proc. Natl. Acad. Sci. U.S.A.* **71:**5022.

Whatley, R., Ng, S. K.-C., Rogers, J., McMurray, W. C., and Sanwal, B. D., 1976, Developmental changes in gangliosides during myogenesis of a rat myoblast cell line and its drug resistant variants, *Biochem. Biophys. Res. Commun.* **70:**180.

Wherrett, J. R., 1976, Gangliosides of the lacto-N-glycaose series (glucosamine containing gangliosides), in: *Glycolipid Methodology* (L. A. Witting, ed.), pp. 215–232, American Oil Chemists' Society Press, Champaign, Illinois.

Whitehead, J. S., Bella, A., Jr., and Kim, Y. S., 1974*a*, An N-acetylgalactosaminyltransferase from human blood group A plasma. I. Purification and agarose binding properties, *J. Biol. Chem.* **249:**3442.

Whitehead, J. S., Bella, A., Jr., and Kim, Y. S., 1974*b*, An N-acetylgalactosaminyltransferase from human blood group A plasma. II. Kinetic and physicochemical properties, *J. Biol. Chem.* **249:**3448.

Whur, P., Herscovics, A., and Leblond, C. P., 1969, Radioautographic visualization of the incorporation of galactose-^{3}H and mannose-^{3}H by rat thyroids *in vitro* in relation to the stages of thyroglobulin synthesis, *J. Cell Biol.* **43:**289.

Wilkinson, F. E., Morré, D. J., and Keenan, T. W., 1976, Ganglioside biosynthesis. Characterization of uridine diphosphate galactose: G_{M2} galactosyltransferase in Golgi apparatus from rat liver, *J. Lipid Res.* **17:**146.

Wilson, J. R., Williams, D., and Schachter, H., 1976, The control of glycoprotein synthesis: N-Acetylglucosamine linkage to a mannose residue as as signal for the attachment of L-fucose to the asparagine-linked N-acetylglucosamine residue of glycopeptide from α_1-acid glycoprotein, *Biochem. Biophys. Res. Commun.* **72:**909.

Witting, L. A., 1975, Separation of complex lipids: Gangliosides, galactosides, sphingolipids, in: *The Analysis of Lipids and Lipoproteins* (E. G. Perkins, ed.), pp. 90–107, American Oil Chemists' Society Press, Champaign, Illinois.

Yu, R. K., and Lee, S. H., 1976, *In vitro* biosynthesis of sialosylgalactosylceramide (G_7) by mouse brain microsomes, *J. Biol. Chem.* **251:**198.

Yu, R. K., and Yen, S. I., 1975, Gangliosides in developing mouse brain myelin, *J. Neurochem.* **25:**229.

Yusuf, H. K. M., Merat, A., and Dickerson, J. W. T., 1977, Effect of development on the gangliosides of human brain, *J. Neurochem.* **28:**1299.

Zanetta, J.-P., Vitiello, F., and Robert, J., 1977, Thin-layer chromatography of gangliosides, *J. Chromatogr.* **137:**481.

Surface Carbohydrate Alterations of Mutant Mammalian Cells Selected for Resistance to Plant Lectins

Pamela Stanley

1. SELECTION OF CELLS WITH ALTERED SURFACE CARBOHYDRATE

In recent years it has become possible to select mutants of mammalian cells which exhibit structural alterations in the carbohydrate moieties of cell surface glycoproteins and glycolipids. The selective agents which have enabled the isolation of these mutants are the cytotoxic plant lectins. These include the phytohemagglutinin from *Phaseolus vulgaris* (PHA), agglutinins from wheat germ (WGA), concanavalin A from jack beans (Con A), toxin from *Ricinus communis* (RIC), and agglutinins from *Lens culinaris* (LCA).

Lectins are proteins or glycoproteins from plants or lower animals which bind specifically to certain sugar residues in carbohydrate moieties (reviewed by Nicolson, 1974). Although lectin binding may usually be inhibited by simple sugars (e.g., lactose, α-D-glucosides) specific for the

Pamela Stanley · Department of Cell Biology, Albert Einstein College of Medicine, Bronx, New York 10461.

particular lectin, there is evidence that lectin receptors of highest affinities reside in more than one sugar residue and may also be dependent on the conformation of sugars in the carbohydrate side chain (Kornfeld *et al.*, 1971; Toyoshima *et al.*, 1972; Kornfeld and Ferris, 1975). Lectins have been shown to bind to the surface of animal cells by a variety of microscopic and biochemical techniques. This interaction can be inhibited or reversed by simple sugars (at relatively high concentrations) and by glycopeptides of known composition. Thus it is assumed that lectins interact with the carbohydrate moieties of glycosylated membrane macromolecules (such as glycoproteins, glycolipids, proteoglycans, collagen), in a manner equivalent to their interaction with oligosaccharides in solution. The fact that lectins are also cytotoxic to many cultured cells suggested their use as selective agents in obtaining membrane mutants of mammalian cells. That is, at least one type of resistance to the toxicity of a particular lectin might be expected to involve a structural membrane alteration such as a decreased ability to bind the lectin at the cell surface. In fact, it is now apparent that this type of lectin-resistant (Lec^R) cell is by far the most common phenotype isolated in lectin selections of mammalian cell populations.

2. BIOCHEMICAL BASIS OF A SURFACE CARBOHYDRATE ALTERATION

Since the first report of Con A-resistant (ConA^R) SV3T3 cells by Ozanne and Sambrook (1971), many different Lec^R cell lines have been isolated from a variety of cell types. However, at the present time, there is only one Lec^R phenotype which has been shown to be always correlated with a specific enzymic lesion. The properties of this phenotype have been extensively studied by a number of laboratories, and thus it represents the best-characterized Lec^R cell line described to date. Consequently, it provides an excellent prototype around which to discuss the properties of other Lec^R cell lines.

This Lec^R phenotype is resistant to and may be selected by at least four lectins: PHA, WGA, Con A, or RIC. It was independently isolated by Gottlieb *et al.* (1974) in a selection for Chinese hamster ovary (CHO) cells resistant to RIC, by Stanley *et al.* (1975a) in a selection for CHO cells resistant to PHA, and by Meager *et al.* (1975) in a selection for BHK cells resistant to RIC. Subsequently, the same phenotype was isolated in eight independent selections from CHO cells using the lectins PHA, WGA, RIC, or LCA (Stanley *et al.*, 1975b). Each of these isolates exhibits a similarly complex pleiotrophic phenotype which appears to stem directly from the loss of a specific N-acetylglucosaminyltransferase (GlcNAc-T) activity (Gottlieb *et al.*, 1974, 1975; Stanley *et al.*, 1975*a,b,c*; Meager *et al.*, 1975, 1976). Thus although there is no formal proof (such as complementation analysis) that the isolates from the different laboratories represent mutations at the same genetic locus, the available evidence suggests that they possess identical properties and may be discussed as a single mutant phenotype.

The GlcNAc-T activity which is missing in these Lec[R] cells appears to be one of at least two activities present in parental CHO cells, both of which are specific for transferring N-acetylglucosamine (GlcNAc) to Man residues in the outer "branches" of asparagine (Asn)-linked carbohydrate moieties (Gottlieb *et al.*, 1975; Stanley *et al.*, 1975*c*). The substrate specificity of the missing GlcNAc-T enzyme activity has been investigated by Narasimhan *et al.* (1977). By comparing the abilities of extracts from parental and GlcNAc-T-deficient Lec[R] CHO cells to transfer GlcNAc to a series of glycopeptide and oligosaccharide acceptors, these authors obtained evidence that the GlcNAc-T activity lacking in the Lec[R] CHO cells is specific for the addition of GlcNAc to terminal Man residues linked α1,3. A GlcNAc-T activity which is *not* diminished in the Lec[R] cells appears to be specific for the transfer of GlcNAc to the terminal Man residue in an IgG glycopeptide of the structure GlcNAc-Man-(Man)-Man-GlcNAc-GlcNAc-Asn. Although structural analyses were not performed in these studies, it was postulated that the terminal Man in this IgG glycopeptide was linked α1,6. Grey *et al.* (1980) have subsequently confirmed this supposition by direct structural analysis using 360-MHz nuclear magnetic resonance (NMR) spectroscopy of the IgG glycopeptides prepared by Narasimhan *et al.* (1977). Thus it appears that the GlcNAc-T activity missing in the Lec[R] CHO cells (termed GlcNAc-T1) is specific for adding GlcNAc to Man residues linked α1,3 while a GlcNAc-T activity which is not altered in these cells (termed GlcNAc-T2) is specific for adding GlcNAc to terminal Man residues linked α1,6. Recently, further evidence in support of this hypothesis has been obtained. Wilson *et al.* (1978) have found that essentially all of the GlcNAc transferred to an IgG glycopeptide with two terminal Man residues (linked α1,3 and α1,6, respectively) by GlcNAc-T1 from liver Golgi preparations is transferred specifically to the α1,3-linked Man. The fact that the Lec[R] cell extracts do not transfer GlcNAc to the latter acceptor (Gottlieb *et al.*, 1975; Stanley *et al.*, 1975*c*; Narasimhan *et al.*, 1977) strongly suggests that the Lec[R] cells are missing the enzyme specific for transferring GlcNAc to α1,3-linked terminal Man residues (GlcNAc-T1).

The various Lec[R] cell lines shown to lack GlcNAc-T1 activity have been designated by a variety of names by different authors (see Table 1).

Table 1. Carbohydrate Lesions of Lec^R Cell Lines

Parental cell line	Selective lectin	Lec^R cell line	Glycoprotein sugar transfer affected	Sugar newly exposed at cell surface	Glycolipids affected	Specifically altered enzyme activities
CHO	RIC	CHO-15B Ric^R₁	GlcNAc → Man (α1,3)	Man	−	↓ GlcNAc-T1
	PHA	Pha^R₁				
	WGA	Wga^R₁				
	LCA	Lca^R₁				
BHK	RIC	Ric^R₁₄				
CHO	WGA	Wga^R₁₁	Sialic acid → Gal/GalNAc	Gal/GalNAc	+	None detected
		Wga^R₁₁₁				None detected
		Clone 1021				Variable Sialyl-T(GP) ↓ Sialyl-T(GL)
		Clone 13	Gal → GlcNAc	GlcNAc	+	None detected

Cell	Lectin	Mutant				Defect
Mouse L	RIC	CL 3	Sialic acid → ? ; GlcNAc → Man	Sialic acid	+	↑ Sialyl-T (GP and GL)
		CL 6		Man	?	↓ GlcNAc-T (GP); ↓ Gal-T (GP); ↑ GlcNAc glycosidase (GP); ↑ Gal glycosidase (GP)
CHO	Con A	B211 ($ConA^{R_I}$)	?	?	+	None detected
		C321 ($ConA^{R_{II}}$)			(Dolichol intermediate)	
CHO Glt$_1$⁻	Con A	$Pha^R ConA^{R_u}$	?	Man	? (Dolichol intermediate)	↓ GlcNAc-T1; No other detected

[a] A summary of Lec^R cell lines which have been shown to exhibit specific glycosylation defects. The appropriate references are provided in the text. Abbreviations: GP, glycoprotein; GL, glycolipid; sialyl-T, Gal-T, and GlcNAc-T, specific glycosyltransferase enzymes.

However, their biochemical properties are essentially identical. Therefore for ease of discussion in this article, it is proposed to refer to Lec^R cells of this phenotype as Glt_1^- cell mutants.

3. GENETIC AND BIOCHEMICAL PROPERTIES OF Glt_1^- CHO CELLS

3.1. Localization of the Mutated Gene Product

The question arises as to whether the GlcNAc-T1 enzyme deficiency provides the basis of the mutation to lectin resistance in Glt_1^- cells. Indirect evidence in favor of this hypothesis has been obtained in genetic studies of the Lec^R phenotype in CHO cells (Stanley *et al.*, 1975*a,b*; Stanley and Siminovitch, 1976, 1977). For these studies, a Glt_1^- mutant was defined as exhibiting decreased GlcNAc-T activity for the transfer of GlcNAc to glycosidase-treated α_1-acid glycoprotein and as possessing the specific lectin-resistance properties which always correlated with the GlcNAc-T1 deficiency (resistance to PHA, WGA, RIC, and LCA and sensitivity to Con A). Briefly, it was shown that: (1) Glt_1^- mutants can be isolated from independent CHO cell auxotrophs via a single-step selection with any of the lectins PHA, WGA, RIC, or LCA; (2) the Glt_1^- phenotype is stable over long periods of continuous culture and clones derived from Glt_1^- cells exhibit all of the characteristics of their parent clones; (3) the frequency of Glt_1^- cells in a CHO cell population is increased markedly by the mutagen ethylmethanesulfonate; and (4) the characteristic Lec^R phenotype of Glt_1^- cells is extinguished in somatic cell hybrids formed between the Glt_1^- cells and parental CHO cells showing that information contributed by the wild-type cell can correct the defect in Glt_1^- cells (i.e., the phenotype behaves recessively in somatic cell hybrids). Consistent with this is the finding that the spontaneous mutation rate to PHA resistance in CHO cells is $\sim 10^{-6}$ per cell per generation (Stanley and Siminovitch, 1976), suggesting that CHO cells are functionally hemizygous for the gene product which gives rise to the Glt_1^- phenotype (see Siminovitch, 1976).

These genetic studies suggest that cells lacking GlcNAc-T1 activity arise as the result of a single mutational event and represent *bona fide* somatic cell mutants. However, direct evidence that the GlcNAc-T1 enzyme represents the mutated gene product has not been forthcoming. Unfortunately, attempts to isolate revertants of this Lec^R phenotype in order to determine whether reversion of lectin resistance correlates with recovery of GlcNAc-T1 activity have not been successful (Stanley and

Siminovitch, 1976). The reason for this may simply be that the frequency of revertants is very low. In fact, the correlation of the phenotypic properties of these cells and the lack of a GlcNAc-T1 activity is so far absolute. It may be assumed therefore that if the mutation has not occurred in the GlcNAc-T1 enzyme, the mutated gene product has a directly inhibitory effect on the expression of GlcNAc-T1 activity in these Lec^R cells.

The modified carbohydrate moieties manufactured in Glt$_1^-$ CHO cells and expressed at the cell surface have been investigated by examining the structure of the carbohydrate of the G glycoprotein of vesicular stomatitis virus (VSV) grown in Glt$_1^-$ compared with parental CHO cells (Schlesinger *et al.*, 1976; Robertson *et al.*, 1978; Tabas *et al.*, 1978) and by direct structural analysis of the mutant cell surface glycopeptides (Li and Kornfeld, 1978). These studies have shown that Glt$_1^-$ CHO cells synthesize Asn-linked carbohydrate chains with the structure Man$_5$GlcNAc$_2$Asn instead of chains of the type [sialic acid-Gal-GlcNAc]$_{2-3}$Man$_3$GlcNAc$_2$(Fuc)Asn. As discussed in detail in Chapters 1 and 3, the Man$_5$GlcNAc$_2$Asn structure appears to be an intermediate in the pathway of oligosaccharide biosynthesis which is not completely processed in Glt$_1^-$ cells due to their GlcNAc-T1 deficiency (Tabas and Kornfeld, 1978).

3.2. Altered Membrane Properties

The description of the biochemical basis of lectin resistance thought to occur in Glt$_1^-$ CHO cells provides a biochemical explanation for the complex pleiotrophic phenotype of Glt$_1^-$ cells. Thus, it would appear that the lack of GlcNAc-Gal-sialic acid sequences and the increased number and exposure of Man residues in Asn-linked carbohydrate side chains of a variety of cell surface glycoproteins in Glt$_1^-$ cells accounts for (1) the high degree of resistance of Glt$_1^-$ cells to PHA, WGA, RIC, and LCA and their decreased ability to bind each of these lectins at the cell surface (Stanley and Carver, 1977*a*); (2) the increased sensitivity of Glt$_1^-$ cells to Con A and their increased ability to bind Con A at the cell surface (Gottlieb *et al.*, 1974; Stanley *et al.*, 1975*a*; Meager *et al.*, 1976; Stanley and Carver, 1977*a*); (3) the fact that Glt$_1^-$ cells exhibit a dramatic reduction in surface labeling via the galactose oxidase:[^{3}H]borohydride technique (Juliano and Stanley, 1975); and (4) the decreased content of sialic acid, Gal, and GlcNAc and increased content of Man in crude membrane preparations from Glt$_1^-$ cells (Gottlieb *et al.*, 1974). These latter findings suggest that the altered carbohydrate moieties synthesized in Glt$_1^-$ cells represent a significant fraction of the total membrane carbohydrate.

4. SURFACE CARBOHYDRATE ALTERATIONS OF OTHER LECTIN-RESISTANT (Lec^R) CELL LINES

The properties of Glt_1^- CHO cells have provided a focus for investigating the biochemical phenotypes and mechanisms of lectin-resistance in other Lec^R cell lines. In particular, the fact that Glt_1^- mutants express their lesion in the carbohydrate moieties of Sindbis and VSV (Schlesinger et al., 1976; Robertson et al., 1978; Tabas et al., 1978) suggested that other mutations leading to lectin resistance might be delineated with the help of enveloped viruses. Although this approach is clearly biased towards describing Lec^R cells which possess structurally altered carbohydrate moieties in membrane glycoproteins, the properties of the relatively large number of Lec^R cell lines isolated to date from different laboratories suggest that this bias is justified. In fact, the majority of Lec^R cell lines possess phenotypes consistent with the expression of distinctive structural alterations in cell surface carbohydrate. Specific glycosylation defects have been localized to certain sugar residues in a number of these isolates. These Lec^R phenotypes will be discussed in some detail since they provide an indication of the diversity of carbohydrate alterations which may lead to lectin resistance in a mammalian cell.

4.1. Ricin-Resistant (Ric^R) Mouse L Cells

Gottlieb and Kornfeld (1976) have isolated two Ric^R cell lines from L cells (designated CL 3 and CL 6) and shown them to exhibit specific and distinct glycosylation defects. CL 3 cells possess increased sialylation of glycoproteins and glycolipids. They exhibit a reduced ability to bind RIC at the cell surface, increased sialic acid in crude membrane preparations, and increased G_{M3} ganglioside content. Also, the glycopeptides of VSV grown in CL 3 cells appear to be oversialylated. In searching for a possible enzymic basis of this Lec^R phenotype, it was found that cell-free extracts from CL 3 cells exhibit an increased transfer rate of sialic acid from CMP-[^{14}C]sialic acid to both desialized fetuin (a glycoprotein) and lactosylceramide (a glycolipid). The authors conclude that the CL 3 cell line possesses an increased ability to sialylate glycoproteins and glycolipids which results in a masking of Gal/GalNAc residues normally involved in RIC binding at the cell surface. Consistent with this hypothesis is their finding that neuraminidase treatment of CL 3 cells partially restores the sensitivity of the cells to RIC.

The question remains, however, as to the molecular basis of lectin resistance in CL 3 cells. Because the activities of two theoretically distinct sialyltransferases (glycoprotein and glycolipid, respectively) are increased in CL 3 cells, the authors suggest that these cells may have arisen as the result of "multiple epigentic or mutational events." Although this possibility exists if the mutants were selected from an uncloned population of L cells which had been continuously cultured over a long period of time, it is probably less likely than the alternative possibility of a single mutational event giving rise to a pleiotrophic phenotype. The precedent for a complex phenotype correlating with lectin resistance is well-established by the properties of the Glt_1^- mutant. Also, accumulated evidence indicates that the majority of somatic cell mutants exhibit specific genetic and biochemical properties consistent with a single mutational event (Siminovitch, 1976). Since glycosyltransferase enzymes are membrane bound and may exist as complexes within the cell (Schachter and Rodén, 1973), it is possible that a mutation in one glycosyltransferase may affect the activity of another. Alternatively a mutation affecting the availability of a cofactor for sialic acid transfer may alter both enzyme activities. Finally, considering the complexities which might arise in measuring glycosyl transfer in detergent-solubilized cell-free extracts, it is conceivable that sialic acid is being transferred to both the desialyzed fetuin and the lactosylceramide exogenous acceptors by the same enzyme activity. Only ~1% or less of the acceptor molecule becomes glycosylated during these reactions (Gottlieb et al., 1975; Stanley et al., 1975c; Narasimhan et al., 1977) and therefore the specificity of the in vitro reactions must be questioned. Thus, in the absence of evidence to the contrary, the increased sialylation of both glycoproteins and lactosylceramide in CL 3 cells could conceivably be caused by a single enzyme lesion or at least by a single factor which is instrumental in activating both enzyme activities.

Similar questions regarding the molecular basis of RIC-resistance also apply to the second Ric^R L cell mutant (CL 6) isolated by Gottlieb and Kornfeld (1976). This mutant exhibits a decreased binding ability for RIC and an altered sugar content of crude membrane preparations (sialic acid, 58%; Gal, 31%; hexosamine, 48%; and increased Man, 123%). A high proportion of the glycopeptides of VSV grown in CL 6 cells possess a molecular weight lower than S3 (Sefton, 1976), suggesting that they lack "branch" sugar residues other than sialic acid. Consistent with these results is the finding that glycoproteins from CL 6 cells labeled by incorporation of [^{14}C]-GlcNAc possess a lower molecular weight, indicating decreased glycosylation compared with parental L cells. Finally, CL 6 cells appear to possess multiple changes in glycosyltransferase and glycosidase activities. Cell-free extracts exhibit decreased transfer of Gal and GlcNAc to the appropriate fetuin acceptors. Also, they exhibit a 1.5- to 2-fold increase in β-galactosidase and β-N-acetylhexosiminidase activities under certain assay conditions. Thus the complex phenotype of CL 6 cells also appears to include changes in enzyme activities having different specificities. However, as mentioned previously, these findings may simply reflect a mutation in a gene product which affects each enzyme activity. Alternatively, they may reflect the problems associated with interpreting the results of *in vitro* glycosyl transfer assays. In fact, the membrane properties reported for CL 6 L cells are very similar to those of the Glt_1^- CHO cells (CHO-15B) described by Gottlieb *et al.* (1974, 1975) and Schlesinger *et al.* (1976). However, the decreased Gal transferase activity and altered glycosidase activities are not consistent with the Glt_1^- phenotype.

Some of the complexities surrounding both the CL 3 and CL 6 phenotypes should be ruled out by further characterization at the genetic level. It is important to determine whether independent isolates expressing phenotypes identical to CL 3 and CL 6 may be obtained from freshly cloned L cell populations via a single-step selection protocol. In addition, it would be of interest to know the effect of mutagens on the frequency of these phenotypes in a cell population. If cell lines of the CL 3 and CL 6 type can be isolated in a single-step selection and their frequency in a population is increased by mutagens, it would be consistent with the possibility that they arise as the result of a single mutational event.

4.2. Wheat Germ Agglutinin-Resistant (Wga^R) CHO Cells

Another example of the difficulties of determining molecular bases for lectin resistance is provided by the two Wga^R CHO cell mutants isolated by Briles *et al.* (1977). Both these mutants exhibit specific glycosylation defects affecting the addition of either sialic acid alone (clone 1021) or sialic acid and Gal residues (clone 13) to membrane macromolecules. However, no specific enzyme defects have been uncovered which might account for their complex phenotypes. Clone 1021 cells exhibit decreased WGA binding and increased binding of *Ricinus communis* agglutinin and soybean agglutinin. They possess a decreased sialic acid content in crude membrane preparations and in a total cellular lipid fraction. Consistent with these findings, they exhibit decreased incorporation (over a 3-day period) of [^{3}H]-N-acetyl-D-mannosamine into the ganglioside G_{M3} and a concomitant increase in the incorporation of [^{14}C]-Gal into lactosylceramide. The glycopeptides released by extensive pronase digestion of crude membranes from [^{3}H]fucose-labeled clone 1021 cells exhibit a decreased molecular weight compared with glycopeptides from parental cells. Finally, clone 1021 cells possess increased intracellular CMP-sialic acid levels and exhibit an 80% reduction in the incorporation of [^{3}H]-N-acetylmannosamine into membrane-bound sialic acid. However, no consistent glycosyltransferase or glycosidase defects appear to correlate with this phenotype. Clone 1021 cells possess a variable level of sialyltransferase activity for desialyzed fetuin and a markedly decreased sialyltransferase activity for lactosylceramide. Despite the reduced G_{M3} synthetase (or sialyltransferase) activity, the primary defect does not appear to be in the enzyme itself since the levels of this enzyme activity for exogenous lactosylceramide may be markedly increased by incubation of cells in the presence of sodium butyrate. However, this does not increase cellular G_{M3} content in either parental or clone 1021 CHO cells. The authors suggest that the properties of this mutant may arise from a compartmentalization defect.

The second Wga^R CHO cell line (clone 13) described by Briles et al. (1977) exhibits decreased incorporation of Gal and sialic acid into membrane glycoproteins and glycolipids. This is reflected in the sugar content of a crude membrane fraction and of a total lipid fraction and in the incorporation pattern of [^{3}H]-N-acetylmannosamine and [^{14}C]-Gal into gangliosides. The glycopeptides released from [^{3}H]fucose-labeled clone 13 cells are of lower molecular weight than those from parental CHO cells. Also, intact clone 13 cells appear to have more exposed GlcNAc residues at the cell surface than parental CHO cells. Clone 13 cells exhibit increased intracellular CMP-sialic acid levels similar to those of clone 1021 cells. However, no enzyme lesion has been observed which might account for the biochemical phenotype of clone 13 cells.

Stanley et al. (1975b) have also isolated Wga^R CHO cells and subsequently shown them to possess sialic acid glycosylation defects in glycoproteins and glycolipids (Robertson et al., 1978; Stanley et al., 1980). These Lec^R phenotypes have been designated Wga^{RII} and Wga^{RIII} and have been shown to be genetically distinct by complementation analysis (Stanley and Siminovitch, 1977). They also exhibit markedly different WGA binding abilities (Stanley and Carver, 1977b) although their lectin-resistant phenotypes are very similar (Stanley et al., 1975b). Wga^{RII} cells have lost particular WGA binding sites from the cell surface whereas Wga^{RIII} cells bind WGA essentially identically to parental CHO cells over a millionfold range of WGA concentrations. Both cell lines exhibit decreased sialylation of membrane glycoproteins and of the ganglioside G_{M3} (Stanley et al., 1980). However, the Wga^{RII} defect appears to be more marked in terms of the extent of sialic acid loss. In fact the Wga^{RII} cell line has similar properties to clone 1021 of Briles et al. (1977) and may represent the same mutation. By contrast, the Wga^{RIII} phenotype appears to be unique. The sialylation defect in Wga^{RIII} cells is reflected in the carbohydrate moieties of VSV, the majority of which do not contain sialic acid when propagated in Wga^{RIII} cells (Robertson et al., 1978). This suggests that possibly the same Asn-linked carbohydrate moieties altered in Glt_1^- CHO cells may be deficient in sialic acid residues in Wga^{RIII} CHO cells. However despite this structural change in cell surface carbohydrate, $Wga^{R_{III}}$ cells do not appear to be altered in their ability to bind WGA at the cell surface. Thus the location of the sialic acid residues lost in these cells would appear to be different from at least one population of sialic residues lost from $Wga^{R_{II}}$, clone 1021, and possibly even clone 13 Wga^R CHO cells (see Briles et al., 1977) which all exhibit decreased WGA binding. No obvious enzyme lesions which correlate with the $Wga^{R_{II}}$ or $Wga^{R_{III}}$ phenotypes have been detected in glycosyltransferase assays using a variety of exogenous glycoprotein acceptors (Stanley, unpublished observations). However, the glycolipids of $Wga^{R_{II}}$ and $Wga^{R_{III}}$ cells are also affected by the mutation to WGA resistance and thus the biochemical lesion might be expected to lie in a step common to the biosynthesis of sialylated glycoproteins and glycolipids.

4.3. Concanavalin A-Resistant (ConA^R) CHO Cells

The Lec^R cell lines discussed so far possess carbohydrate lesions affecting the more terminal sugars in the carbohydrate moieties of glycoproteins and glycolipids. All of them appear to affect sugars in the "branches" of typical Asn-linked carbohydrate moieties, producing structures which terminate prematurely at Man (Glt_1^- and CL 6 cells), GlcNAc (clone 13 cells), or Gal/GalNAc (clone 1021, $Wga^{R_{II}}$ and $Wga^{R_{III}}$ cells). However, there are two Lec^R phenotypes in which the glycosylation defects appear to be localized to sugars in the "core" region of Asn-linked carbohydrate side chains. Both these mutants were selected with the lectin Con A.

The ConA^{R_I} mutants described by Stanley et al. (1975b), Stanley and Siminovitch (1977), and Stanley and Carver (1977a) were originally isolated from CHO cells by R. M. Baker (Cifone and Baker, 1976). They have been shown to exhibit a defect in the incorporation of Man from GDP-Man into endogenous oligosaccharide-lipid intermediates and also into a protein fraction (Krag et al., 1977). The remaining endogenous activity for the transfer of Man to oligosaccharide–lipid is not inhibited by tunicamycin in these mutants. Recent experiments suggest that perhaps earlier steps in the addition of Man and Glc to dolichol phosphate intermediates may be impaired in these cell lines (Krag, 1979). However, no specific enzyme lesion has as yet been described which might represent the mutated gene product in ConA^{R_I} cells.

Wright and co-workers (Wright, 1973, 1975; Wright and Ceri, 1977) have independently isolated and partially characterized a number of ConA^R CHO cells. The properties of these isolates are similar to those of ConA^{R_I} CHO cells. A characteristic of both phenotypes is a decreased ability to bind Con A at the cell surface (Wright and Ceri, 1977; Stanley and Carver, 1977a). The residual Con A binding activity of ConA^R mutants appears to be noncooperative whereas parental CHO cells exhibit positively cooperative binding of Con A over a certain concentration range (Wright and Ceri, 1977). However, this latter finding may reflect the fact that these binding data were obtained over a relatively small range of Con A concentrations due to the low specific activity of the [^{3}H]-Con A utilized. Binding studies with [^{125}I]-Con A of high specific activity may reveal additional Con A binding sites (cooperative or non-cooperative) in the ConA^R mutants.

Consistent with the observation that ConA^R CHO cells possess a glycosylation defect and exhibit reduced Con A binding at the cell surface is the finding that a number of the cell membrane glycoproteins in ConA^R cells exhibit a reduced molecular weight in SDS gels (Ceri and Wright, 1978, Cifone *et al.*, 1979). Ceri and Wright (1978) also observed a glycoprotein of molecular weight ~155 K in membranes from ConA^R cells which is absent from membranes of parental CHO cells. The amount of this glycoprotein appears to be diminished in a revertant cell line. In addition, ConA^R cells exhibit a decreased ability to form caps in the presence of Con A whereas a revertant exhibits a similar cap-forming ability to parental CHO cells. Colchicine-resistant CHO cells have also been shown to express a "new" glycoprotein (mol. wt. ~170 K) at the cell surface which is diminished in a revertant cell line (Juliano *et al.*, 1976). However, these cells appear to possess an ability similar to parental cells to form "caps" in the presence of Con A (Aubin *et al.*, 1975). Thus the relationship of the high-molecular-weight glycoprotein to the resistance phenotype is not clear in the case of either ConA^R or colchicine-resistant cells.

A third property of ConA^R CHO cell lines (which is so far unique amongst Lec^R isolates), is that they are usually unable to grow at a temperature of 39°C (Wright, 1973, 1975; Stanley *et al.*, 1975*b*; Cifone and Baker, 1976). However, the exact relationship between the temperature sensitive (ts) phenotype and the biochemical lesion leading to Con A resistance is not clear. For example, the ts phenotype is essentially abolished if ConA^{R_1} CHO cells are plated at high cell density and, under these culture conditions, their decreased ability to transfer Man to endogenous oligosaccharide–lipid is also lost (Krag *et al.*, 1977). In addition, selection of ConA^{R_1} cell lines which can grow at 39°C ("ts revertants") gives rise to cell lines which may or may not have reverted for Con A resistance (Krag *et al.*, 1977; Cifone *et al.*, 1979). In fact, Krag *et al.* (1977) have obtained evidence that the ts phenotype is unrelated to Con A resistance since they have isolated a ts revertant which retains its resistance to Con A and continues to exhibit decreased incorporation of Man into oligosaccharide–lipid. On the other hand, the ConA^R revertant studied by Ceri and Wright (1978) was also selected by its ability to grow at 39°C. Therefore if the altered membrane properties of ConA^R CHO cells arise as a direct consequence of the mutation to Con A resistance (as might be predicted), the fact that the ts revertant of Ceri and Wright (1978) exhibits membrane properties similar to parental CHO cells suggests that the ts and Con A-resistant phenotypes may arise from a single mutation in some ConA^R cell lines.

The relationship between the ts phenotype and resistance to Con A in ConA^R CHO cells should be clarified by genetic studies utilizing somatic cell hybrids. The ConA^{R_I} phenotype behaves recessively in somatic cell hybrids (Stanley *et al.*, 1975b; Wright and Ceri, 1977; Cifone *et al.*, 1979). Therefore, if the ts and Con A-resistant phenotypes are linked, they should segregate together in a selection for Con A resistance or in a selection for cells with a ts phenotype from a hybrid cell population. In summary, the ConA^R CHO cell lines are clearly mutants defective in the glycosylation of glycoproteins but the molecular lesion associated with their complex phenotype is not defined.

The second ConA^R phenotype for which a specific "core"-related glycosylation defect has been observed is the Pha^{R_I}ConA$^{R_{II}}$ CHO cell line (Stanley *et al.*, 1975b). This "double" mutant was selected from a Glt$_1^-$ CHO cell population for resistance to Con A. Pha^{R_I}ConA$^{R_{II}}$ cells are highly resistant to each of the lectins PHA, WGA, RIC, LCA, and Con A and also exhibit decreased binding abilities for these lectins at the cell surface (Stanley and Carver, 1977a). Pha^{R_I}ConA$^{R_{II}}$ cells are not ts for growth. They exhibit complementation for resistance to Con A in hybrids formed with ConA^{R_I} cells (Stanley *et al.*, 1975b) and therefore represent a distinct ConA^R genotype. The membrane glycoproteins of this mutant are altered to the same extent as Glt$_1^-$ glycoproteins (Sudo and Stanley, unpublished observations) as would be expected from the GlcNAc-T1 lesion possessed by these cells (Stanley *et al.*, 1975b).

The carbohydrate alteration which distinguishes Pha^{R_I}ConA$^{R_{II}}$ cells from Glt$_1^-$(Pha^{R_I}) and from parental CHO cells has been delineated via structural studies of the carbohydrate moieties of VSV grown in Pha^{R_I}ConA$^{R_{II}}$ cells (Robertson *et al.*, 1978). In contrast to G from VSV grown in parental CHO cells (which possesses three Man residues in the "core" portion) and to G from VSV grown in Glt$_1^-$ cells (which possesses five Man residues in the "core" portion), G from VSV grown in Pha^{R_I}ConA$^{R_{II}}$ cells possesses a "core" containing four Man residues— (Man)$_4$(GlcNAc)$_2$Asn. Therefore, in addition to the Glt$_1^-$ mutation, these cells appear to have acquired a second mutation which affects the structure of the "core" portion of Asn-linked carbohydrate side chains. There is some evidence that the linkage of these Man residues may differ from those synthesized in Glt$_1^-$ or parental CHO cells. Thus there may also be a structural difference between the "cores" synthesized by the different cell lines (Robertson *et al.*, 1978). No enzyme activity has yet been described which could account for this Lec^R phenotype. However, it is clear that both the Pha^{R_I}ConA$^{R_{II}}$ and ConA^{R_I} cell lines should prove invaluable in dissecting the biochemical reactions involved in carbohydrate biosynthesis in mammalian cells. For example, it will be important to determine whether the lesions expressed in the glycoproteins made by these cells are also reflected in their dolichol phosphate intermediates. If so, it may be possible to use these mutants to further examine the question of whether there is a common oligosaccharide precursor for Asn-linked carbohydrate moieties (see P. W. Robbins *et al.*, 1977; Chapters 2 and 3).

5. MEMBRANE PROPERTIES OF Lec^R CELLS POSSESSING SPECIFIC CARBOHYDRATE ALTERATIONS

A summary of the Lec^R cell lines which have been shown to exhibit specific glycosylation defects is given in Table 1. Each of these defects gives rise to a variety of altered membrane properties. In addition to the mutants described in Table 1, there is a spectrum of Lec^R cell lines that are less well-studied. The specific biochemical changes associated with their phenotypes have not yet been defined at the molecular level but in all cases they have been shown to possess distinctive membrane alterations. In particular, each isolate has been tested for alterations in the ability to be agglutinated by and/or to bind the selective lectin at the cell surface. Therefore, by comparing these properties with the membrane properties of Lec^R lines which are known to possess specific biochemical lesions (described in Table 1), it is possible to interpret the data for the less well-characterized Lec^R lines and to arrive at some general conclusions. Before describing these conclusions, however, it is necessary to summarize the general membrane properties of the Lec^R lines presented in Table 1.

5.1. Lectin Binding and Cytotoxicity

All but one of the Lec^R mutants described in Table 1 exhibit reduced binding of the selective lectin at the cell surface. Since the highest affinity lectin receptors at the cell surface are almost certainly located in the carbohydrate moieties of glycoproteins and/or glycolipids, a reduction in binding of the selective lectin is a positive indication of a surface carbohydrate alteration. However, it is clear that reduced lectin binding is not necessarily a correlate of altered surface carbohydrate. Wga$^{R_{III}}$ CHO cells have been shown to bind WGA similarly to parental CHO cells over a millionfold range of WGA concentrations (Stanley and Carver, 1977b). Therefore it seems likely that the mutant possesses essentially all of the WGA binding sites present in parental CHO cells. However, Wga$^{R_{III}}$ cells exhibit altered surface binding of lectins of different specificity to WGA (Stanley and Carver, 1977a). They also are altered in their sensitivity to the toxicity of lectins other than WGA (Stanley et al., 1975b). This suggests that the decreased sialylation known to affect the surface carbohydrate of Wga$^{R_{III}}$ cells, also changes the interaction of Wga$^{R_{III}}$ cells with other lectins. For example, Wga$^{R_{III}}$ cells are more sensitive to and also bind increased amounts of PHA and RIC as might be expected from the increased exposure of Gal/GalNAc residues at the cell surface. Thus, for Lec^R cells which do not exhibit decreased binding of the selective lectin, altered interaction with lectins of different sugar specificities is indicative of structural changes in surface carbohydrate. In fact, each of the Lec^R mutants described in Table 1 has been shown to interact differentially (compared with parental cells) with lectins other than the selective lectin. Therefore, a consistent correlate of altered surface carbohydrate in a Lec^R cell line is a change in sensitivity to and/or binding ability for lectins of different specificity than the one used to select the cell line.

5.2. Lectin Agglutination and Capping

Many of the Lec^R isolates described to date have been characterized by an altered ability to be agglutinated by the selective lectin. Three of the Wga^R CHO cell mutants described in Table 1 have recently been examined for their abilities to be agglutinated by WGA over a millionfold range of WGA concentrations (Stanley and Carver, unpublished observations). Although two of these mutants [Wga^{R_I}(Glt$_1$$^-$) cells and Wga$^{R_{II}}$ cells] exhibit markedly decreased WGA binding abilities, neither of them appear to be differentially agglutinated by WGA compared with parental CHO cells. In addition, neither of them exhibit an altered ability to be capped by fluorescein-conjugated WGA. Thus altered lectin agglutination or capping does not appear to necessarily correlate with reduced lectin binding.

By contrast, the ConA^R cells of Wright (1975) which exhibit reduced Con A binding abilities (Wright and Ceri, 1977), also exhibit reduced abilities to be agglutinated by and to be capped by Con A (Ceri and Wright, 1978). Since these mutants appear to be essentially identical to those described by Krag et al. (1977), it would seem that the carbohydrate defect which affects many of the glycoproteins in ConA^{R_I} CHO cells may also give rise to the reduced mobility of membrane components. The alternative possibility, that these cells possess a primary lesion which alters membrane fluidity, cannot be ruled out but is less attractive because of the demonstration by Krag *et al.* (1977) of the glycosylation defect in ConA^{R_I} CHO cells. In summary, it would seem that altered agglutination or capping properties may or may not be correlated with the known surface carbohydrate alterations in particular Lec^R cell lines.

5.3. *Membrane Glycoproteins and Glycolipids*

Many membrane glycoproteins are affected by each of the carbohydrate lesions described in Table 1. This has been demonstrated by SDS gel electrophoresis of the radiolabeled membrane glycoproteins. Thus, Lec^R cells which possess decreased surface carbohydrate (such as Glt$_1^-$ CHO cells, CL 6 mouse L cells, Wga^R CHO cells and ConA^R CHO cells) give rise to SDS gels in which many membrane glycoproteins migrate faster than their parental cell counterparts. Also, Lec^R cells which possess decreased sialic acid or Gal/GalNAc residues on surface glycoproteins may be observed in SDS gels via the galactose oxidase:[^{3}H]borohydride technique (see Juliano and Stanley, 1975). For example, cells which have only lost sialic acid from the cell surface (such as clone 1021, Wga$^{R_{II}}$ and Wga$^{R_{III}}$ CHO cells) have an increased number of Gal/GalNAc residues available for labeling in comparison with parental cells. By contrast, cells which have lost sialic acid and Gal residues from the cell surface (such as Glt$_1^-$ and clone 13 CHO cells and CL 6 mouse L cells) exhibit reduced labeling via galactose oxidase:[^{3}H]borohydride compared with parental cells. In summary, if a Lec^R isolate is shown to possess glycoproteins of increased mobility in SDS gels or to exhibit altered labeling via the galactose oxidase:[^{3}H]borohydride technique compared with parental cells, it most probably possesses structurally altered cell surface carbohydrate. However, the converse does not necessarily follow. For example, Hyman and Trowbridge (1978) and Trowbridge *et al.* (1978) have recently obtained evidence that certain Thy-1$^-$ mutants of S49 lymphoma cells possess glycosylation defects which appear to specifically affect the Thy-1 molecule while not affecting other glycoproteins made by the cells.

Other membrane molecules which have been shown to be affected by glycosylation defects in Lec^R cells are the glycolipids (see Chapter 3). Alterations in membrane glycolipid content have been demonstrated using thin-layer chromatography of lipid extracts. In this manner it was shown that Ric^R CL 3 mouse L cells exhibit increased sialylation of G$_{M3}$ (Gottlieb and Kornfeld, 1976); clone 1021, Wga$^{R_{II}}$, and Wga$^{R_{III}}$ CHO cells exhibit decreased sialylation of G$_{M3}$ accompanied by a concomitant increase in lactosylceramide (Briles *et al.*, 1977; Stanley *et al.*, 1980); and clone 13 Wga^R CHO cells exhibit decreased sialic acid and Gal on G$_{M3}$ with a concomitant increase in glucosylceramide (Briles *et al.*, 1977). So far, however, no Lec[R] cell lines have been reported which possess carbohydrate alterations in glycolipids in the absence of a similar lesion affecting glycoproteins.

6. PROPERTIES OF OTHER Lec[R] CELL LINES

As mentioned previously many of the Lec[R] cell lines described in the literature have not been specifically investigated for surface carbohydrate defects (see review by Baker and Ling, 1978). However, all of them have been examined for altered membrane properties by one or more of the techniques described in the previous section. Therefore it is now possible to interpret their altered membrane properties in light of the changes known to occur in the membranes of Lec[R] cell lines which possess specific glycosylation defects. The outcome of this analysis indicates that the majority (if not all) of the Lec[R] cell lines isolated to date might possess structural alterations in surface carbohydrate indicative of glycosylation defects.

For example, Hyman et al. (1974) describe Wga[R] and Ric[R] S49 lymphoma cell lines which exhibit reduced binding of the selective lectin (~50–70% compared with parental cells) and a reduced ability to be agglutinated by the selective lectin. Both these changes are consistent with a surface carbohydrate defect. J. C. Robbins et al. (1977) have also isolated a Ric[R] S49 lymphoma line which exhibits a 30% reduction in RIC binding ability. Membranes from these cells which have been labeled via the lactoperoxidase technique exhibit a number of bands in SDS gels which migrate faster than the equivalent bands from parental membranes. This suggests that the Ric[R] S49 lymphoma cells may possess altered membrane glycoproteins with a reduced carbohydrate content. Similar results were observed by Juliano and Stanley (1975) for lactoperoxidase-iodinated Glt_1^- cells compared with parental CHO cells.

A number of Lec[R] isolates have been examined for altered interactions with lectins other than the selective lectin and in these cases the evidence is even more suggestive of altered surface carbohydrate. Stanley et al. (1975b) have described CHO Lec[R] mutants selected with PHA(Pha[RII]), RIC(Ric[RII]), or WGA and RIC(Wga[RII]Ric[RIII]) which exhibit decreased binding of the selective lectin as well as altered sensitivities to and binding for lectins of different carbohydrate specificity (Stanley and Carver, 1977a). Also, Meager et al. (1975, 1976) have described a variety of BHK cells selected for resistance to RIC which exhibit decreased binding of the selective lectin and, in many cases, altered interactions with other lectins. Since all of the Lec[R] cell lines described in Table 1 exhibit altered interactions with lectins other than the selective lectin, it seems likely that Lec[R] cell lines expressing such properties probably possess altered cell surface carbohydrate moieties. One of the Ric[R] BHK lines (Ric[R] 21) of Meager *et al.* (1975) appears to possess increased sialylation at the cell surface, since neuraminidase treatment of these cells increases their sensitivity to RIC as well as their RIC binding ability (Rosen and Hughes, 1977). This phenotype is similar to the CL 3 phenotype described by Gottlieb and Kornfeld (1976) and to a Ric[R] HeLa cell line isolated by Sandvig *et al.* (1978).

Some of the Ric[R] cell lines have been shown to exhibit a reduced uptake of RIC, and it has been suggested that these mutants are "endocytosis" rather than "receptor" (or carbohydrate) mutants. These include the Ric[R] S49 lymphoma cell line of J. C. Robbins *et al.* (1977), which was studied by Nicolson *et al.* (1976), the Ric[R] HeLa cell line of Sandvig *et al.* (1978), and the Ric[R] BHK cell line (Ric[R] 14) of Meager *et al.* (1975) which was studied by Sandvig *et al.* (1978). The molecular basis of the altered membrane properties in the Ric[R] S49 lymphoma cells is not clear. However, it is known that Ric[R] 14 BHK cells possess a GlcNAc-transferase lesion, and it appears that the Ric[R] HeLa cells exhibit increased sialylation of RIC "receptors." Thus both the latter cell lines possess altered surface carbohydrate and presumably arise as the result of a biochemical lesion in carbohydrate biosynthesis. It is perhaps unlikely, therefore, that the mutation to lectin resistance in these cells resides in a molecule specifically concerned with endocytosis. Rather, it seems probable that the reduced rate of RIC uptake observed in these cells results from the fact that their altered RIC receptors are functionally less efficient at endocytosis compared with the normally glycosylated RIC receptor glycoproteins in parental cells.

Another group of partially characterized Lec[R] lines which probably also possesses altered surface carbohydrate has been isolated from a variety of cell lines using Con A as the selective agent. Wollman and Sachs (1972) described SV40-transformed hamster cells which exhibit a reduced ability to be agglutinated by Con A; Culp and Black (1972) described ConA[R] SV40-transformed 3T3 cells which exhibit certain nontransformed growth properties and an increased sialic acid content similar to 3T3 cells; Ozanne (1973) also described ConA[R] SV40-3T3 cells which exhibit properties of "reversion" for transformation (a reduced ability to be agglutinated by Con A and WGA) and which bind Con A similarly to 3T3 and SV3T3 cells; Wright (1973) described ConA[R] CHO cells which exhibit altered sensitivity to certain "membrane-reactive" agents but which appear to bind Con A similarly to parental cells and are not to be cross-resistant to PHA; and finally, Guérin *et al.* (1974) described ConA[R] MOPC-173 plasmocytoma cells which exhibit a reduced ability to be agglutinated by Con A, no apparent reduction in Con A binding ability, and intramembranous particles which are not induced to aggregate in the presence of Con A.

Essentially, all of these ConA[R] cell lines exhibit a reduced ability to be agglutinated by Con A. However, no concomitant reduction in Con A binding abilities appears to correlate with this property in those ConA[R] lines which were examined for Con A binding (Ozanne, 1973; Wright, 1973; Guérin *et al.*, 1974). The explantation for this may be the fact that these binding studies were performed over relatively narrow ranges of Con A concentrations in the presence of rather low concentrations of bovine serum albumin (which inhibits the high background of nonspecific Con A binding). Thus the authors' conclusions that no differences exist between the Con A binding abilities of the respective ConA[R] and parental cell lines are open to some question. In fact, the agglutination and the freeze-fracture electron-microscopic results of Guérin *et al.* (1974) suggest that their ConA[R] plasmocytoma mutants are analogous to the ConA[R] CHO cells described by Wright and Ceri (1977) and Ceri and Wright (1978). Thus they might be expected to possess the glycosylation defect described by Krag *et al.* for ConA[R1] CHO cells. Alternatively, they might possess an increased sialic acid content as found by Culp and Black (1972) for their ConA[R] SV40-3T3 cells. In summary, the available evidence is consistent with the probability that the majority of the ConA[R] cell lines which have been described also arise as the result of a mutation affecting cellular glycosylation mechanisms.

7. MECHANISMS OF LECTIN CYTOTOXICITY

The properties of the Lec[R] cell lines isolated from a variety of cell types suggest that the overwhelming majority possess mutations in pathways related to the synthesis of the carbohydrate moieties of glycoproteins and glycolipids. Many of them have been shown to express specifically altered carbohydrate structures at the cell surface (see Table 1), and the remainder may probably be classified as "lectin-receptor" mutants. No Lec[R] isolates have yet been described as exhibiting an intracellular mechanism of lectin resistance. This may be due to the fact that little is known about mechanisms of lectin toxicity or to the fact that cell surface mutants occur more frequently than those affected intracellularly.

Of the lectins used most commonly to isolate Lec[R] cell lines, the mechanism leading to the toxicity of RIC is best understood. Briefly, RIC is composed of two, nonidentical, disulfide-bonded subunits, A and B. Subunit B binds specifically to the cell surface, the RIC molecule is transported into the cell by endocytosis, and subunit A interacts catalytically with 60S ribosomal subunits to inhibit protein synthesis (Olsnes *et al.*, 1974). However, none of the Ric^R mutants isolated so far appear to be resistant to the toxic effects of RIC at the level of protein synthesis (Gottlieb *et al.*, 1974; Hyman *et al.*, 1974; Meager *et al.*, 1976; J. C. Robbins *et al.*, 1977; Gottlieb and Kornfeld, 1976; Sandvig *et al.*, 1978). By contrast, the mechanisms of cytotoxicity of the other lectins commonly used as selective agents are not clear. It is possible that they exert their toxicity at the cell surface by inhibiting membrane mobility or the transport of specific molecules. For example, WGA has been shown to inhibit the transport of α-aminoisobutryric acid, cycloleucine, and thymidine in parental CHO cells. The WGA-resistant CHO mutants of Briles *et al.* (1977) appear to be resistant to this inhibitory effect (Li and Kornfeld, 1977). However, the specificity of this toxic mechanism is not clear.

8. GLYCOSYLATION MUTANTS SELECTED WITHOUT THE USE OF LECTINS

Mutant cell lines in which the primary defect appears to be an affect on carbohydrate biosynthesis have been isolated by at least three types of selection protocols which do not involve the use of lectins. Hyman (1973) used an immunoselective protocol (anti-Thy-1 antibody with complement) to isolate S49 lymphoma cells which had lost Thy-1 antigenic reactivity at the cell surface. Some of the isolates were subsequently shown to express a molecule at the cell surface which cross-reacts with anti-Thy-1 antiserum. These molecules migrated slightly faster than Thy-1 in SDS gels and also incorporated less [2-³H]-Man into the altered molecule, suggesting that the alteration in the Thy-1 molecules might reside in the carbohydrate moieties (Trowbridge and Hyman, 1975). Evidence in favor of this hypothesis has since been obtained by Trowbridge *et al.* (1978) and Hyman and Trowbridge (1978). They have isolated a number of Thy-1 S49 lymphoma cell lines and shown them to fall into five distinct complementation groups designated A, B, C, D, and E (Hyman and Trowbridge, 1978). Four of the genotypes appear to have arisen from different mutations affecting the glycosylation of Thy-1 (Trowbridge *et al.*, 1978). Each of these isolates synthesizes a molecule with Thy-1 antigenic reactivity which migrates faster than Thy-1 in SDS gels. Mutants in groups A, C, and E produce a "Thy-1 molecule" which does not contain Gal, indicating that they possess prematurely terminated carbohydrate moieties. In two of the mutants (A and C) the glycosylation lesion affecting Thy-1 molecules does not appear to affect other membrane glycoproteins. By contrast, mutants in complementation group E exhibit many membrane glycoproteins which are affected by the glycosylation defect. Buxbaum *et al.* (1977) have also reported a Thy-1 variant of S49 lymphoma cells which might fall into complementation group E since it possesses a number of membrane glycoproteins (including Thy-1) which migrate faster than their parental counterparts in SDS gels.

The glycosylation defects exhibited by the Thy-1 mutants of Trowbridge *et al.* (1978) suggested that these cells might be resistant to and therefore be selectable via certain plant lectins. In fact, Hyman and Trowbridge (1978) have been able to isolate complementation group E mutants by direct selection with Con A. They have also obtained other Lec^R S49 lymphoma cells which express glycosylation defects but which remain to be further characterized at the molecular level.

The second type of selection which has given rise to mutants affected in glycoprotein synthesis was designed to select for 3T3 cells with reduced adhesiveness for a substratum (Pouysségur and Pastan, 1976). Two clones were obtained following several cycles of selection and these were subsequently shown to possess membrane glycoproteins with apparently reduced molecular weights compared with similar glycoprotein classes in parental cells. In addition, the mutant cells exhibited reduced incorporation of glucosamine into acid-precipitable material (Pouysségur and Pastan, 1977). When fed glucosamine, the mutants were shown to accumulate glucosamine-6-phosphate in a soluble intracellular pool rather than the UDP-GlcNAc which accumulates in parental cells under the same incubation conditions (Pouysségur and Pastan, 1977). Since the phenotypic characteristics of these mutants can be reversed by 10 mM exogenously added N-acetylglucosamine (Pouysségur *et al.*, 1977), they were thought to be blocked in their ability to acetylate glucosamine-6-phosphate (Pouysségur and Pastan, 1977). In fact, the glucosamine phosphate acetyltransferase activity in mutant cell extracts is negligible compared with parental 3T3 extracts, suggesting that this enzyme defect may provide the basis of the mutant phenotype (Neufeld and Pastan, 1978). The mutants are probably "leaky" however, since a complete blockage in the acetylation of glucosamine-6-phosphate would inhibit the biosynthesis of UDP-GlcNAc, UDP-GalNAc and CMP-N-acetylneuraminic acid which would presumably be lethal (Schachter and Rodén, 1973). Also it is not clear whether the two mutants, AD6 and AD8, arise from the same or different mutations since they appear to exhibit some phenotypic differences (Pouysségur and Pastan, 1976).

The third type of glycosylation mutant selected without the use of lectins was isolated from CHO cells by Atherly *et al.* (1977). The mutant was obtained from a mutagenized population of CHO cells following ten trypsinization selection steps in which cells with *increased* adherence to a substraum (i.e., resistance to trypsinization) were sought. The cell line which emerged adheres more strongly to substrates than parental CHO cells and has been shown to exhibit broad changes in the synthesis of cell surface glycoproteins and mucopolysaccharides (Atherly *et al.*, 1977). The most markedly deficient molecule is hyaluronic acid, which is hardly synthesized at all in these mutants. However, they also exhibit decreased levels of the sugars associated with glycoproteins (30–76% of parental levels depending on the sugar) and of heparin sulfate and chondroitin sulfate (approximately 65% and 20% of the respective parental levels). There is some evidence suggesting that the mucopolysaccharides of the mutant cells turnover more rapidly than those of parental cells since they incorporate relatively more glucosamine into mucopolysaccharides during a 21-hr label. However, there is no evidence pertaining to a specific biochemical basis for the mutation(s) which have occurred in this line.

There is one other Chinese hamster cell mutant in which glycosylation mechanisms are affected. This is a temperature-sensitive mutant in which one of the earliest consequences of shift-up to the nonpermissive temperature is a marked decrease in the incorporation of labeled sugars (Fuc, Man, and GlcNAc) into acid-precipitable material (Tenner *et al.*, 1977). Recent evidence suggests that this mutant may be defective in the transfer of oligosaccharide moieties from dolichol phosphate to protein (Tenner and Scheffler, 1979). However the phenotype of this mutant is complex since it also exhibits an alteration in the induction of ornithine decarboxylase activity at the nonpermissive temperature (Landy-Otsuka and Scheffler, 1978). Thus the primary defect in this mutant is not clear. A lesion affecting internal membrane structure may be a likely candidate.

9. CONCLUDING REMARKS

Mammalian cultured cell lines possessing structural alterations in surface carbohydrate moieties may be selected with relative ease from a variety of different cell populations. The agents which appear to select specifically for this type of mutant are the cytotoxic plant lectins. For some cell lines it may be necessary to mutagenize to obtain Lec^R cells at a reasonable frequency (see Meager *et al.*, 1975). However, in many cases, the frequency of Lec^R cells in a population is $\sim 10^{-5} - 10^{-6}$. Among these, there might be expected to be a variety of distinct genotypes resistant to any one lectin. Thus, it is crucial to establish the clonal purity of a Lec^R isolate prior to characterizing its biochemical phenotype. To diminish the number of mutants surviving a particular lectin, specific selection protocols may be designed to select against unwanted phenotypes. For example, Glt_1^- mutants are highly sensitive to Con A and therefore may be specifically eliminated from a Lec^R population by the inclusion of Con A in the selective medium (Stanley, unpublished observations).

Lec^R cell lines are clearly very valuable biological tools with which to investigate the biosynthesis of glycoproteins and glycolipids in mammalian cells (see Chapters 2 and 3). They should also contribute to our understanding of the function of the carbohydrate moieties in these macromolecules and of the structure/function relationships which exist in the plasma membrane. An indication of some of the applications of Lec^R cell lines to these areas of membrane biology and carbohydrate biochemistry may best be exemplified by a consideration of the Glt$_1^-$ mutant phenotype. Through this mutant, it has been possible to: (1) define the specificities of two GlcNAc-T activities in CHO cells (Narasimhan et al., 1977), (2) observe the accumulation of an apparent intermediate in the "processing" of Asn-linked carbohydrate side chains (Robertson et al., 1978; Tabas et al., 1978; Li and Kornfeld, 1978), (3) obtain Sindbis and VSV with specifically altered carbohydrate moieties and thus study the function of the carbohydrate of these viruses in relation to their biological activities (Schlesinger et al., 1976; Robertson et al., 1978; Tabas et al., 1978), (4) investigate the properties of a particular class of WGA binding site at the CHO cell surface (Stanley and Carver, 1977b), and (5) observe the biological effect upon the cell of possessing a large number of specifically altered membrane glycoproteins.

Clearly, Glt$_1^-$ mutant cell lines will be useful for many further studies related to membrane biology. The other Lec^R cell lines also possess enormous potential in this area. These studies should be greatly facilitated by the major advances which have recently been made in the technology of glycopeptide isolation and structural analysis. For example, as discussed in Chapter 5, endoglycosidases are available to release specific oligosaccharides from the cell surface; lectin columns (Ogata et al., 1975; Krusius, 1976) and high-pressure liquid chromatography may be employed to separate glycopeptides of different carbohydrate structure; and nuclear magnetic resonance spectroscopic techniques have recently been developed for rapid and nondestructive compositional and linkage analysis of glycopeptides (Strecker et al., 1977; Grey et al., 1980). These techniques should ultimately enable the precise definition of specific carbohydrate structures at the cell surface of parental and Lec^R cells.

However, the full potential of Lec^R cells will not be realized until the mutated gene product leading to lectin resistance is identified in each different Lec^R mutant line. The interpretation of the biochemical properties and biological consequences of a particular Lec^R phenotype rests entirely on knowing the site of the original mutation. For example, the site of the mutation in Glt$_1^-$ cells is thought to be the GlcNAc-T1 enzyme (see Chapters 1 and 3). However, the primary defect could be another gene product which somehow alters the activity of the GlcNAc-T1 enzyme. In order to understand the biochemistry of carbohydrate biosynthesis, it is clearly important to know exactly which protein in Glt_1^- cells is responsible for the mutant phenotype. For example, a "processing" defect may alter both simple and complex Asn-linked carbohydrate side chains whereas a GlcNAc-T1 defect might be expected to affect only the "complex" type. Thus one of the major challenges facing the future development of Lec^R cell lines is to determine the mutated gene product which provides the basis for each Lec^R phenotype.

The first task is to locate a putative mutant gene product. It is clear from Table 1 that the mechanisms leading to lectin resistance are many and varied, as might be expected from what is known of the mechanisms of carbohydrate biosynthesis and maturation in mammalian cells. Alterations in lipid, nucleic acid, amino acid, or sugar metabolism may conceivably give rise to altered carbohydrate structures at the cell surface. Thus, maybe it is not surprising that a molecular basis for lectin resistance appears to have been characterized in only one Lec^R mutant (Glt_1^- cells). The development of *in vitro* assays for enzymes involved in glycosylation mechanisms (other than glycosyltransferases and glycosidases) may be required before progress is made in this area. Meanwhile, as mentioned previously, it is important to ensure the consistent correlation of a particular Lec^R phenotype with a unique biochemical lesion via genetic studies hypothesizing that an altered enzyme activity forms the molecular basis of lectin resistance.

Having localized a "putative mutant gene product," the second task is to obtain evidence to support the theory. The most direct approach is to isolate the enzyme and compare its amino acid sequence (tryptic peptides) with the same enzyme obtained from parental cells. However, this approach may be very difficult for glycosylation-related enzymes which are membrane bound and present in minute amounts. An alternative approach is the genetic one. For example, the demonstration that revertant cell lines exhibit parental enzyme activity in an *in vitro* assay or that a temperature-sensitive (ts) cell line exhibits ts enzyme activity in cell-free extracts would provide some evidence for the location of the mutation in the enzyme under examination.

In summary, the present data suggest that lectins are relatively specific selective agents for obtaining membrane mutants of mammalian cells which possess defects in the glycosylation of macromolecules. In addition, many different Lec^R phenotypes may be selected with a given lectin. In fact, the number of unique Lec^R mutants which might exist is potentially enormous. Such mutants should provide novel biological material with which to study membrane structure/function relationships and the various roles of membrane-bound carbohydrate.

ACKNOWLEDGMENT

This work was supported by a grant from the National Science Foundation PCM-76-84293.

10. REFERENCES

Atherly, A. G., Barnhart, B. J., and Kraemer, P. M., 1977, Growth and biological characteristics of a detachment variant of CHO cells, *J. Cell Physiol.* **89**:375.

Aubin, J. E., Carlsen, S. A., and Ling, V., 1975, Colchicine permeation is required for inhibition of concanavalin A capping in Chinese hamster ovary cells, *Proc. Natl. Acad. Sci. U.S.A.* **72**:4516.

Baker, R. M., and Ling, V., 1978, Membrane mutants of mammalian cells in culture, *Methods Membr. Biol.* **9**:337.

Briles, E. B., Li, E., and Kornfeld, S., 1977, Isolation of wheat germ agglutinin-resistant clones of Chinese hamster ovary cells deficient in membrane sialic acid and galactose, *J. Biol. Chem.* **252**:1107.

Buxbaum, J. N., Basch, R. S., and Szabadi, R. R., 1977, Analysis of Thy-1 variants of murine lymphoma cells, *Somat. Cell Genet.* **3**:1.

Ceri, H., and Wright, J. A., 1978, Mammalian concanavalin A-resistant cells contain altered cell surface glycoproteins, *Exp. Cell Res.* **115**:15.

Cifone, M. A., and Baker, R. M., 1976, Concanavalin A-resistant temperature-sensitive CHO cells, *J. Cell Biol.* **70**:77a.

Cifone, M. A., Hynes, R. O., and Baker, R. M., 1979, Characteristics of concanavalin A-resistant Chinese hamster ovary cells and certain revertants, *J. Cell. Physiol.* **100**:39.

Culp, L. A., and Black, P. H., 1972, Contact-inhibited revertant cell lines isolated from simian virus 40-transformed cells. III. Concanavalin A-selected revertant cells, *J. Virol.* **9**:611.

Gottlieb, C., and Kornfeld, S., 1976, Isolation and characterization of two mouse L cell lines resistant to the toxic lectin ricin, *J. Biol. Chem.* **251**:7761.

Gottlieb, C., Skinner, A. M., and Kornfeld, S., 1974, Isolation of a clone of Chinese hamster ovary cells deficient in plant lectin-binding sites, *Proc. Natl. Acad. Sci. U.S.A.* **71**:1078.

Gottlieb, C., Baenziger, J., and Kornfeld, S., 1975, Deficient uridine diphosphate-*N*-acetylglucosamine: Glycoprotein *N*-acetylglucosaminyltransferase activity in a clone of Chinese hamster ovary cells with altered surface glycoproteins, *J. Biol. Chem.* **250**:3303.

Grey, A., Narasimhan, S., Schachter, H., and Carver, J., 1980, The complete structure of the monosialylated glycopeptide of human αIgG myeloma protein as determined by highfield proton nuclear magnetic resonance spectroscopy, submitted for publication.

Guérin, C., Zachowski, A., Prigent, B., Paraf, A., Dunia, I., Diawara, M.-A., and Benedetti, E. L., 1974, Correlation between the mobility of inner plasma membrane structure and agglutination by concanavalin A in two cell lines of MOPC 173 plasmocytoma cells, *Proc. Natl. Acad. Sci. U.S.A.* **71**:114.

Hyman, R., 1973, Studies on surface antigen variants. Isolation of two complementary variants for Thy 1.2, *J. Natl. Cancer Inst.* **50**:415.

Hyman, R., and Trowbridge, I., 1978, Analysis of the biosynthesis of T25 (Thy-1) in mutant lymphoma cells: A model for plasma membrane glycoprotein biosynthesis, *Cold Spring Harbor Conference on Cell Proliferation* **5**:49.

Hyman, R., Lacorbière, M., Stavarek, S., and Nicolson, G., 1974, Derivation of lymphoma variants with reduced sensitivity to plant lectins, *J. Natl. Cancer Inst.* **52**:963.

Juliano, R., and Stanley, P., 1975, Altered cell surface glycoproteins in phytohemagglutinin-resistant mutants of Chinese hamster ovary cells, *Biochim. Biophys. Acta* **389**:401.

Juliano, R., Ling, V., and Graves, J., 1976, Drug resistant mutants of CHO cells possess an altered cell surface carbohydrate component, *J. Supramol. Struct.* **4**:521.

Kornfeld, R., and Ferris, C., 1975, Interaction of immunoglobulin glycopeptides with concanavalin A, *J. Biol. Chem.* **250**:2614.

Kornfeld, S., Rogers, J., and Gregory, W., 1971, The nature of the cell surface receptor site for *Lens culinaris* phytohemagglutinin, *J. Biol. Chem.* **246**:6581.

Krag, S. S., 1979, A concanavalin A-resistant Chinese hamster ovary cell line is deficient in the synthesis of [^{3}H]glucosyl oligosaccharide–lipid, *J. Biol. Chem.* **254**:9167.

Krag, S. S., Cifone, M., Robbins, P. W., and Baker, R. M., 1977, Reduced synthesis of [^{14}C]mannosyl oligosaccharide-lipid by membranes prepared from concanavalin A-resistant Chinese hamster ovary cells, *J. Biol. Chem.* **252**:963.

Krusius, T., 1976, A simple method for the isolation of neutral glycopeptides by affinity chromatography, *FEBS Lett.* **66**:86.

Landy-Otsuka, R., and Scheffler, I. E., 1978, Induction of ornithine decarboxylase activity in a temperature-sensitive cell cycle mutant of Chinese hamster cells, *Proc. Natl. Acad. Sci. U.S.A.* **75**:5005.

Li, E., and Kornfeld, S., 1977, Effects of wheat germ agglutin on membrane transport, *Biochim. Biophys. Acta* **469**:202.

Li, E., and Kornfeld, S., 1978, Structure of the altered oligosaccharides present in glycoproteins from a clone of Chinese hamster ovary cells deficient in N-acetylglucosaminyltransferase activity, *J. Biol. Chem.* **253**:6426.

Meager, A., Ungkitchanukit, A., Nairn, R., and Hughes, R. C., 1975, Ricin resistance in baby hamster kidney cells, *Nature* **257**:137.

Meager, A., Ungkitchanukit, A., and Hughes, R. C., 1976, Variants of hamster fibroblasts resistant to *Ricinus communis* toxin (ricin), *Biochem. J.* **154**:113.

Narasimhan, S., Stanley, P., and Schachter, H., 1977, Control of glycoprotein synthesis. Lectin-resistant mutant containing only one of the two distinct N-acetylglucosaminyltransferase activities present in wild type Chinese hamster ovary cells, *J. Biol. Chem.* **252**:3926.

Neufeld, E. J., and Pastan, I., 1978, A mutant fibroblast cell line defective in glycoprotein synthesis due to a deficiency of glucosamine phosphate acetyltransferase, *Arch. Biochem. Biophys.* **188**:323.

Nicolson, G. L., 1974, The interactions of lectins with animal cell surfaces, *Int. Rev. Cytol.* **39**:89.

Nicolson, G. L., Robbins, J. C., and Hyman, R., 1976, Cell surface receptors and their dynamics on toxin-treated malignant cells, *J. Supramol. Struct.* **4**:15.

Ogata, S., Muramatsu, T., and Kobata, A., 1975, Fractionation of glycopeptides by affinity column chromatography on concanavalin A-Sepharose, *J. Biochem.* **78**:687.

Olsnes, S., Refsnes, K., Pihl, A., 1974, Mechanisms of action of the toxic lectins abrin and ricin, *Nature* **249**:627.

Ozanne, B., 1973, Variants of simian virus 40-transformed 3T3 cells that are resistant to concanavalin A, *J. Virol.* **12**:79.

Ozanne, B., and Sambrook, J., 1971, Isolation of lines of cells resistant to agglutination by concanavalin A from 3T3 cells transformed by SV40, in: *The Biology of Oncogenic Viruses* (L. G. Silverstri, ed.), pp. 248–257, North-Holland, Amsterdam.

Pouysségur, J. M., and Pastan, I., 1976, Mutants of BALB/c 3T3 fibroblasts defective in adhesiveness. Evidence for an alteration in cell surface proteins, *Proc. Natl. Acad. Sci. U.S.A.* **73**:544.

Pouysségur, J., and Pastan, I., 1977, Mutants of mouse fibroblasts altered in the synthesis of cell surface glycoproteins, *J. Biol. Chem.* **252**:1639.

Pouysségur, J., Willingham, M., and Pastan, I., 1977, Role of cell surface carbohydrates and proteins in cell behavior: Studies on the biochemical reversion of an *N*-acetylglucosamine-deficient fibroblast mutant, *Proc. Natl. Acad. Sci. U.S.A.* **74**:243.

Robbins, J. C., Hyman, R., Stallings, V., and Nicolson, G. L., 1977, Cell-surface changes in a *Ricin communis* toxin (ricin)-resistant variant of a murine lymphoma, *J. Natl. Cancer Inst.* **58**:1027.

Robbins, P. W., Hubbard, S. C., Turco, S. J., and Wirth, D. F., 1977, Proposal for a common oligosaccharide intermediate in the synthesis of membrane glycoproteins, *Cell* **12**:893.

Robertson, M. A., Etchison, J. R., Robertson, J. S., Summers, D. F., and Stanley, P., 1978, Specific changes in oligosaccharide moieties of VSV grown in different lectin-resistant CHO cells, *Cell* **13**:515.

Rosen, S. W., and Hughes, R. C., 1977, Effects of neuraminidase on lectin binding by wild-type and ricin-resistant strains of hamster fibroblasts, *Biochemistry,* **16**:4908.

Sandvig, K., Olsnes, S., and Phil, A., 1978, Binding, uptake and degradation of the toxic proteins abrin and ricin by toxin-resistant cell variants, *Eur. J. Biochem.* **82**:13.

Schachter, H., and Rodén, L., 1973, The biosynthesis of animal glycoproteins, in: *Metabolic Conjugation and Metabolic Hydrolysis* (W. H. Fishman, ed.), pp. 1–149, Academic Press, New York.

Schlesinger, S., Gottlieb, C., Feil, P., Gelb, N., and Kornfeld, S., 1976, Growth of enveloped RNA viruses in a line of Chinese hamster ovary cells with deficient *N*-acetylglucosaminyltransferase activity, *J. Virol.* **17**:239.

Siminovitch, L., 1976, On the nature of hereditable variation in cultured somatic cells, *Cell* **7**:1.

Stanley, P., and Carver, J. P., 1977a, Lectin receptors and lectin resistance in Chinese hamster ovary cells, *Adv. Exp. Med. Biol.* **84**:265.

Stanley, P., and Carver, J. P., 1977b, Selective loss of wheat germ agglutinin binding to agglutinin-resistant mutants of Chinese hamster ovary cells, *Proc. Natl. Acad. Sci. U.S.A.* **74**:5056.

Stanley, P., and Siminovitch, L., 1976, Selection and characterization of Chinese hamster ovary cells resistant to the cytotoxicity of lectins, *In Vitro* **12**:208.

Stanley, P., and Siminovitch, L., 1977, Complementation between mutants of CHO cells resistant to a variety of plant lectins, *Somat. Cell Genet.* **3**:391.

Stanley, P., Caillibot, V., and Siminovitch, L., 1975a, Stable alterations at the cell membrane of Chinese hamster ovary cells resistant to the cytotoxicity of phytohemagglutinin, *Somat. Cell Genet.* **1**:3.

Stanley, P., Caillibot, V., and Siminovitch, L., 1975b, Selection and characterization of eight phenotypically distinct lines of lectin-resistant Chinese hamster ovary cells, *Cell* **6**:121.

Stanley, P., Narasimhan, S., Siminovitch, L., and Schachter, H., 1975c, Chinese hamster ovary cells selected for resistance to the cytotoxicity of phytohemagglutinin are deficient in a UDP-*N*-acetylglucosaminyl-glycoprotein *N*-acetylglucosaminyltransferase activity, *Proc. Natl. Acad. Sci. U.S.A.* **72**:3323.

Stanley, P., Sudo, T., and Carver, J. P., 1980, Differential involvement of cell surface sialic acid residues in WGA binding to parental and WGA-resistant CHO cells, submitted for publication.

Strecker, G., Herlant-Peers, M.-C., Fournet, B., Montreuil, J., Dorland, L., Haverkamp, J., Vliengenthart, J. F. G., and Farieux, J.-P., 1977, Structures of seven oligosaccharides excreted in the urine of a patient with Sandhoff's Disease (GM$_2$ gangliosidosis—varient 0), *Eur. J. Biochem.* **81**:165.

Tabas, I., and Kornfeld, S., 1978, The synthesis of complex-type oligosaccharides. III. Identification of an α-mannosidase activity involved in a late stage of processing of complex-type oligosaccharides, *J. Biol. Chem.* **253:**7779.

Tabas, I., Schlesinger, S., and Kornfeld, S., 1978, Processing of high mannose oligosaccharides to form complex type oligosaccharides on the newly synthesized polypeptides of the vesicular stomatitis virus G protein and the IgG heavy chain, *J. Biol. Chem.* **253:**716.

Tenner, A., Zeig, J., Scheffler, I. E., 1977, Glycoprotein synthesis in a temperature-sensitive Chinese hamster cell cycle mutant, *J. Cell Physiol.* **90:**145.

Tenner, A. J., and Scheffler, I. E., 1979, Lipid–saccharide intermediates and glycoprotein biosynthesis in a temperature-sensitive Chinese hamster cell mutant, *J. Cell Physiol.* **98:**251.

Toyoshima, S., Fukuda, M., and Osawa, T., 1972, Chemical nature of the receptor site for various phytomitogens, *Biochemistry* **11:**4000.

Trowbridge, I. S., and Hyman, R., 1975, Thy-1 variants of mouse lymphomas. Biochemical characterization of the genetic defect, *Cell* **6:**279.

Trowbridge, I. S., Hyman, R., and Mazauskas, C., 1978, The synthesis and properties of T25 glycoprotein in Thy-1 negative mutant lymphoma cells, *Cell* **14:**21.

Wilson, J., Narasimhan, S., and Schachter, H., 1978, Structural analysis of the products of liver golgi glycosyltransferases using *Cl. perfringens* endo-β-*N*-acetylglucosaminidase C1 (endo-C1), *Can. Fed. Biol. Soc.* **21**(abst. 233):XV.

Wollman, Y., and Sachs, L., 1972, Mapping of sites on the surface membrane of mammalian cells. II. Relationship of sites for concanavalin A and an ornithine, leucine polymer, *J. Membrane Biol.* **10:**1.

Wright, J. A., 1973, Evidence for pleiotropic changes in lines of Chinese hamster ovary cells resistant to concanavalin A and phytohemagglutinin-P, *J. Cell Biol.* **56:**666.

Wright, J. A., 1975, Concanavalin A as a selective agent in tissue culture for temperature-sensitive hamster cell lines, *Can. J. Microbiol.* **21:**1650.

Wright, J. A., and Ceri, H., 1977, The concanavalin A binding properties of concanavalin A-resistant and -sensitive hamster cell lines, *Biochim. Biophys. Acta* **469:**123.

Alterations in Glycoproteins of the Cell Surface

Paul H. Atkinson and John Hakimi

1. INTRODUCTION

It has been a widely held view over the last several years that the surface membrane of mammalian cells can initiate growth regulatory events when presented with specific external stimuli. This idea has stemmed mainly from the effect of various substances which are presumed not to enter the target cell and elicit a specific response in the cell.

It has been suggested (Kalckar, 1965) and is now generally held (see Todaro and Green, 1963; Levine *et al.*, 1965; Stoker and Rubin, 1967; and also, for example, Dulbecco, 1970) that a specific event at the cell surface is a primary step in the sequence of events resulting in a marked decrease in the rate of DNA, RNA, and protein synthesis (Levine *et al.*, 1965) and stabilization of cell population in contact-inhibited cells. Transformed cells have altered surface membranes, as demonstrated by the occurrence to tumor-specific surface antigens (Old and Boyse, 1964); by a decrease in cellular adhesiveness (Coman, 1960); by an increase in agglutinability by plant lectins (Burger and Goldberg, 1967; Inbar and Sachs, 1969; Pollack and Burger, 1969); by alterations in the patterns of glycolipids and glycoproteins (see below); and by alterations of contact inhibition of cell growth (Todaro and Green, 1963; Levine *et al.*, 1965; Stoker and Rubin, 1967). Such alterations in the surface membrane of transformed cells are possibly related and may be manifestations of an

Paul H. Atkinson and John Hakimi • Departments of Pathology and Developmental Biology and Cancer, Albert Einstein College of Medicine, Bronx, New York 10461.

overall process in which regulation by the cell surface of internal meta-bolic events in contact-inhibited cells is altered by transformation.

Many of the oligosaccharide moieties of glycoproteins are accessible in the medium, as demonstrated by removal of large proportions of bound surface sugars by proteases or glycosidases (probably in the manner depicted by Winzler, 1970). Two-thirds of the cell-bound sialic acid can be removed from intact cells by neuraminidase in CHO, L, HeLa and BHK variant cells (Kraemer, 1967); up to 66% of various bound sugars (40% of the fucose in HeLa Spinner cells) were released from the surface of HeLa cells by mild trypsin (Shen and Ginsberg, 1968); up to 80% of the bound glucosamine was removed from the surface of 3T3 cells by trypsin (Onodera and Sheinin, 1970). Fucosyl glycopeptides have been obtained from baby hamster kidney cells (Buck *et al.*, 1971*a*; Warren *et al.*, 1972*b*), chick cells (Buck *et al.*, 1971*b*), and also human fibroblasts (Muramatsu *et al.*, 1973) by protease treatment of the intact cultured cells. These data are supported by an earlier observation, utilizing elec-tron-microscope histochemical techniques, that much of the sialic acid of liver cells is on the outer surface of the plasma membrane (Benedetti and Emmelot, 1967). Data derived in more recent approaches also sup-port an exterior location of the oligosaccharides of cell surface glycopro-teins (Hirano *et al.*, 1972; Gahmberg and Hakomori, 1973; Steck and Dawson, 1974; see also lectin binding studies of Oseroff *et al.*, 1973). Thus, consistent with thermodynamic strictures (Singer and Nicolson, 1972; Singer, 1971), the polar hydrophilic sugar moieties of these cell surface macromolecules are probably oriented in the exterior medium. Furthermore, at least in some cases, the C terminus of the protein of these glycoproteins is embedded in, if not completely penetrating, the lipid bilayer (Segrest *et al.*, 1973; Gahmberg *et al.*, 1972; Bretscher, 1971; Winzler, 1970). Glycolipids are concentrated, if not exclusively located, in the plasma membranes (Winzler, 1970; Klenk and Choppin, 1970; Hakomori *et al.*, 1971) with their terminal nonreducing sugars exposed to the environment (Winzler, 1970). The red cell major glycoprotein (PAS-1, glycophorin) has recently been sequenced (Tomita and Marchesi, 1975) leading to a direct demonstration (Cotmore *et al.*, 1977) of the transmembrane disposition (N terminus exterior to the cell, C terminus on the cytoplasmic side) and that the sialic acid containing moieties are exterior to the cell. Thus, sugars are undoubtedly available for interaction with external factors.

The theoretical suitability of these sugars as specific structures, perhaps receptors, was pointed out by Shen and Ginsburg (1968) in that four different monosaccharides forming a heterosaccharide can be ar-ranged in more combinations than can 20 amino acids in a polypeptide of similar size. The actual function of these sugars as receptors or specific sites is now being documented and is discussed in detail in Chapter 6.

That surface sugar receptors can also be involved in control functions was suggested by Kalckar (1965) and by Burger and Noonan (1970) (reviewed, Burger, 1973). Growth control (similar to density inhibition of growth) over virally transformed 3T3 cells apparently could be imposed by the surface-acting lectin concanavalin A and was reversible by competition with such sugars as α-methyl glucose (Burger and Noonan, 1970). This implies that the surface receptors which interact with the lectin to produce contact inhibition are carbohydrates (Burger and Noonan, 1970). This work was subsequently confirmed when SV40-transformed 3T3 cells were shown to stop growth and movement in time-lapse cinematography on the addition of succinyl concanavalin A. Such growth inhibition of these transformed cells was reversed by the addition of α-methyl mannoside (Mannino et al., 1977). Burger has shown (1969) that trypsinization also releases cells from contact inhibition of growth with the exposure of "agglutination sites" on the membrane similar to those observed in mitotic cells (Fox et al., 1971) and virus-transformed cells (Burger and Goldberg, 1967). Consistent with this, Reich and his co-workers observed the appearance of a fibrinolytic system associated with the oncogenic transformation of culture chick cells (Unkeless et al., 1973) and later showed that an active fibrinolytic system was a necessary, although not sufficient, requirement for morphological changes in transformation of hamster cells (Ossowski et al., 1974). Involvement of surface sugars in control functions has further been suggested in that neuraminidase alone has been found sufficient to stimulate cell division in density-inhibited cultures (Vaheri et al., 1972). However, BHK cells were found to grow at their usual rate in the presence of sufficient neuraminidase, which would very rapidly remove sialic acid even in newly inserted molecules (Hughes and Clark, 1974). However, these authors did see a small (5%) increase in confluent density, thus these observations do not necessarily conflict. Moreover, lymphocyte transformation can be induced and reversed by specific sodium periodate oxidation and sodium borohydride reduction of surface sialic residues (Zatz et al., 1972). Similarly, periodate increases the permeability of chicken erythrocytes to radioactive uridine and also its incorporation into RNA (Kent and Pogo, 1974).

In a different manifestation of the possible involvement of surface sugars in control functions, density inhibition of growth itself may be the result of specific intercellular glycosylation of exposed oligosaccharides (Roseman, 1970; Roth and White, 1972). This latter subject has now been extensively reviewed by Roth (1973). A finding of major interest is that loss of anchorage dependence of cells in culture is a characteristic strongly correlated with the tumorogenecity of such cells in nude mice (Freedman and Shin, 1974). The chemical basis of anchorage dependence should be a very interesting property to study and may well be related to particular types of complex surface oligosaccharides. It should be noted that 3T3 cells do not have the ability to form colonies in soft agar or viscous media and by themselves are not tumorigenic in nude mice. 3T3 cells can be tumorigenic if injected while adhered to the surface of glass beads (Boone, 1975), hence anchorage and adhesion phenomena are clearly complex.

Recently, a new high-molecular-weight (~250,000) class of cell surface-associated glycoproteins appear related because of their cross-reactivity to common antisera. These glycoproteins known variously as "LETS protein" (large external transformation-sensitive) (Hynes, 1973; Hynes and Bye, 1974), "galactoprotein" (Gahmberg et al., 1974), "fibroblast surface antigen" (Ruoslahti et al., 1973), "cell surface protein" (Yamada and Weston, 1974), and "250K protein" (Stone et al., 1974). In general these proteins appear to be lost from their cell surface association on transformation (Hynes, 1976). However, there are exceptions to this statement depending on the exact cell-transforming agent combination. Also, the loss of this protein in transformed cells is not a specific in vitro marker for malignancy (Kahn and Shin, 1979). The protein(s) has been purified and when added back to various transformed chick, mouse, rat, and hamster (but not human) cells in culture restored normal morphology, adhesiveness, cytoskeletal structures, and density-dependent inhibition of movement (Ali et al., 1977; Yamada et al., 1976). The role of the carbohydrates in the adhesive functioning of this molecule (Yamada and Pastan, 1976) is not yet known but their presence in light of the preceding discussion is suggestive. The relationship of LETS-type proteins, if any, to proteoglycans of hyaline cartilage-like aggregates which show alterations as chick limb mesenchyme cells develop into chondrocytes (DeLuca et al., 1977; Hascall et al., 1976) remains to be investigated. Both seem to be related to cell adhesive properties but in what way remains to be determined. It is worth noting that the former (Heinegard and Axelsson, 1977) does show a qualitatively similar monosaccharide content to CSP, another LETS-type protein (Yamada and Pastan, 1976).

Kalckar (1965) noted that cancer cells seem to have lost the important link in the sequence of events which would begin with cell contact in normal cells. The nature of this "link" or "receptor" is still unknown, but from the foregoing discussion would seem that surface sugars at carbohydrate chain termini are likely candidates. The assumption that the terminally located sugars such as galactose and sialic acid have much to do with receptors involved in the control of various cell functions, including density-dependent inhibition of growth, finds support in structural studies of cell surface components. Transformed or growing cells have consistently been shown to have perturbations in these terminal residues of cell surface oligosaccharide chains when compared to normal or density-inhibited cells (for reviews see Oseroff et al., 1973). It has been shown, for example, that the glycolipids of transformed cells contain new carbohydrate chains and perhaps a greater proportion of chains which are shortened, lacking the nonreducing sugar terminals compared with the corresponding compounds in normal cells (Hakomori and Murakami, 1968; Mora et al., 1969; Brady et al., 1969; Hakomori et al., 1971). Certain enzymes involved in the addition of further sugar residues to an existing acceptor sequence are either greatly reduced (Den et al., 1971; Grimes, 1970) or entirely absent in transformed cells compared to normal cells (Cumar et al., 1970). In this latter report it was shown that the absence or inactivity of the enzyme catalyzing the transfer of a GalNAc from UDP-GalNAc to hematoside (N-Gly-sialic acid-Cer and sialic acid-Gal Glu-Cer) resulted in a profound decrease in the higher gangliosides (see also Itaya et al., 1976). Changes in the oligosaccharides of glycolipids are reviewed in Chapter 3.

This review is confined in scope to studies on growth- and transformation-dependent alterations in the oligosaccharides of animal cell membrane glycoproteins and in mucopolysaccharides. The coverage is not intended to include changes related to developmental or embryonic systems. Neither does it deal with alterations in the glycoprotein changes associated with lectin-resistant mutants discussed in Chapter 4.

For the purpose of this chapter oligosaccharide changes can be categorized into three major classes (see Chapter 1). Class A changes are those in oligomannosyl asparagine-linked glycopeptides. Class B changes are alterations that have been detected in sialoglycopeptides, generally considered to be of asparagine-linked type because they contain fucose. It should be noted sialoglycopeptides from human erythrocyte membrane glycoproteins are significantly different in that one structure does not have a dichitobiosyl core and has distal terminal fucose (Thomas and Winzler, 1971). The other structure has a side chain (not fucose) linked to the N-acetylglucosamine of the protein carbohydrate linkage (Kornfeld and Kornfeld, 1970). Structures of calf thymocyte plasma membrane glycopeptides have also been determined. Several unusual oligosaccharide structures of the general class B type have been demonstrated to lack sialic acid (Kornfeld, 1978). Class C changes are those found in glycopeptides of the mucin type which are O-glycosidically linked to serine or threonine.

A primary focus in this review is on the structural underpinnings of the various alterations observed, and the literature is discussed mainly from this viewpoint. Finally, it should be noted that, despite the weight of circumstantial evidence summarized in the foregoing introduction, there is to date no direct evidence that changes in various oligosaccharides are of primary importance in growth regulation. We still await definitive demonstration that such alterations are not secondary consequences of a more fundamental change in metabolism.

2. GROWTH-DEPENDENT CHANGES IN ASPARAGINE-LINKED OLIGOSACCHARIDES OF MEMBRANE GLYCOPROTEINS

Growth-dependent alterations in asparagine-linked glycopeptides are often studied by comparisons of the sizes of such molecules labeled with fucose or mannose. The probable cell surface origin of these glycopeptides is indicated when the glycopeptides are cleaved from the intact cells by mild protease treatments which do not rupture the cells. Codington *et al.* (1970) were among the first to show that such treatment released glycocalyx or "sialomucin" (Gasic and Gasic, 1962) material composed of mannose, galactose, glucose, *N*-acetylgalactosamine, *N*-acetylglucosamine, and sialic acid as the major constituents. Alternatively, isolated surface membranes are sometimes used as starting material. Other studies have used total membranes in which no attempt is made to segregate the subcellular distribution of these glycopeptides. Whether obtained by a prior step of proteolytic cleavage from intact cells or from membranes, glycopeptides are then extensively degraded with pronase which removes all but a few (often only 1–4) amino acids contained in the peptide to which the oligosaccharide is attached. In many studies, use of labeled monosaccharides such as mannose or fucose is the only evidence that the glycopeptides being studied fall into class A or class B. Few studies have directed efforts toward rigorously excluding class C glycopeptides or mixtures of subpopulations of all classes. It should be clearly pointed out that this general approach used in a number of different investigations does not distinguish whether alterations occur on the same or different glycoprotein backbones. Exhaustive proteolytic digestion would, of course, obscure the nature of the protein to which the various glycopeptide oligosaccharides were attached. This problem, general to the field, is in part caused by the difficulty of purifying membrane glycoproteins. This difficulty has been skirted by the use of membrane maturing viruses which often contain only one glycoprotein—a subject which is reviewed more fully below. Also, more recently, use of an affinity column technique has allowed purification of a liver cell membrane glycoprotein solubilized from membranes by nonionic detergents (e.g., Hudgin *et al.*, 1974) which may form a prototype approach to future studies on growth-dependent alterations (see Section 3.7.2).

2.1. Glycopeptides of Class B

In the first of a series of studies, Buck *et al.* (1971a) showed that fucose-labeled glycopeptides derived from purified surface membranes or from intact baby hamster kidney cells (BHK) by mild trypsin treatment were resolvable into two broad size classes in gel filtration: a bulk species and a higher-molecular-weight shoulder all within the range of 2000–5000 daltons. Rapidly growing cells expressed more high-molecular-weight glycopeptides than did "plateau" or density-inhibited cells. The same increases in higher-molecular-weight glycopeptides are observed in BHK_{21}/C_{13} cells which were in metaphase of mitosis when compared with control cells not in mitosis (Glick and Buck, 1973), underscoring that the phenomenon was growth related. Ceccarini (1975) found similar results in rapidly growing and density-inhibited WI38 cells. Muramatsu *et al.* (1973) confirmed these results with fucose-labeled glycopeptides removed from the surfaces of rapidly growing and density-inhibited human diploid fibroblasts. In that study, however, the use of endo-β-N-acetylglucosaminidase D (Koide and Muramatsu, 1974) in the presence of neuraminidase, β-galactosidase, and exo-β-N-acetylglucosaminidase to further probe the basis of the size alteration led to the conclusion that the core or protein linkage region of the oligosaccharides was altered by growth status (see Section 6). Warren and his co-workers had shown the high-molecular-weight shoulder could be explained by increased sialylation of terminal portions of the oligosaccharides. The change observed by Muramatsu *et al.* (1973) is not related to this type of alteration, since the glycosidase-digested fragments analyzed were only small fragments of the intact glycopeptides containing sialic acids. The differences may have been in the peptide portion of the glycopeptides, possibly indicating that different glycoproteins were involved. Van Nest and Grimes (1977) compared the sizes of glucosamine-labeled glycopeptides from crude membranes of rapidly growing normal Balb/c mouse cells or of confluent density-inhibited cells and did not see a differential emphasis in the high-molecular-weight glycopeptides as in the previously mentioned studies. Sakiyama and Burge (1972) extracted glucosamine-labeled glycoproteins from confluent and rapidly growing 3T3 cells with EDTA. Comparison of pronase-digested glycopeptides prepared from the extracted proteins indicated an increase in high-molecular-weight glycopeptides from the rapidly growing cells. The same general observation was true when glycopeptides from membrane fractions were compared. The increase in high-molecular-weight glycopeptides was due to sialic acid, since its removal with neuraminidase gave identical glycopeptide size profiles from rapidly growing and confluent cells. Glycopeptides from a nuclear fraction of growing 3T3 cells, labeled with glucosamine, showed a similar overall shift to the high-molecular-weight region when compared to those from confluent cells (Meezan *et al.*, 1969).

In these various studies, it is possible that alterations are most pronounced in material removed from the exterior of the cells with protease and therefore most likely represents a mature or completed molecule. Crude membranes (Van Nest and Grimes, 1977), on the other hand, must contain oligosaccharides, a proportion of which may be incomplete oligosaccharides, especially if labeled with mannose or glucosamine, and might represent precursor species yet to be processed or completed (see Chapter 3). Furthermore, glucosamine also labels class C glycopeptides and glycosaminoglycans, which in general yield pronase fragments found in the void volume upon gel filtration. However, pronase-digested class C glycopeptides may also be similar in size to asparagine-linked glycopeptides, since they contain tetra- and trisaccharides (see Bhavanandan *et al.*, 1977; Section 4 below) of the type found in fetuin (Spiro and Bhoyroo, 1974). Thus it is clear that care must be taken in interpretation of glucosamine-labeled glycopeptide elution profiles since an enrichment in higher-molecular-weight class B glycopeptides (also labeled with glucosamine) in gel filtration may be obscured by either of these class C-type glycopeptides.

Another factor which must be taken into account in comparing various studies is feeding regimen. Normal cells in culture stop dividing at a characteristic density per culture vessel. This behavior can be termed "contact inhibition" when it is shown nutrient supply is not limiting (Dulbecco, 1970; Ceccarini and Eagle, 1971). Confluent cultures are not necessarily contact inhibited, and even though they are nongrowing, they may only be starved. Comparison of results from different laboratories may be difficult unless growth conditions are the same, since physiologic status (growing, nongrowing) is known to affect the turnover rates of radioactively labeled carbohydrate (Warren and Glick, 1968; see Section 3). Nevertheless, it does seem warranted to conclude that rapidly growing cultured cells do display an enrichment in higher-molecular-weight glycopeptides of the class B type.

2.2. Glycopeptides of Class A

To probe the generality of whether other growth-dependent oligosaccharide changes (besides enrichment sialic acid in fucosyl glycopeptides) occurred in glycopeptides, Ceccarini *et al.* (1975) labeled glycopeptides from the surfaces of rapidly growing and density-inhibited human diploid fibroblasts with mannose. Mannose, in contrast to fucose, is universally found in asparagine-linked carbohydrates (Spiro, 1973). Mannose-labeled glycopeptides from rapidly growing human diploid fibroblast cells (KL2) or in material removed from cell surfaces by mild protease treatment were also enriched in high-molecular-weight glycopeptides compared to density-inhibited cells. Neutral glycopeptides isolated from the mixture displayed a differential susceptibility to endo-β-N-acetylglucosaminidase D (see Section 6, Table 2). This enzyme is specific towards the number and configuration of mannoses in the core and in general hydrolyzes mannose-terminating glycopeptides with three to five mannose residues. Glycopeptides containing more than five mannoses or lacking a particular α-linked terminal mannose are resistant to the enzyme. Greater quantities of resistant material were observed in the rapidly growing cells (Ceccarini *et al.*, 1975). Use of α-mannosidase showed that these endo-β-N-acetylglucosaminidase D resistant glycopeptides terminated in mannose. Endo-β-N-acetylglucosaminidase H (see Section 6) usually does not digest the small (3) core glycopeptides, which seem to form the core of the asparagine-linked heterosaccharides containing sialic acid, but does hydrolyze species containing dichitobiose with multiple mannoses. This enzyme converted the glycopeptides to an array of mannose-containing oligosaccharides; the largest contained 8–9 monosaccharides and had a strong affinity for concanavalin A–Sepharose (Muramatsu *et al.*, 1976a), and the smallest contained ~6 residues. Rapidly growing cells expressed much more of the large mannosyl oligosaccharides than did the density-inhibited nongrowing cells (see class A above). This type of growth-dependent alteration is fundamentally different from the types reviewed above since the increased level of sialic acids implied in the former cannot be involved in the changes seen in class A glycopeptides. These mannosyl glycopeptides are similar to thyroglobulin in unit A (Arima and Spiro, 1972; Tarentino *et al.*, 1973) in that they contain mannose and glucosamine as the major component. It is not known whether these class A species are a major cellular species chemically, but the labeling data would place them as 65% of the total in rapidly growing cells and 41% in density-inhibited cells (Muramatsu *et al.*, 1976a). Even allowing for the fact that class A glycopeptides may contain three times the number of mannoses as the heterosaccharide class B species, they would still represent a considerable proportion of the total.

3. TRANSFORMATION-DEPENDENT CHANGES IN ASPARAGINE-LINKED GLYCOPEPTIDES

3.1. Comparison Systems

A problem in comparing cultured transformed cells with cultured control cells is that the control cell is assumed to be a "normal" or a noncancer cell when, in fact, it often is not. Both 3T3 and BHK_{21} cells, for example, are a "line" and do not have the limited life span of genuinely normal cell. Furthermore, they are aneuploid, not diploid. 3T3 and also C3H/10T1/2, apparently incapable of causing tumors in syngeneic mice, do so when the cells are first attached to glass beads and then injected into the animal (Boone, 1975; Boone and Jacobs, 1976). BHK_{21} clone C_{13} cells are frankly tumorigenic in animals (Defendi *et al.*, 1963). Thus, the data in this field need to be extended; the normal cell must be carefully examined, defined, and used as the control for whatever *in vitro* cancer cell system is being studied (spontaneous tumors, virus transformants, chemical transformant, aggressive metastasizers, regressing tumors, and so on). It would seem primary fibroblasts and their transformants are generally more appropriate.

3.2. Difference in Sialoglycopeptides (Class B Glycopeptides)

Much of the work cited in the previous section was stimulated by the initial observation (Buck *et al.*, 1970; Meezan *et al.*, 1969) that virus-transformed cells had an enrichment in higher-molecular-weight glycopeptides associated with cell surfaces compared to their normal counterparts. The alteration was found temperature sensitive in cells reversibly transformed by temperature-sensitive Rous sarcoma virus mutants (Warren *et al.*, 1972*a*). This difference was seen in isolated surface membranes (Meezan *et al.*, 1969; Buck *et al.*, 1970) and in trypsinates from intact cells (Buck *et al.*, 1970). In the latter studies (Buck *et al.*, 1970) the specific radioactivity of the fucose tracer in the normal cell glycoprotein (BHK_{21}/C_{13}) was similar to that in the Rous sarcoma virus-transformed counterpart ($C_{13}B_4$). The labeling period with radioactive fucose was 72–78 hr, thus it is possible in their experiments that macromolecular fucose would have equilibrated with the precursor GDP-fucose specific radioactivity, if BHK cells were like HeLa cells (Kaufman and Ginsburg, 1968; Yurchenco and Atkinson, 1977) in utilization of exogenous fucose. It follows that radioactivity incorporated per unit time under equilibration conditions should allow calculation of the chemical amounts of the fucose in fucose-containing oligosaccharides. It also follows that a differential rate of synthesis from an equilibrated radioactive precursor pool, without subsequent accumulation of chemical amount, would not show up as a separate peak. The fact that Buck *et al.* (1970) did see separate peaks under these conditions would point to greater chemical quantities of these peaks in the transformed versus the normal cells if the precursor pools were equilibrated. There are few data on the rates at which nucleotide sugar pools equilibrate with isotopically labeled precursor and product glycoprotein in different cell types (Yurchenco *et al.*, 1978). This measurement would be very useful in determining whether differences in synthetic rate or accumulated amounts were involved in growth-altered glycopeptides, since without it, use of radioactive monosaccharide precursors obscures the distinction. By direct chemical determinations there was 0.44 nmole/10^6 cells of sialic acid in untransformed BHK_{21}/C_{13} cells whereas there was 0.85 nmole/10^6 cells in the BHK_{21}-C_{13}/B_4 transformed cells (Glick, 1974); there was twice as much sialic acid in the high-molecular-weight shoulder as in the bulk lower-molecular-weight glycopeptides. These results are not consistent with those from other laboratories for Rous sarcoma virus-transformed BHK cells (Table 1; Hartman et al., 1972) where overall little difference in sialic acid content was seen. In some cases (see, for example, Buck et al., 1970, especially Figure 3) the peak of glycopeptides enriched in sialic acid and labeled with radioactive fucose from transformed cells is around 30% of the total, which should be readily detectable as an increase in sialic acid per cell. The fact that it is not (Hartman et al., 1972) argues that in some studies cells were not at equilibrium with the radioactive fucose precursor and the higher apparent content of sialoglycopeptides might at least, in part, reflect a higher turnover rate of such fucose-labeled sialoglycopeptides.

Warren et al. (1972b) demonstrated an alteration in sialyltransferase activity in transformed cells. It was found that there was three times more transferase activity in the transformed cells than in the control nontransformed cells; the activity appeared specific for transformed cell glycopeptide acceptor since neither the endogenous activity nor the transferase activity with desialylated fetuin was significantly different in extracts of transformed and control cells. The asialoglycopeptides from the transformed cells resembled, in size distribution, those from the surface of the untransformed control cells. As noted in the previous section, Warren, Buck, and their collaborators also observed an increment in higher-molecular-weight glycopeptides in rapidly growing cells or cells in mitosis compared with nongrowing cells or cells not in mitosis. That the transformation-induced change was an addition to the growth-dependent change was most clearly demonstrated where surface glycopeptides from untransformed cells in mitosis were compared with those from transformed cells. In the latter, the high-molecular-weight glycopeptides were even more emphasized (Glick and Buck, 1973). The relationship between growth rate and presentation of higher-molecular-weight sialoglycopeptides remains to be clearly defined; it is entirely possible all cell surface oligosaccharide changes are related to growth rate alone, since the growth rate in transformed cells is generally greater than that of controls.

Buck et al. (1974) showed that the alterations accompanying transformation seen in surface membrane glycopeptides were also found in the rough endoplasmic reticulum, nuclei, mitochondrial, and smooth endoplasmic reticulum fractions. This study convincingly showed that the appearance of the glycopeptides enriched in sialic acid in the various subcellular fractions could not have been due to cross-contamination from surface membrane vesicles, since prior treatment of intact cells with neuraminidase abolished the high-molecular-weight peak from surface membrane glycopeptides but made no difference to its appearance in the various subcellular fractions obtained from these cells.

Grimes (1973), in an examination of the generality of differences in sialyltransferase activity in various cells and their transformed counterparts, found that the sialic acid content of cells was directly related to sialyltransferase activity. There was no absolutely consistent relationship between either one of these properties and the transformed state except an overall lower sialic acid (and therefore transferase activity) content in transformed cells which grew to higher saturation densities in culture. Grimes' (1973) study did not include comparisons of Rous sarcoma virus-transformed chick cells and, as can be seen in Table 1, these cells have elevated content of sialic acid. Predictably the transformed chick cells would have elevated transferase activity, based on Grimes' (1973) work. It seems possible that in this study and that of Warren *et al.* (1972a) different enzymes were involved since the enzyme studied by Grimes did not transfer sialic acid to a asialofetuin acceptor, whereas that of Warren *et al.* (1972a) did. Grimes (1970) had earlier studied another sialyltransferase which did transfer sialic acid to asialofetuin. That enzyme activity was lower in both virus (SV40 transformed) and spontaneously transformed 3T3 cells than the corresponding activity in untransformed 3T3 cells. Again the overall result does not seem to agree with the results of Warren *et al.* (1972a), but it must be pointed out that chick cells may not behave like 3T3 cells. The studies of Warren *et al.* (1972a) and those of Grimes (1973) are also not directly comparable because the former group used a specific asialoglycopeptide acceptor obtained from the transformed cells. Bosmann (1972) found a 1.5- to 6-fold increase in sialyltransferase activity in polyoma, murine sarcoma, and Rous sarcoma virus-transformed 3T3 cells using acceptors prepared from transformed and untransformed cells. The acceptor from transformed cells was 1.5-fold more active, confirming the work of Warren *et al.* (1972a).

An alteration to higher-molecular-weight glycopeptides was not seen in a comparison of fucose-labeled cell surface glycopeptides from the human diploid fibroblast WI38 and an SV40 transformant WI 18VA (Ceccarini, 1975) or in a comparison of glycopeptides from 3T3 and SV40 transformed 3T3 cells (Smets *et al.*, 1976). However, glycopeptides labeled with fucose derived from tumor cells developed in syngeneic mice by subcutaneous injection of SV3T3 cells did show an enrichment in high-molecular-weight glycopeptides. This suggests that SV40 transfor-

Table 1. Sialic Acid Content of Normal Cells and Their Transformed Counterparts

Cell	Sialic acid (nmole/mg protein)		Oncogenic	Reference
	Normal	Transformed		
Chinese hamster embryo (CHE)	19.9		—	Hartmann et al., 1972
SV-CHE		12.1	ND	Hartmann et al., 1972
Py-CHE		21.0	ND	Hartmann et al., 1972
Chick embryo fibroblasts (CEF)	24.4		—	Hartmann et al., 1972
RSV-CEF		32.6	ND	Hartmann et al., 1972
CEF	106.0		No	Perdue et al., 1971
ALV-CEF		114.0	No	Perdue et al., 1971
RSV-CEF (flat)		77.0	Yes	Perdue et al., 1971
RSV-CEF (round)		67.0	Yes	Perdue et al., 1971
ASV-CEF		79.0	Yes	Perdue et al., 1971
BHK_{21}/C_{13}	6.7		—	Hartmann et al., 1972
$RSV(Bry)-BHK_{21}/C_{13}$		6.8	ND	Hartmann et al., 1972
$RSV(S-R)-BHK_{21}/C_{13}$		6.9	ND	Hartmann et al., 1972
$BHK_{21}/C_{13}-A1$		10.0	—	Meager et al., 1975
$SV-BHK_{21}/C_{13}$		9.4	ND	Meager et al., 1975
$Py-BHK_{21}/C_{12}$		15.9	ND	Meager et al., 1975
BHK	8.1		—	Ohta et al., 1968
BHK_{21}	8.8		ND	Ohta et al., 1968
Py-BHK		6.7	ND	Ohta et al., 1968
BHK	8.1		—	Grimes, 1973
Py-BHK		5.8	ND	Grimes, 1973
BHK	35.4		—	Nigam and Cantero, 1973
Cl_2TSV_5-S		41.1	Yes	Nigam and Cantero, 1973
Cl_2TSV_5-R		65.8	Yes	Nigam and Cantero, 1973
3T3	More		—	Meezan et al., 1969
SV3T3		Less	ND	Meezan et al., 1969
3T3	16.5		—	Grimes, 1970
SV3T3		10.0	ND	Grimes, 1970
3T3	16.2		—	Culp et al., 1971
SV3T3		9.7	ND	Culp et al., 1971
ST3T3		11.0	ND	Culp et al., 1971
3T3A31	19.0		—	Grimes and Greegor, 1976

(Continued)

Table 1. (Continued)

Cell	Normal	Transformed	Oncogenic	Reference
		Sialic acid (nmole/mg protein)		
3T12		10.8	Yes	Grimes and Greegor, 1976
3T3-A31-C15		4.9	Yes	Grimes and Greegor, 1976
3T3	14.0		—	Ohta *et al.*, 1968
Py-3T3		5.5	ND	Ohta *et al.*, 1968
Nil-b	15.9		—	Grimes, 1973
HSV-Nil-b		12.6	ND	Grimes, 1973
Py-Nil-b		7.8	ND	Grimes, 1973
Rat liver	31.0		—	Emmelot and Bos, 1972
Rat liver hepatoma-484		45.0	Yes	Emmelot and Bos, 1972
Rat liver hepatoma-484A		61.0	Yes	Emmelot and Bos, 1972
AGMK	2.8[a]	—	—	Makita and Shimojo, 1973
SV-AGMK	—	2.3[a]	No	Makita and Shimojo, 1973

[a] nmole/mg dry cells.

mation of cultured cells may result in only weakly tumorigenic cells; selection of more highly tumorigenic cells by passage through animals thus may enhance the alteration in class B glycopeptides.

3.3. Carbohydrate Content

In general an overall lower sialic acid content was characteristic of transformed cells grown *in vitro*. In those few cases where an increase was observed, nontumorigenic cells were used as controls (chick embryo fibroblasts, normal rat liver) rather than cell lines such as 3T3 and BHK. Table 1 summarizes the observations in various comparison systems. Transformation of cells with Rous sarcoma virus tends to elevate the content of sialic acid (Hartman *et al.*, 1972) as is the case in chemically induced rat liver hepatomas (Emmelot and Bos, 1972; and see also Glick, 1974, above). However, in most other cases, transformation of cultured cells with SV40, polyoma virus, avian sarcoma virus, and by spontaneous means (Table 1) caused a reduction in cellular sialic acid content. In one study, transformation by different strains of Rous sarcoma virus, avian leukosis virus, or avian sarcoma virus did not lead to an increase of plasma membrane-associated sialic acid in chick cells; rather a decrease was observed (Perdue *et al.*, 1971). One possible explanation of these differences derives from a study by Perdue *et al.* (1972). The quantity of sialic acid in purified plasma membranes made from various uninfected and avian sarcoma virus-transformed clones showed decreasing sialic acid correlated with cell shape. Cells that were more fusiform in shape had more sialic acid. Cells that were more rounded and less well anchored had less sialic acid. Thus, anchorage, adhesiveness, cell shape, and motility may also play a role in the quantity of sialic acid per cell, and the appearance of correlation to transformation may instead be a reflection of one or all of these changes.

There are few studies on the relative quantities of neutral sugars in transformed and control cells. In one study Makita and Shimojo (1973) found overall less fucose, mannose, galactose, hexosamines, and sialic acid in African green monkey kidney cells (AGMK) transformed with SV40 than their nontransformed counterparts. In another study, Perdue *et al.* (1972) observed a 40% increase in neutral sugar content on oncogenically transformed cells; this may be related to the marked increase in neutral mannose and *N*-acetylglucosamine-containing glycopeptides seen in SV40-transformed human diploid fibroblasts (Ceccarini and Atkinson, 1977), although in the latter study, the absolute content of various monosaccharides was not studied. An increase in neutral sugars and hexosamines, however, was not observed in a series of Balb/c mouse cells either spontaneously transformed or virus transformed (Hartman *et al.*, 1972). In fact, these cell lines, capable of producing progressing or regressing tumors in mice, had a lower overall neutral and hexosamine carbohydrate content than the normal 3T3 (A31 clone) counterpart. Similarly, SV40 and polyoma virus-transformed Chinese hamster cells had a slightly lower overall neutral sugar and hexosamine content than the control cell counterpart (Hartman *et al.*, 1972), but Rous sarcoma virus-transformed chick embryo fibroblasts had a slightly higher neutral sugar content. Normal colonic mucosa had overall higher fucose, mannose, galactose, *N*-acetylglucosamine, *N*-acetylgalactosamine, and *N*-acetyl-neuraminic acid than did cancerous colonic mucosa from humans (Kim and Isaacs, 1975). Whether such overall carbohydrate analyses would, in fact, detect changes in subpopulations of the class B type or of the neutral class A type glycopeptide species described above is problematic.

3.4. Alterations and Relation to in Vitro and in Vivo Tumor Cells

The observation that transformed cells have an increase in fucose-labeled glycopeptides containing sialic acid compared with the nontransformed counterparts was confirmed and extended by Van Beek *et al.*

(1973), who studied four pairs of cells. These were hamster fibroblasts (BHK_{21}/C_{13} and polyoma virus-transformed counterparts), mouse fibroblasts (3T3 and 3T3-f, a spontaneous transformant), mouse lymphoblasts (MBIII, a nontransplantable tumor cell and the malignant lymphoblasts MBVIA), and rat epitheloid cancer cells (RLC and Novikoff hepatoma cells). Of these cell pairs, the best defined in terms of the normal vs. the transformed phenotype was the 3T3 and 3T3-f pair. The latter had low serum requirements compared to the "normal" 3T3, grew to approximately three times the saturation density of 3T3, could be grown in soft agar, and was more readily agglutinable by concanavalin A. The 3T3-f is called a cancer cell (tumorigenic) by these authors (Van Beek *et al.*, 1973), although it is not specifically stated whether they were capable of forming tumors in athymic, syngeneic, or whole-body irradiated mice.

The latter comment pertains to much of the work in this field since comparatively few studies have rigorously combined tissue culture criteria for transformation with tumor-forming ability in studies on biochemical alterations such as are under review here. These criteria have included low serum requirement, high saturation density, altered morphology and growth rate, agglutinability with lectins at lower concentrations, and the ability to grow in suspension in methylcellulose or soft agar (see Shin *et al.*, 1975; for a concise review of these various parameters in cell biological studies see Kahn and Shin, 1979). Correlates such as increased sialoglycopeptides should be combined with tests of ability to form tumors and/or metastases in animals with a range of cell types and species (as done by Van Beek *et al.*, 1973). In a recent study (Smets *et al.*, 1978) 3T3 cells did form tumors in some injected animals. These tumors explanted into culture gave rise to cells ("3T3-T") which, unlike the original 3T3 cells, could grow in methylcellulose and showed the alteration in fucose-labeled high-molecular-weight glycopeptides. Consistent with earlier observations from this laboratory, SV3T3 cells showed neither the ability to grow in methylcellulose nor the higher-molecular-weight glycopeptides. But like the 3T3/3T3-T comparison, cultured explants of tumors derived from SV3T3 (cell line "SV-T") showed both. Van Nest and Grimes (1977) partially purified glucosamine-labeled glycoproteins by separation on SDS-polyacrylamide gel electrophoresis. The gel was divided into three regions ranging from large to small glycoproteins which were then digested with pronase and used for comparisons of glycopeptide (oligosaccharide) size in the usual manner. In these studies the control "normal" cell was cloned line A31, derived from Balb/c mice, which was compared to (1) C5, a transformed derivative which produces regressing tumors in mice; (2) MSC, a Moloney sarcoma virus transformant to Balb/c mouse cells and also a regressor; (3) C5T, a variant of C5 which produces progressing killer tumors; and (4) 3T12T, a variant isolated from 3T12 tumors and also a progressor. With the exception of C5, all comparisons of the control with the tumorigenic cells in mice showed an enrichment in the high-molecular-weight glycopeptides. This enrichment was confined to the high-molecular-weight glycoproteins, as fractionated from SDS–polyacrylamide gels. Fascinatingly, the smaller glycoproteins, although containing similar sized oligosaccharides, whether transformed or not, had overall larger glycopeptides than the higher-molecular-weight glycoproteins. Radioactive glucosamine would, of course, label both class B glycopeptides and the oligomannosyl-N-acetylglucosamine-containing class A species. As shown by Ceccarini (1975) and Muramatsu et al. (1976a) these mannose-labeled glycopeptides are smaller and are strongly emphasized in the SV40-transformed human fibroblast (Ceccarini and Atkinson, 1977). It is possible that class A glycopeptides may be more confined to low-molecular-weight proteins. It would be interesting to know if mannose labeling shows alterations in low-molecular-weight glycoprotein oligosaccharides not visible with glucosamine labeling. This supposition presupposes that Balb/c mouse membrane glycoproteins contain both class A and class B glycopeptides, although this has not yet been demonstrated. Neuraminidase digestion did reduce the size of the higher-molecular-weight glycopeptides in the transformed cells, but these glycopeptides were still larger than the bulk lower-molecular-weight glycopeptides characteristic of the normal cells. Neuraminidase treatment did not, however, erase the overall inverse size relationships between glycopeptides derived from high-molecular-weight glycoproteins compared to low-molecular-weight glycoproteins. This is consistent with the suggestion above and the idea of Van Nest and Grimes (1977) that different carbohydrate structures may be involved (see also Ogata et al., 1976; below). In these studies the cells that produced regressing tumors showed no definite correlation with the absence of the high-molecular-weight glycopeptides, even though one regressor did, i.e., that produced by C5. It is quite likely that the "Peak C" glycopeptides described by Buck et al. (1974) correspond to the oligomannosyl (Class A) glycopeptides herein described. This work did not, however, demonstrate a growth-related enhancement possibly because of the use of glucosamine label. Mannose labeling may better show changes in the oligomannosyl glycopeptides.

Glick (1974) showed that the appearance of high-molecular-weight glycopeptides in hamster embryo cells increases with the number of days after infection (transformation) by polyoma virus. The latter time was also correlated with the number of cells required to form tumors in adult hamsters. The efficiency of transformation in culture was also observed in that more of the high-molecular-weight glycopeptides appeared as the number of passages through tumors in hamsters of chemically transformed cells increased (Glick, 1974). This work was an extension of earlier work (Glick et al., 1973) in which the relationship of appearance of the high-molecular-weight fucose-labeled glycopeptides to tumorigenecity was examined: Hamster embryo cells do not form tumors in hamsters; (1) dimethyl nitrosamine transformed cells do; (2) a variant of these, variant 11, does not form tumors at the same dosage as the parent cells; (3) variant 11 will form tumors when innoculated with 10–100 times more cells; (4) a revertant of variant 11 forms tumors at relatively low doses from which (5) cells can be obtained, labeled with fucose and glycopeptides analyzed as in all the previous cases (1–4). The transformed phenotype in culture was monitored by saturation density and ability to form colonies in soft agar. The enhancement of the high-molecular-weight fucosyl glycopeptides was clear-cut in cells derived from tumors from (3), (4), or (5) above. This enhancement was not observed in the original chemically transformed cells in culture without passage through animals (Glick et al., 1973). Thus, several studies are in agreement that passage of transformed cells as tumors in animals, followed by reculturing, results in enhancement of the higher-molecular-weight glycopeptides.

It is appropriate to comment here that, to our knowledge, there was no study among the many earlier studies where the presence of altered glycopeptides had been shown associated with in vivo labeled tumor cells. It could not be unequivocally known, therefore, that the appearance of the high-molecular-weight glycopeptides was not an artifact of cell culturing. This possibility was made less likely (Warren et al., 1975) when the same increase in sialic-acid-rich glycopeptides labeled with fucose was observed in the glycopeptides of various solid melanotic tumors of mice (liver and lung) as compared with normal tissues of the same organ. It should be noted, however, that in vivo tumors can be mixtures of cells, some cancerous, some not, (Farber, 1973), and the exact cellular origin of the biochemical alteration should be directly demonstrated. Similar to the results of Warren's group, solid Novikoff hepatomas, when compared to late embryonic rat liver, both labeled in vivo, also showed the enrichment in sialoglycopeptides (Van Beck et al., 1977). Explants of mouse mammary tumors labeled in vitro also showed the increase in sialoglycopeptides compared with normal mammary tissue (Smets et al., 1977). In a series of papers, this group of workers (Van Beck et al., 1975) has long held the view that such an increase in sialoglycopeptides is the only presently known biochemical characteristic that distinguishes cancer and normal cells. However, characterization of the oligosaccharide structures involved, the number of glycoproteins involved, and their rates of turnover under the conditions of fucose labeling used to detect the sialoglycopeptide enrichment will be necessary before this hypothesis can be termed substantiated. Also as noted, the cellular origin of the biochemical lesion in the animal systems will have to be rigorously demonstrated.

3.5. Microheterogeneity in Oligosaccharides

Microheterogeneity remains a problem. Classical structural analysis degrades starting material to single sugars and is inherently an averaging technique. There is no independent way to know that a deduced structure represents anything more than an average of the complex in the original mixture. An example of the complexity of definite structures present in seemingly homogeneous glycopeptides from ovalbumin is instructive. These glycopeptides are mostly two-sugar species (mannose, N-acetyl-glucosamine), that predominantly contain only one amino acid (aspara-gine) and look very homogeneous in gel filtration. However, when subfractionated on Dowex 50X (Huang et al., 1970) at least five distinct species of glycopeptide (I, II, III, IV, V) are obtained which have differ-ing proportions of mannose and N-acetylglucosamine. Subsequent work (Tai et al., 1977a; Yamashita et al., 1978) has led to the deductions that glycopeptides I and II contain galactose, II consists of two subspecies, III consists of three subspecies, and IV consists of two subspecies. Microheterogeneity can be nondestructively assayed by proton magnetic resonance (PMR) spectroscopy. In addition this technique can simulta-neously provide an analysis of anomeric configuration and in many cases can unambiguously assign linkage and quantity of monosaccharide in oligosaccharides or glycopeptides. PMR spectroscopy thus provides strong corroboration of structures assigned by classical approaches and assays homogeneity of start material.

This is so because resonances appearing in less than molar (frac-tional) quantities usually indicate mixture of compounds. The equivoca-tion derives from incomplete assignments of known oligomannosyl gly-copeptides. In 360-MHz PMR analyses of ovalbumin series we can confirm the deductions of Kobata's group (Carver et al., 1979), although there may be even more subspecies than suspected. This assay can be used in concert with glycopeptide purification techniques (gel filtration, ion exchange, paper electrophoresis) to obtain homogeneous prepara-tions. The efficacy of high-resolution proton magnetic spectroscopy of complex carbohydrates has been demonstrated in structural work on yeast mannans (Gorin and Spencer, 1968; Gorin et al., 1968, 1969), oligosaccharides derived from the liver of patients with G_{M1} gangliosidosis

(Wolfe *et al.*, 1974), in the urine of patients with Sandhoff's disease, G_{M2} gangliosidosis-variant O (Strecker *et al.*, 1977), human serotransferrin glycopeptide (Dorland *et al.*, 1977), human IgG (Grey and Carver, 1978), and sialyl oligosaccharides from patients with sialidosis (Dorland *et al.*, 1978). As an adjunct to classical structural analysis and as a tool in its own right, the power of this technique cannot be overemphasized.

3.6. Class A Glycopeptides

When normal and transformed glycopeptides labeled with mannose rather than fucose were compared, the normal cell had higher-molecular-weight glycopeptides (Ceccarini, 1975), which is the converse of the situation with fucose-labeled glycopeptides. The key to understanding why this occurs is that cellular and cell surface N-linked glycopeptides fall into at least two distinct classes (classes A and B), and both these classes show growth-dependent changes. The class of change studied initially in Warren's laboratory is detected by size alterations of fucose-labeled glycopeptides which in general are sailoglycopeptides containing five to six different monosaccharides. Not visible with a fucose label is a second broad class, namely the neutral glycopeptides containing only *N*-acetylglucosamine and mannose in different proportions and superficially resembling the ovalbumin glycopeptides (Tai *et al.*, 1977a) or thyroglobulin unit A (Arima and Spiro, 1972). These glycopeptide species from human diploid fibroblasts, electrophoretically neutral at ~pH 6.5 because they do not contain sialic acids, are generally smaller (average mol. wt. 1500) than the heterosaccharide acidic species (Ceccarini *et al.*, 1975). The large fucosyl glycopeptides of class B cannot be converted to the small neutral species of class A by treatment with neuraminidase. Also, it is apparent from the work of Muramatsu *et al.* (1976a) and Ceccarini (1975) that oligomannosyl glycopeptides on the cell surface are a major asparagine-linked class of glycopeptides in human diploid fibroblasts and are present in the membrane glycoproteins of SV40-transformed rat fibroblasts (Muramatsu *et al.*, 1975) as well as HeLa cells (Hunt *et al.*, 1978, 1980). Neutral glycopeptides containing only mannose and *N*-acetylglucosamine have also been observed in membrane glycopeptides of cultured rat hepatocytes (Debray and Montreuil, 1977), in rabbit fat cell membranes (Kawai and Spiro, 1977), and in Sindbis virus E1 and E2 glycoproteins (Burke and Keegstra, 1979). Endo-β-acetylglucosaminidase D (see below) will digest glycopeptides of the mannose–*N*-acetylglucosamine–asparagine type efficiently when the number of mannose residues is five or less and when the α-linked mannose attached to the β-linked mannose of the core is not 2-O-substituted with α-mannose (see Section 6). It was observed (Ceccarini and Atkinson, 1977) that mannose-labeled glycopeptides from the surface of human diploid fibrob-last cells (WI38) and from a SV40-transformed close relative of WI38 (WI 18VA), separated by paper electrophoresis into neutral glycopeptides and digested with endo-β-N-acetylglucosaminidase D (in the presence of neuraminidase, β-galactosidase, exo-β-N-acetylglucosaminidase), formed two peaks of digested products. One peak, hydrolyzed by the endogly-cosidase was an oligosaccharide of probable composition Man$_3$-GlcNAc. The other peak was glycopeptide not cleaved by the endoglycosidase activity. Similar endoglycosidase D-resistant mannose-labeled glycopep-tides were seen in membrane glycopeptides of SV40-transformed fibro-blasts (Muramatsu et al., 1975). In this study, such resistant glyco-peptides were hydrolyzed by endo-β-N-acetylglucosaminidase H, demonstrating that there are oligomannosyl glycopeptides present in the membrane glycoproteins of these cells as well. When the amount of these products in glycopeptides from the normal and transformed cells was normalized to those in the probable Man$_3$-GlcNAc peak, there was ap-proximately twice as much resistant mannose–N-acetylglucosamine-con-taining glycopeptide in the transformed as in the normal cells (Ceccarini and Atkinson, 1977). These several papers (Ceccarini et al., 1975; Cec-carini, 1975; Muramatsu et al., 1976b; Ceccarini and Atkinson, 1977) demonstrated that general oligomannosyl neutral glycopeptides from human fibroblast cells are markedly growth dependent; rapidly growing and transformed cells express these mannose-labeled oligosaccharides much more than nongrowing (density inhibited) cells. Also, different sizes of oligomannosyl residues seem to be expressed such that rapidly growing cells have more of the overall larger residues (7–9 mannose species). This difference is emphasized in transformed and rapidly grow-ing cells compared with the density-inhibited cells. In our recent unpub-lished studies chick embryo fibroblasts display similar growth-dependent changes in the oligomannosyl glycopeptides (Atkinson and Ceccarini, unpublished data). The "S4" (oligomannosyl) glycopeptides purifed from the E1 and E2 glycoproteins of Sindbis virus grown in these cells also show these alterations (Hakimi and Atkinson, unpublished data). It will be important to know how the various different oligomannosyl glycopep-tides are synthesized and differentially expressed at the cell surface. Current work on processing (see Chapters 1 and 2) of high-mannose (oligomannosyl) precursors is beginning to give insight into these ques-tions.

3.7. Changes in Oligosaccharides Derived from Single Glycoproteins

In most of the foregoing studies, it was generally assumed that the glycopeptides whose growth-dependent oligosaccharide alterations were being studied were asparagine linked. However, there is very little direct evidence that this is indeed the case. In fact, there are few studies on the complete structure (monosaccharide composition, sequence, and linkage) of animal cell membrane glycopeptides derived from purified membrane glycoproteins.

One way of obtaining relatively large quantities of a pure membrane glycoprotein is to grow a membrane maturing virus in the cells of interest. Many such viruses bud out into the cell milieu from which they can be highly purified. Depending on the specific virus, they only contain one or sometimes two glycoproteins. Vesicular stomatitis virus (VSV), a bovine lytic virus having a very broad host range, contains only five structural proteins. The virion lipid bilayer envelope can be dissociated from the virus at pH 1.5. It contains one glycoprotein (G) and a matrix protein (M) (Mudd, 1974), both of which are known to associate strongly with plasma membranes during infection and maturation (Wagner *et al.*, 1970, 1972; Cohen *et al.*, 1971; David, 1973, 1977; Atkinson *et al.*, 1976; Hunt and Summers, 1976; Morrison and McQuain, 1977; Knipe *et al.*, 1977; Atkinson, 1978). The structure of the oligosaccharide of VSV G-protein in BHK cells has been almost completely elucidated (Etchison *et al.*, 1977) and shown to be modified by prior transformation of its host cell with polyoma virus (Moyer and Summers, 1974). The structure of the oligosaccharide chain of the G-protein in untransformed host cell is shown below (Etchison *et al.*, 1977):

$$\text{Sialic acid} \xrightarrow[(\pm)]{\alpha 2,3} \text{Gal} \xrightarrow{\beta 1,4(3)} \text{GlcNAc}$$
$$\text{Sialic acid} \xrightarrow[(\pm)]{\alpha 2,3} \text{Gal} \xrightarrow{\beta 1,4(3)} \text{GlcNAc} \xrightarrow[(\pm)]{\beta} [\text{Man}]_3 \xrightarrow{\alpha}$$
$$\text{Sialic acid} \xrightarrow{\alpha 2,3} \text{Gal} \xrightarrow{\beta 1,4(3)} \text{GlcNAc}$$

$$\text{Man} \rightarrow \text{GlcNAc} \xrightarrow{\beta} \text{GlcNAc} \rightarrow \text{Asn}$$
$$\uparrow$$
$$\text{Fuc}^{(\pm)}$$

A similar structure containing only three mannose moieties has been reported by Reading *et al.* (1978).

To compare the oligosaccharide chain of the G-protein in normal and transformed cells, mannose- or fucose-labeled glycopeptides were prepared by pronase digestion. The glycopeptides were then tested for their susceptibility to digestion by the endo-β-N-acetylglucosaminidase D. With either label the transformed cell viral glycopeptide showed an increase of glycopeptides not hydrolyzed by this glycosidase. At the time this study was published, the detailed specificity of the enzyme was not known. However, based on current knowledge of the specificity of the

endo-glycosidase, it seems likely that transformation can cause alterations in the mannose core of this glycopeptide, perhaps by adding extra
mannoses. It is not likely the resistant glycopeptides are of the neutral
class A type (Muramatsu *et al.*, 1975, 1976a), seen greatly increased in
SV40-transformed human fibroblasts (Ceccarini and Atkinson, 1977).
This is because VSV glycoprotein in virus grown in BHK cells contains
two oligosaccharides per molecule, both asparagine-linked and both of
the class B type (Etchison *et al.*, 1977). In addition, the endoglycosidase-
resistant glycopeptide was labeled with fucose (Moyer and Summers,
1974), which also points to an alteration in the class B glycopeptide.

   In summary, these observations do raise the interesting possibility
that growth- or transformation-dependent changes may fall into four
distinct types: those in which extra sialic acid residues are present (Warren and co-workers); those in which extra sialic acids and outer terminal
sugars (galactose and *N*-acetylglucosamine) are present (Ogata *et al.*,
1976); those in which the proportion of the neutral mannose- and *N*-
acetylglucosamine-containing glycopeptides is increased (Muramatsu *et
al.*, 1975, 1976a); and those in which subtle changes occur in monosaccharide composition or linkage in the core region of the heterosaccharide
class B type of glycopeptide. What seems clear is the necessity of analyzing growth-dependent structural changes in homogeneous glycopeptides from purified single species of glycoprotein in order to detect subtle
and possibly important changes in growth-dependent oligosaccharides of
the cell surface.

4. CHANGES IN SERINE- (THREONINE-) LINKED OLIGOSACCHARIDES OF MEMBRANE GLYCOPROTEINS

   The chemistry and biosynthesis of mucin-type glycoproteins, of
which O-glycosidically linked oligosaccharides are a major class, have
been reviewed by Carlson (1977). Studies on growth-dependent changes
in class C (mucin-type) membrane glycopeptides are very few, although
in general much more detailed. Jeanloz and Codington (1974) obtained a
high-molecular-weight membrane glycoprotein from the surface of mouse
ascites cells which strongly binds the agglutinin of *Vicia graminea*. This
lectin binds to β-D-galactopyranosyl residues and agglutinates erythrocytes containing N-blood group substances. The glycoprotein strongly
inhibits this agglutination. Interestingly, like the asparagine-linked oligosaccharides, the oligosaccharides are of two general types—acidic
sialic acid containing glycopeptides and neutral glycopeptides:

Sialic acid → Gal → (Gal, GalcNAc) → Gal → GalNAc → Ser (Thr)

Gal → GalNAc → Ser (Thr)

It is possible that a sialoglycopeptide from Novikoff ascites cells was also of these compositions (Walborg et al., 1969), although in addition to sialic acid, galactosamine, and galactose, this fraction contained glucosamine, mannose, glucose, and uronic acid. Therefore, it seems likely that the fraction was a mixture of asparagine-linked (classes A and B), C-type glycopeptides, and possibly glycoaminoglycans in addition (see Section 5). Class C glycopeptides were probably also present in glucosamine-labeled glycopeptides released extracellularly by Ehrlich ascites carcinoma cells (Molnar et al., 1965b). In these latter studies, released glycoprotein was not digested with pronase and could be separated into several fractions by DEAE-cellulose ion-exchange chromatography. The released material contained glucosamine, galactosamine, neutral hexoses, sialic acid, and uronic acid, again pointing to a complexity of glycopeptide types from ascites cells.

Davidson and his co-workers have studied the growth dependency of oligosaccharides of class C glycopeptides in melanoma and control pigmented (iris) cells. In an elegant series of papers, this group has established a fractionation scheme (Bhavanandan et al., 1977) which enables the chemical separation of class C glycopeptides labeled with glucosamine from glycosaminoglycans and from serum class A and B glycopeptides. Based on the results of glycosidase digestion and periodate oxidation, it was suggested the oligosaccharides of melanoma class C glycopeptides may be tetrasaccharides (and some trisaccharides) similar to those found O-glycosidically linked to fetuin, namely (Spiro and Bhoyroo, 1974):

$$\text{Sialic acid} \xrightarrow{\alpha2,3} \text{Gal} \xrightarrow{\beta1,3} \text{GalNAc} \xrightarrow{\alpha1,3} \text{Ser}$$

Since the molecular weight of the pronase-resistant glycopeptides appeared to be 12,000–15,000, 7–9 such units could account for this molecular weight. It is of interest to note that pronase resistance of the peptide moiety is perhaps due to the presence of clusters of such oligosaccharides. Class C glycopeptides in these studies (Bhavanandan et al., 1977) were strongly bound to immobilized WGA (on a column). The binding ligand is probably sialic acid (Sharon and Lis, 1972) since the molecule does not contain N-acetylglucosamine; free class C oligosaccharides do not bind, leading these authors to suggest that several adjacent sialic acids (on a peptide backbone) are necessary for binding. This conclusion is not unlike that reached by Muramatsu and his co-workers on the numbers of necessary concurrent mannose residues (two) necessary for binding to immobilized concanavalin A (Ogata et al., 1975).

Satoh et al. (1974) had previously compared the production of class C glycopeptides in normal (iris) and transformed (B16 tumor cells) melanocytes of mice. Although characterization of this material was then not as extensive, it was clear that the glucosamine-labeled material was of the class C type (97–100% of the label was found in galactosamine and sialic acid). Glycopeptide, extracted from cetylpyridinium chloride precipitates and chromatographed on glass beads (called Peak II in this study) was practically absent in the normal iris melanocytes. There was six times the quantity of class C sialoglycopeptides in the B16 tumor cells compared to the normal iris melanocytes and 22 times as much in an amelanotic clone of B16. In the neutral asialoglycopeptides the numbers were, respectively, 2.5- and 8-fold. In apparent contrast to these results, Kim (1977) found that blood group A glycoproteins were markedly decreased in cancer tissues.

It should be noted that with the class C O-glycosidically linked glycopeptides an enrichment in sialic acids was also observed similar to that in the asparagine-linked glycopeptides. There is no reported counterpart to an enrichment in neutral O-glycosidic glycopeptides of the type where numbers of extra galactoses might possibly be added, analogous to the high mannose type in asparagine-linked molecules described above.

Gacto and Steiner (1976) examined the content of two "fucolipids," (named FL3 and FL4) as a function of cell density. These compounds, initially thought to be sphingosine fucosides, were actually glycopeptides containing fucose and threonine (FL3), and glucose, fucose, and threonine (FL4) (Larriba et al., 1977). In murine sarcoma virus-transformed rat cells, the quantity of FL3 and FL4 increased markedly due to an increased rate of synthesis. There was an apparent increase in FL3 and FL4 in highly tumorigenic Balb/c mouse cell lines but not weakly tumorigenic lines (SV3T3) as compared to the control 3T3 (Steiner and Steiner, 1976). This was measured by the amount of radioactive fucose incorporated, and it is not known if chemical amounts increased. The membrane location (cell surface? membrane bound at all?) of these novel compounds is unknown. Also, given their low molecular weight, it would be important to know that they are not degradation products.

5. ALTERATIONS IN GLYCOSAMINOGLYCANS

The subject of glycosaminoglycans in animal cells, as well as coverage of growth-dependent alterations of these molecules has been authoritatively and extensively reviewed by Kraemer (1978). The biosynthesis and catabolism of glycosaminoglycans is discussed in Chapter 7. The purpose of the present discussion is to review recent papers and to place alterations of glycosaminoglycans in context with those of the other complex glycan components associated with the cell surface. In reiteration of the earlier reviews by Kraemer (1971, 1978) glycosaminoglycans fall into one of the major classes characterized by the repeat units of the polymer (keratosulfates may be a distinct class of glycosaminoglycans but also resemble in composition the class C glycopeptides; see Kraemer, 1971).

Hyaluronic acid (HA)*

$$\xrightarrow{,4} GlcUA \xrightarrow{\beta 1,} [\xrightarrow{3} GlcNAc \xrightarrow{\beta 1,4} GlcUA \xrightarrow{\beta 1,}] \xrightarrow{3} GlcNAc \xrightarrow{\beta 1,}$$

Chondroitin sulfates (CS)

$$\xrightarrow{,4} GlcUA \xrightarrow{\beta 1,} [\xrightarrow{3} GalNAc \xrightarrow{\beta 1,4} GlcUA \xrightarrow{\beta 1,}]_n \xrightarrow{3} GalNAc \xrightarrow{\beta 1,}$$

CSA: 4-O-sulfoGalNAc substituted
CSB: (dermatan SO$_4$) IdUA for GlcUA and 4-O-sulfoGalNAc
CSC: 6-O-sulfoGalNAc

Heparan sulfate (HS)

$$\xrightarrow{4} GlcUA \xrightarrow{\alpha 1,} [\xrightarrow{4} GlcNSO_4 \xrightarrow{\alpha 1,4} GlcUA(IdUA) \xrightarrow{\alpha 1,}]_n \xrightarrow{4} GlcNSO_4 \xrightarrow{\alpha 1,}$$

In heparan sulfates, the glucosaminidic and iduronic linkages may be β, but most apparently are α (Kraemer, 1978).

Keratan sulfates are a somewhat indistinct group of complex carbohydrates, containing galactose and glucosamine. Both N-linked glucosamine and O-linked N-acetylglucosamine protein carbohydrate linkages are present. Some purified preparations contain sialic acid, mannose, and fucose (Kraemer, 1978). The latter author prefers the term "sulfated glycoprotein" to keratan sulfate due to the apparent heterogeneity of types.

The protein carbohydrate linkages for chondroitin sulfates and heparan sulfates involves the sequence (Kraemer, 1971):

$$GlcUA \xrightarrow{\beta 1,3} Gal \xrightarrow{\beta 1,3} Gal \xrightarrow{\beta 1,4} Xyl \xrightarrow{\beta} O\text{-}Ser$$

The protein carbohydrate linkage for HA is not known; it is not even completely clear that there is a covalent linkage to proteins (reviewed by Kraemer, 1978).

---

* HA, hyaluronic acid; CSA, -B, -C, chondroitin sulfates A, B, C; HS, heparan sulfate; BS (=CSB), dermatan sulfate; KS, keratan sulfate.

Interest in the relationship of glycosaminoglycan alterations to alteration in cellular growth status derives from knowledge that the bulk of this material is associated with the cell surface (see Kraemer, 1978). In one case, for example, 63–84% of the cellular sulfated GAG could be removed by mild trypsin treatment of the cells (Roblin et al., 1975b). Among the first to document growth-dependent alterations of this nature were Hamerman et al. (1965), who measured hyaluronic acid (testicular hyaluronidase susceptible) levels in 3T3 and several clones of polyoma virus and SV40-transformed 3T3. In all the transformed cells, there was much less hyaluronate and in one clone it was one-twentieth that of the untransformed 3T3 cells. A similar magnitude of reduction in HA was seen in human fibroblasts transformed with SV40. However, all measurements in this study were made on dense, nongrowing cultures (in the case of normal cells) or in dense confluent cultures (in the case of transformed cells). As discussed below, Hopwood and Dorfman (1977) showed that not only do normal human fibroblasts synthesize much more HA and sulfated GAGs at low densities while rapidly growing, but that the effect of cell density on such synthesis in transformed cells is insignificant. In other studies, an adhesive rat fibrosarcoma had ~10-fold the quantity of HA and ~6-fold the quantity of CS than did nonadhesive mouse ascites cells (Danishefsky et al., 1966), but the validity of detecting alterations by such a cross species/cell type of comparison is doubtful. Saito and Uzman (1971) compared acid mucopolysaccharide production by labeling with [^{35}S]sulfate in a variety of cell lines: lymphoid L-1210, P-1534; Ehrlich ascites; HeLa, KB, S-180, L-929, DON, 3T3, and SV3T3. In the latter pair, 3T3 had more CSC, CSA, and DS than SV3T3 and also had more chondroitinase ABC-resistant material (KS and HS). Unlike the previous study, these comparisons were made of cultures in log phase growth. Similar results were obtained by Roblin et al. (1975b), who found that 3T3 had more CSA, CSC, and DS than did SV3T3. In contrast to the above results, these workers found less material resistant to chondroitinase ABC in 3T3 cells. It is not stated in this paper what the relative growth phases of 3T3 and SV3T3 cells were, i.e., whether they were confluent or sparse and rapidly growing. However, one figure does record the number of cells per plate whose values for 3T3, SV3T3, and Con A selected revertants of SV3T3 imply the cells were in the rapidly growing phase. Goggins et al. (1972) also found a reduction in acid mucopolysaccharides in subconfluent growing SV3T3 compared to subconfluent growing 3T3 cells. These studies utilized [^{35}S]sulfate label which Fratantoni et al. (1968) had noted was incorporated in general only into acid mucopolysaccharides in mammalian cells. Under these conditions HA would not, of course, be "visible." Goggins et al. (1972) found a reduction in the SV3T3 cells in CSA and CSC and also in DS.

In agreement with the work of Saito and Uzman (1971), Roblin *et al.* (1975*b*) found more material resistant to chondroitinase ABC (i.e., HS and KS) in the 3T3 cells. In this latter study the identity of the sulfated material was determined by sulfate labeling, use of specific degradative enzymes (chondroitinase ABC, testicular hyaluronidase), precipitation after enzyme digestion of polyanions with cetylpyridinium chloride, and identification of various glycosaminoglycans by the use of DEAE-cellulose chromatography and cellulose acetate electrophoresis. Few of these steps were taken in another study (Underhill and Keller, 1975), where cell surface material labeled with [^{35}S]sulfate and [^{3}H]- or [^{14}C]glucosamine was removed by mild trypsinization, pronased, and directly chromatographed on DEAE-cellulose. These authors stated that no differences between 3T3 and SV3T3 appeared in CS. However, the data in this paper were not sufficient to draw this conclusion because the observed galactosamine-containing material, thought to be CS and eluting from the DEAE column with HS, may have been obscured by class C-type glycopeptide material (see Section 4) incompletely removed prior to the chromatography (see the work of Davidson and co-workers, and Codington *et al.* in Section 4). Class C glycopeptides can be both anionic and high-molecular-weight. There was a difference in the amount of sulfated material in SV3T3 and 3T3 cells eluting in the region of HS. The transformed cell material eluted significantly before that from 3T3 and it seems possible that the observed change may have depended on the degree of sulfation. HA, although reportedly not different in amount between 3T3 and SV3T3 cells, did seem somewhat lower. However, the growth phase of the cells in which the comparisons were made was not stated and hence interpretation of the results is difficult (see below). Vannucci and Chiarugi (1977) compared glycosaminoglycan synthesis in rapidly growing 3T3, confluent nongrowing 3T3, SV40-transformed 3T3, and polyoma virus-transformed 3T3. Radioactive glucosamine-labeled HA appeared greater in trypsinates from Py3T3, but alterations based on label incorporation may have only reflected differences in UDP-GlcNAc pool sizes. Many papers have depended on estimating the relative concentrations of various glycosaminoglycans by comparing amounts of labeled monosaccharide incorporated into a particular substance. This leads to some ambiguity in interpretation because labeling of kinetics of products and precursor pool equilibration rates may be different (cf. Hopwood and Dorfman, 1977, below) under different physiologic conditions (Kim and Conrad, 1976). No apparent differences were seen in CS, although this finding is subject to the same reservation. Less equivocal was their observation that the transformed cells (Py3T3) had more material labeled with sulfate and degradable by hyaluronidase than the confluent nongrowing 3T3 cells. This material presumably could only be CSA and CSC. These findings do not agree with work discussed above from other laboratories. The reason for this discrepancy with the previous publications is not clear but presumably does not involve the growth state in which the untransformed cells were measured. This is because only slightly less hyaluronidase-degradable material was observed in Py3T3 than even in rapidly growing 3T3 cells (Vannucci and Chiarugi, 1977). Although Roblin et al. (1975b) did not describe the growth status of the transformed and control 3T3 being compared, the growth status must have been between the two extreme conditions observed by Vannucci and Chiarugi (1977). The explanation may be that SV40 transformation was used in the former studies whereas the latter comparisons reviewed here are with polyoma virus transformation.

By direct determination, Makita and Simojo (1973) saw a slight decrease in KS and HS, an increase in CS and a decrease in HA in African green monkey kidney cells (AGMK) compared to their SV40-transformed counterparts. This observation also does not fit the general pattern of observations reviewed above from a number of laboratories in comparisons of mouse 3T3 vs. SV3T3 cells, where overall decrease of glycosaminoglycans occurred on transformation. However, consistent with the latter studies, infection of chick embryo fibroblasts with avian sarcoma viruses caused a fivefold increase in hyaluronic acid synthetase activity and an increased HA (Ishimoto et al., 1966) when compared to the growing uninfected cells. Satoh et al. (1973) observed a large increase in HA with no corresponding increase in the sulfated glycosaminoglycans in African green monkey cells transformed with SV40 (BSC1 cells) and also in hamster embryo fibroblasts transformed with Herpes simplex virus type 2. The control untransformed cells were presumably rapidly growing since they were passaged every 3–4 days, although this is not stated. Terry and Culp (1974), comparing SV3T3 with 3T3, showed an increase in HA, labeled with glucosamine and digested with hyaluronidase in substrate-attached material, from SV3T3 compared to 3T3 cells. The phase of growth of the cells from which the comparisons were made was not stated; however, the object of this study was mostly the identity of the substrate-attached material (SAM, found to be mainly HA). The authors did note that there were very few qualitative differences found in SAM in SV3T3 compared to 3T3 or in their metabolic properties, synthesis, and turnover (Culp et al., 1975).

It is important to establish the growth phase of the cells when making comparisons between transformed and normal cells in the above papers, because it has been long known that the rate of GAG synthesis is greater during log phase of growth at low cell densities (Morris, 1960; Davidson, 1963; Nameroff and Holtzer, 1967; Bischoff, 1971; Tomida et al., 1975; Lembach, 1976). This point for HA was again confirmed by Hopwood and Dorfman (1977), who in addition noted that the rate of synthesis of HA was several-fold greater in sparse, rapidly growing normal cells than their dense, nongrowing counterparts. Transformed cells did not display this alteration whether sparse or dense. They conclude that virus transformation of human fibroblasts results in a marked increase in HA synthesis with a moderate decrease in DS synthesis compared to untransformed fibroblasts. However, based on their data, this appears to be true only when the latter cells are confluent and nongrowing. When normal cells at low densities and presumably rapidly growing are compared, the virus-transformed cells do not display this increase in HA synthesis.

With growth status of the cells notwithstanding and making allowances for it in the various studies, it is still not easy to reconcile the data from various laboratories. Presumably explanations must be sought with other usually uncontrolled factors such as pH (Lie et al., 1972); the quantity of secreted/released GAGs already in the cell milieu (since it is known that GAGs and polyanions can affect the rate of cell proliferation); the presence and concentration of monosaccharides in the medium (for example, β-xyloxides which cause the stimulation of HA synthesis in normal but not transformed fibroblasts, Hopwood and Dorfman, 1977); the presence of specific serum glycoprotein factors which stimulate hyaluronic acid synthesis (Tomida et al., 1977); or starvation for essential amino acids (Morris, 1960).

6. USE OF ENDOGLYCOSIDASES FOR DETECTING ALTERATIONS IN CELL SURFACE GLYCOPEPTIDES

6.1. Class A and B Glycopeptides

In further distinguishing the major types of animal cell oligosaccharide glycopeptides, studies have focused on the determination of the oligosaccharide structures. These include (1) sugar composition, (2) size of the oligosaccharide, (3) nature of the oligosaccharide linkage to the peptide chain, (4) sequence of the sugars in the oligosaccharide chain, and (5) anomeric configuration of each glycosidic bond. The methods used have been discussed in Chapter 1. In addition to these techniques, a new class of degradative endoglycosidases has been isolated which have proven to be valuable tools for structural analysis of the carbohydrate units of glycoproteins. The endoglycosidases described here are the endo-β-N-acetylglucosaminidases (Muramatsu, 1971; Koide and Muramatsu, 1974; Arakawa and Muramatsu, 1974; Ito et al., 1975; Tai et al., 1977a,b; Nishigaki et al., 1974; Tarentino and Maley, 1974, 1975; Tarentino et al., 1972, 1973).

In 1971, Muramatsu demonstrated the first example of an endoglycosidase activity capable of acting on animal membrane glycoproteins. This enzyme was demonstrated in the culture fluid of Diplococcus pneu-

moniae (Muramatsu, 1971). In the presence of β-galactosidase and β-*N*-acetylglucosaminidase, the enzyme cleaved a glycopeptide isolated from IgG mouse myeloma MPC-11 cells into two products. One product contained mannose and *N*-acetylglucosamine with glucosamine in the reducing end, and the second product contained fucose, *N*-acetylglucosamine, and amino acids. The enzyme released these products without any detectable cleavage of the peptide linkages; therefore, the endoglycosidase-released material was not due to cleavage of the glycopeptide bond.

Prior to fully characterizing this new enzyme, the endoglycosidase was used to investigate gross size differences in the carbohydrate–peptide linkage region of KL-2 cell surface glycopeptides from growing and nongrowing cells (Muramatsu *et al.*, 1973). Similar oligosaccharide fragments were released with the endoglycosidase as found with enzyme treatment of mouse myeloma IgG glycopeptide (Muramatsu, 1971). It was found in these studies that the release of fucose-labeled fragments from surface KL-2 glycopeptide was much more complete in the presence of β-galactosidase, neuraminidase, and β-*N*-acetylglucosaminidase (all of which are found in the *Diplococcus* culture fluid) (Koide and Muramatsu, 1974). Following purification, the endoglycosidase from *Diplococcus pneumoniae* was further characterized with respect to its specificity in the cleavage of di-*N*-acetylchitobiose structures (Koide and Muramatsu, 1974). Using a purified [^{14}C]acetylated ovalbumin glycopeptide, $Asn(GlcNAc)_2(Man)_5$, as a substrate, the purified endoglycosidase (referred to as endo-β-*N*-acetylglucosaminidase D), containing practically no other exoglycosidases, cleaved the chitobiose structure of this glycopeptide yielding equimolar amounts of [^{14}C]acetyl-GlcNAc-Asn and $(Man)_5GlcNAc$. Using unfractionated ovalbumin glycopeptides, only 20% of the glycopeptides were hydrolyzed by the enzyme. It was suggested that since the di-*N*-acetylchitobiose structure is considered to be common to ovalbumin glycopeptides (Lee and Scocca, 1972), the enzyme specificity is determined not only by the di-*N*-acetylchitobiose structure itself but also by the surrounding monosaccharide configuration (Koide and Muramatsu, 1974). The enzyme, in conjunction with β-galactosidase and β-*N*-acetylglucosaminidase, hydrolyzed mouse myeloma IgG and bovine IgG glycopeptides, releasing a fucose-containing glycopeptide fragment and a mannose–*N*-acetylglucosamine oligosaccharide. With the addition of neuraminidase to the enzyme mixture, porcine thyroglobulin glycopeptide unit B (see Kornfeld and Kornfeld, 1976, for structure) was similarly hydrolyzed. The attachment of fucose on glucosamine on the inner core region did not hinder the action of the endoglycosidase. These heterosaccharide glycopeptides were resistant to the endoglycosidase action in the absence of the exoglycosidases. The ovalbumin glycopeptides $(Man)_6(GlcNAc)_2 Asn$ and $(Man)_4(GlcNAc)_4 Asn$ (average composition of ovalbumin glycopeptides III) as well as unit A glycopeptide from calf thyroglobulin, all considered high-mannose-containing structures, were resistant to endo-β-N-acetylglucosaminidase D hydrolysis.

In 1972, Tarentino and colleagues partially purified another endoglycosidase which acts on animal cell glycoproteins from the culture fluid *Streptomyces griseus* grown on chitin. Enzyme hydrolysis of ovalbumin $(Man)_6(GlcNAc)_4Asn$ released GlcNAc-Asn and $(Man)_6(GlcNAc)_3$, thus demonstrating cleavage of the di-N-acetylchitobiose unit of the ovalbumin glycopeptide. This endoglycosidase was also used to release a disaccharide Man-GlcNAc from $Man_1(GlcNAc)_2Asn$ which was prepared by treating ovalbumin glycopeptide $Man_5(GlcNAc)_2Asn$ with α-D-mannosidase. This disaccharide was shown to possess a β-mannosidic linkage. Similar studies using the *S. griseus* enzyme on thyroglobulin unit A glycopeptide also demonstrated a β-mannosidase linkage in the core structure of the glycopeptide (Tarentino *et al.*, 1973). It was noted at this time that the endoglycosidase from *S. griseus* cultural filtrate contained two enzymes. One enzyme, an endoglycosidase which hydrolyzed the di-N-acetylchitobiose structure of $Man_1(GlcNAc)Asn$, was later referred to as endo-β-N-acetylglucosaminidase L (L for low-molecular-weight oligosaccharides) (Tarentino and Maley, 1974). The second enzyme, an endo-β-N-acetylglucosaminidase H (H for high-molecular-weight oligosaccharides), hydrolyzed $Asn(GlcNAc)_2(Man)_5$ and higher mannose-containing oligosaccharides.

Purification and properties of endo-β-N-acetylglucosaminidase H were further elucidated (Tarentino and Maley, 1974; Tarentino *et al.*, 1974). A complete separation of endo-β-N acetylglucosaminidase L and H was achieved by Sephadex G-100 gel chromatography, and their substrate specificity was determined. Removal of the asparagine moiety from the glycopeptide $(Man)_6(GlcNAc)_2(GlcNAc)_2Asn$ or the addition of a dansyl group to the asparagine did not effect the hydrolysis rate of endoglycosidase H to any great degree. Proteins containing both neutral and acidic oligosaccharides, such as thyroglobulin, immunoglobin M, and/or porcine ribonuclease, released only neutral oligosaccharides when treated with endoglycosidase H. Acidic glycoproteins, DNase, RNase, transferrin, fibrinogin, α-acid glycoprotein, and IgG were resistant to endoglycosidase H treatment. Removal of sialic acid from these complex glycopeptides did not alter the endoglycosidase H activity. Neutral oligosaccharides attached to proteins via asparagine having the general structure $(Man)_x(GlcNAc)_y(GlcNAc)_2Asn$ were susceptible to endoglycosidase H hydrolysis. These glycopeptides included sulfitolyzed ovalbumin, bovine pancreatic deoxyribonuclease A, ribonuclease B, and an invertase from *Saccharomyces cerevisiae*.

Comparative studies on the substrate specificities of endoglycosidase D and H have been reported (Arakawa and Muramatsu, 1974; Tarentino and Maley, 1975). The findings of these studies are summarized in Table 2. Muramatsu's group did not find any endoglycoside L activity in their enzyme preparation of endoglycosidase H, possibly because bacterial strains and/or culture conditions they used differed from those of Tarentino's. Muramatsu also found that side chain-free IgG and thyroglobulin unit B glycopeptides were resistant to their endoglycosidase H whether the structures were defücosylated or not. Tarentino found that the defucosylated structures were susceptible to endoglycosidase H if longer hydrolysis times and higher enzyme concentrations were used.

Using both endoglycosidase D and H, the structural differences between the ovalbumin glycopeptides $(Man)_5(GlcNAc)_2 Asn$ and $(Man)_6(GlcNAc)_2 Asn$ were determined; this was useful in elucidating structural specificities of these two enzymes. The structures were determined by the combination of methylation analysis, acetolysis, $[^3H]$-$NaBH_4$ reduction and α-mannosidase digestions (Tai et al., 1975a). Table 2 gives the proposed structures of the two ovalbumin glycopeptides. The presence of the mannose linked $\alpha1,3$ to the core β-mannose was demonstrated to be essential for endoglycosidase D activity in a nonreducing terminal configuration. If this mannose residue was removed or substituted by N-acetylglucosamine, the resulting structures were not susceptible to endoglycosidase D. In addition to these results, it was established with methylation studies that the di-N-chitobiose structure of these ovalbumin glycopeptides contained no substitution on the N-acetylglucosamines other than the β-mannose linkage to the C-4 position of the distal N-acetylglucosamine. Therefore, no α-mannose linkages are attached to the di-N-acetylchitobiose structure, but are in fact linked to the β-mannose residue of the core structure (Tai et al., 1975a).

In 1975 two new endo-β-N-acetylglucosaminidases (C_I and C_{II}) were isolated and purified from the culture fluid of Clostridium perfringens (Ito et al., 1975). C_I was demonstrated to have substrate specificities indistinguishable from endo-β-N-acetylglucosaminidase D. The C_{II} enzyme was similar to the activity of endo-β-N-acetylglucosaminidase H but distinct with respect to relative activity towards ovalbumin glycopeptides and unit A glycopeptide of thyroglobulin. The relative activities of C_{II} are $(Man)_6(GlcNAc)_2 Asn \geq$ unit A $> (Man)_5(GlcNAc)_2 Asn$ while those of endoglycosidase H are $(Man)_5(GlcNAc)_2Asn >$ $(Man)_6(GlcNAc)_2 Asn >$ unit A. This suggested that the two enzymes, C_{II} and H, recognized different portions of the oligosaccharide core structures. In addition the C_{II} enzyme cannot hydrolyze ovalbumin glycopeptides I and II which are longer sugar chain glycopeptides (Tai et al., 1977a). In contrast, the H enzyme can cleave all these glycopeptides (Tai et al., 1975a). Comparative studies of the substrate specificities of endo-β-N-acetylglucosaminidases C_{II} and H helped elucidate structures of

Table 2. Differential Specificities of

$N[^{14}C]$-Acetylated substrates		Clostridium perfringens C_I	C_{II}	Diplococcus pneumoniae Endo D	Streptomyces griseus Endo H
I. Ovalbumin glycopeptides	Gp				
a. $(Man)_5$-$(GlcNAc)_4$-Asn	III A	R	S	R	S
b. $(Man)_7$-$(GlcNAc)_2$-Asn	III B	R	S	R	S
c. $(Man)_6$-$(GlcNAc)_5$-Asn	III C	R	R	R	S
d. $(Man)_6$-$(GlcNAc)_2$-Asn	IV	R	S	R	S
e. $(Man)_5$-$(GlcNAc)_2$-Asn	V	S	S	S	S
f. $(Man)_4$-$(GlcNAc)_2$-Asn	VI	—	S	—	S
g. Man -$(GlcNAc)_2$-Asn	*	R	R	R	R**
II. Thyroglobulin glycopeptides					
a. Unit A		R	S	R	S
b. Unit B		R	R	R	R
c. Unit B side chain free		S	R	S	R
d. Unit B defucosyl		—	—	S	R***
III. IgG glycopeptides					
a. Intact		R	R	R	R
b. Intact (side chain free)		S	R	S	R
c. Defucosylated		—	—	R	R
d. Degalactosylated		—	—	R	R
e. Defucosylated (side chain free)		—	—	S	R

---

[a] Data from Muramatsu, 1971; Koide and Muramatsu, 1974; Arakawa and Muramatsu, 1974; Ito et al., 1975; Tai et al., 1977a,b; Nishigaki et al., 1974; Tarentino and Maley, 1974, 1975; Tarentino et al., 1972, 1973.

[b] R = resistant to enzyme hydrolysis; S = susceptible to enzyme hydrolysis; * Glycopeptide V digested with α-mannosidase; ** sensitive to endoglycosidase L; *** in nanogram quantities.

Endo-β-N-Acetylglucosaminidases[a,b]

Oligomannosyl and complex oligosaccharide structures

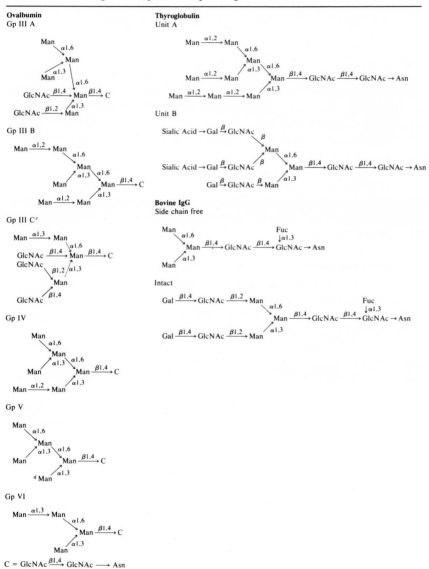

C = GlcNAc $\xrightarrow{\beta1,4}$ GlcNAc $\longrightarrow$ Asn

---

[c] Possible recognition point of C_{II} enzyme GlcNAcβ1 $\longrightarrow$ 4 substitution on the mannose results in a resistant substrate enzyme.

[d] This mannose must be terminal for the Endo D enzyme activity.

ovalbumin glycopeptides III and VI (Table 2) (Tai *et al.*, 1977*a,b*). Endoglycosidase C_{II} only hydrolyzed 75% of ovalbumin glycopeptide III
while endoglycosidase H hydrolyzed it completely. The partial susceptibility of this glycopeptide to endoglycosidase H suggested that ovalbumin glycopeptide III was a mixture of glycopeptides. This was in fact
shown to be the case. The ovalbumin structures and the specificities of
the two enzymes are shown in Table 2. Examining the structural differences in Gp III A and Gp III C, it is apparent that, if the mannose linked
$\alpha 1,3$ to the β-mannose is substituted at the C-4 position by another sugar,
the glycopeptide is resistant to the C_{II} enzyme but not to endoglycosidase
H. Apparently, this is the only structural specificity difference in these
two similar enzymes thus far.

Two distinct endo-β-N-acetylglucosaminidases have also been partially purified from fig latex and are denoted F-I and F-II (Chien
et al., 1977). F-I hydrolyzed the di-N-acetylchitobiose linkage in
$(Man)_3(GlcNAc)Asn$ prepared from IgG at a faster rate than
$(Man)_5(GlcNAc)Asn$ and $(Man)_6(GlcNAc)_2 Asn$ ovalbumin glycopeptides.
F-II hydrolyzed the two ovalbumin glycopeptides similarly but not
$(Man)_3(GlcNAc)_2 Asn$. In addition, endo-β-N-acetylglucosaminidases
have been isolated from rat liver, spleen, and kidney, and porcine liver
and kidney that hydrolyze N-acetylated $(Man)_6(GlcNAc)_2 Asn$ to GlcNAc-
$[^{14}C]$acetyl-Asn and $(Man)_6GlcNAc$. The activities of the porcine liver
endoglycosidase enzyme were studied in more detail by Nishigaki *et al.*
(1974). The role of these endoglycosidases in mammalian tissues is unknown at this time, and additional studies to determine their occurrence
and function in different sources are essential in the understanding of
glycoprotein metabolism. Whether carbohydrate moieties of glycopeptides are degraded by a concerted action of endo- and exoglycosidases,
as suggested by Nishigaki *et al.* (1974), remains to be demonstrated.

It seems likely that combining such enzyme techniques with both
the classical approaches to carbohydrate structural analysis and the more
recent nuclear magnetic resonance spectroscopy technique will lead to
rapid advances. The latter techniques have been described in structural
work on yeast mannans (Gorin and Spencer, 1968; Gorin *et al.*, 1968,
1969), in oligosaccharides derived from the liver of patients with G_{M1}
gangliosidosis (Wolfe *et al.*, 1974), in the urine of patients with Sandhoff's
disease, in G_{M2} gangliosidosis-variant O (Strecker *et al.*, 1977), in the
structure of human serotransferring glycopeptide (Dorland *et al.*, 1977),
and in human IgG (Grey and Carver, 1978). These techniques provide a
nondestructive analysis of anomeric configuration, linkage, sequence,
and even quantity of monosaccharide in oligosaccharides or glycopeptides. NMR spectroscopy thus provides strong corroboration of struc-

tures assigned by classical approaches, prediction of contaminants, and altogether unsuspected new structures.

6.2. Endoglycosidases Useful in Structural Analysis of Class C Glycopeptides

Another group of endoglycosidases have also been isolated and are proving to be useful in structural characterization of class C glycopeptides. Reissig *et al.* (1975) purified an endogalactosaminidase from the cultural filtrate of *Streptomyces griseus*. This enzyme was active towards cleaving galactosamine–galactosamine linkages in a galatosamine-rich oligogalactosaminoglycan isolated from the culture filtrate of a *Neurospora* mutant. The endogalactosaminidase catalyzed the release of *Neurospora* sporelings attached to glass surfaces. This enzyme was inactive towards *N*-acetyloligogalactosaminoglycan molecules and chitosan. In conjunction with these studies the previously discussed endoglycosidase H, which was also isolated from *Streptomyces griseus* culture filtrates, did not contain any endogalactosaminidase activity.

Endo-β-galactosidase activity has been isolated from the culture filtrate of *Diplococcus pneumoniae*, which also is the source of endo-β-*N*-acetylglucosaminidase D (Takasaki and Kobata, 1976). This enzyme was found to release trisaccharides from blood group A and B active mucins purified from ovarian cyst fluid. The enzyme was shown to catalyze the following reaction:

$$
\begin{array}{ccc}
\text{GalNAc} & & \text{GalNAc} \\
\text{or} \xrightarrow{\alpha 1,3} \text{Gal} \xrightarrow{\beta 1,4} \text{R} + \text{H}_2\text{O} \longrightarrow & \text{or} \xrightarrow{\alpha 1,3} \text{Gal} + \text{R} \\
\text{Gal} \quad \uparrow {\scriptstyle \alpha 1,2} & & \text{Gal} \quad \uparrow {\scriptstyle \alpha 1,2} \\
\text{Fuc} & & \text{Fuc}
\end{array}
$$

where R is Glc or GlcNAc. Removal of the α-fucosyl residue linked at the C-2 hydroxyl group of the galactose moiety of blood group A and B substances resulted in inactive substrates. This suggested that the enzyme specificity requires Fuc $\xrightarrow{\alpha 1,2}$ Gal for its substrate. Other sources of endo-β-galatosidases have been reported with similar specificities as those described above. An endo-β-galactosidase has been purified from *Escherichia freundii* (Fukuda and Matsumura, 1975, 1976), *Actinobacillus* sp. (Nakazawa and Suzuki, 1975), and *Coccobacillus* (Hirano and Meyer, 1971).

In 1976 two independent laboratories partially purified and characterized an endo-β-*N*-acetylgalactosaminidase enzyme from the culture filtrates of *Diplococcus pneumonia* (Endo and Kobata, 1976; Bhavanandan *et al.*, 1976). These studies demonstrated that this endoglycosidase is capable of acting on the O-glycosidic linkage between *N*-acetylgalactosamine and serine/threonine residues. The purified enzyme released a disaccharide, Gal $\xrightarrow{\beta 1,3}$ GalNAc, from mouse melanoma, fetuin, and pig submaxillary mucin (Umemoto *et al.*, 1977; Bhavanandan *et al.*, 1976). The same dissaccharide was released from human erythrocyte membrane glycopeptides (Endo and Kobata, 1976). The enzyme was unable to hydrolyze the following substrates: Sialic acid → Gal → GalNAc-Ser/Thr, GalNAc → Gal → GalNAc → Ser/Thr, GalNAc → (Fuc) → GalNAc → Ser/Thr, Sialic acid → GalNAc-Ser/Thr, and galactose acted as an inhibitor of the enzyme. It was suggested (Umemoto *et al.*, 1977) that a nonreducing galactose terminus is necessary for the recognition of the substrate by the enzyme. With further characterization of the substrate specificities of the endo-β-galactosidases and endo-β-*N*-acetylgalactosaminidases, these enzymes should prove to be useful tools for structural analysis of complex oligosaccharide chains linked O-glycosidically to protein.

7. REFERENCES

Ali, I. U., Mautner, V., Lanza, R., and Hynes, R. O., 1977, Restoration of normal morphology, adhesion and cytoskeleton in transformed cells by addition of a transformation-sensitive surface protein, *Cell* **11**:115.

Arakawa, M., and Muramatsu, E. T., 1974, Endo-β-*N*-acetyl glucosaminidases cuting on the carbohydrate moieties of glycoproteins, *J. Biochem.* **76**:307.

Arima, T., and Spiro, R. G., 1972, Studies on the carbohydrate units of thyroglobulin. Structure of the mannose-*N*-acetylglucosamine unit (unit A) of the human and calf proteins, *J. Biol. Chem.* **247**:1836.

Atkinson, P. H., 1978, Glycoprotein and protein precursors to plasma membrane in vesicular stomatitis virus infected cells, *J. Supramol. Struct.* **8**:89.

Atkinson, P. H., and Summers, D. F., 1971, Purification and properties of HeLa plasma membranes, *J. Biol. Chem.* **246**:5162.

Atkinson, P. H., Moyer, S. A., and Summers, D. F., 1976, Assembly of vesicular stomatitis virus glycoprotein and matrix protein into HeLa cell plasma membrane, *J. Mol. Biol.* **102**:613.

Benedetti, E. L., and Emmelot, P., 1967, Studies on plasma membranes. IV. The ultrastructure localization and content of sialic acid in plasma membranes isolated from rat liver and hepatoma, *J. Cell Sci.* **2**:499.

Bhavanandan, V. P., Umemoto, J., and Davidson, E. A., 1976, Characterization of an endo-α-*N*-acetyl galactosaminidase from *Diplococcus pneumoniae, Biochem. Biophys. Res. Commun.* **70**:738.

Bhavanandan, V. P., Umemoto, J., Banks, J. R., and Davidson, E. A., 1977, Isolation and partial characterization of sialoglycopeptides produced by a murine melanoma, *Biochemistry* **16**:4426.

Bischoff, R., 1971, Acid mucopolysaccharide synthesis by chick amnion cell cultures, *Exp. Cell Res.* **66**:224.

Boone, C. W., 1975, Malignant hemangioendotheliomas produced by subcutaneous inoculation of Balb/3T3 cells attached to glass beads, *Science* **188**:68.

Boone, C. W., and Jacobs, J. B., 1976, Sarcomas routinely produced from putatively nontumoigenic Balb/3T3 and C3H/10T 1/2 cells by subcutaneous inoculation attached to plastic platelets, *J. Supramol. Struct.* **5**:131.

Bosmann, H. B., 1972, Sialyltransferase activity in normal and RNA- and DNA-virus transformed cells utilizing desialyzed trypsinized cell plasma membrane external surface glycoproteins. *Biochem. Biophys. Res. Commun.* **49**:1256.

Brady, R. O., Borek, C., and Bradley, R. M., 1969, Composition and synthesis of gangliosides in rat hepatocyte and hepatoma cell lines, *J. Biol. Chem.* **244**:6552.

Bretscher, M. S., 1971, A major problem which spans the human erythrocyte membrane, *J. Mol. Biol.* **59**:351.

Buck, C. A., Glick, M. C., and Warren, L., 1970, A comparative study of glycoproteins from the surface of control and Rous sarcoma virus transformed hamster cells, *Biochemistry* **9**:4567.

Buck, C. A., Glick, M. C., and Warren, L., 1971a, Effect of growth on the glycoproteins from the surface of control and Rous sarcoma virus transformed hamster cells, *Biochemistry* **10**:2176.

Buck, C. A., Glick, M. C., and Warren, L., 1971b, Glycopeptides from the surface of control and virus transformed cells, *Science* **172**:169.

Buck, C. A., Fuhrer, J. P., Soslau, G., and Warren, L., 1974, Membrane glycopeptides from subcellular fractions of control and virus transformed cells, *J. Biol. Chem.* **249**:1541.

Burger, M. M., 1969, A difference in the architecture of the surface membrane of normal and virally transformed cells, *Proc. Natl. Acad. Sci. U.S.A.* **62**:994.

Burger, M., 1973, Surface changes in transformed cells detected by lectins, *Fed. Proc.* **32**:93.

Burger, M. M., and Goldberg, A. R., 1967, Identification of a tumor-specific determinant on neoplastic cell surfaces, *Proc. Natl. Acad. Sci. U.S.A.* **57**:359.

Burger, M. M., and Noonan, K. D., 1970, Restoration of normal growth by covering agglutinin sites on tumor cell surface, *Nature* **228**:512.

Burke, D., and Keegstra, K., 1979, Carbohydrate structure of Sindbis virus glycoprotein E_2 from virus grown in hamster and chicken cells, *J. Virol.* **29**:546.

Carlson, D. M., 1977, Chemistry and biosynthesis of mucin glycoproteins, *Adv. Exp. Med. Biol.* **89**:251.

Carver, J. P., Atkinson, P. H., Hakimi, J., Grey, A., and Ceccarini, C.,1979, in preparation.

Ceccarini, C., 1975, Appearance of smaller mannosyl-glycopeptides on the surface of a human cell transformed by simian virus 40, *Proc. Natl. Acad. Sci. U.S.A.* **72**:2687.

Ceccarini, C., and Atkinson, P. H., 1977, Studies on mannose-containing glycopeptides from a normal and an SV40 transformed human cell, *Biochim. Biophys. Acta* **500**:197.

Ceccarini, C., and Eagle, H., 1971, pH as a determinant of cellular growth and contact inhibition, *Proc. Natl. Acad. Sci. U.S.A.* **68**:229.

Ceccarini, C., Muramatsu, T., Tsang, J., and Atkinson, P. H., 1975, Growth-dependent alterations in oligomannosyl cores of glycopeptides, *Proc. Natl. Acad. Sci. U.S.A.* **72**:3139.

Chien, S., Weinburg, R., Li, S., and Li, Y., 1977, Endo-β-N-acetylglucosaminidase from pig latex, *Biochem. Biophys. Res. Commun.* **76**:317.

Codington, J. F., Sanford, B. H., and Jeanloz, R. W., 1970, Glycoprotein coat of the TA_3 cell. I. Removal of carbohydrate and protein material from viable cells, *J. Natl. Cancer Inst.* **45**:637.

Cohen, G. H., Atkinson, P. H., and Summers, D. F., 1971, Interactions of vesicular stomatitis virus structural proteins with HeLa plasma membranes, *Nature (London) New Biol.* **231**:121.

Coman, D. R., 1960, Reduction in cellular adhesiveness upon contact with a carcinogen, *Cancer Res.* **20:**1202.

Cotmore, S. F., Furthmay, H., and Marchesi, V., 1977, Immunochemical evidence for the transmembrane orientation of glycophorin A. Location of ferritin-antibody conjugates in intact cells, *J. Mol. Biol.* **113:**539.

Culp, L. A., Grimes, W. J., and Black, P. H., 1971, Contact inhibited revertant cell lines isolated from SV40-transformed cells, *J. Cell Biol.* **50:**682.

Culp, L. A., Terry, A. H., and Buniel, J. F., 1975, Metabolic properties of substrate-attached glycoproteins from normal and virus-transformed cells, *Biochemistry* **14:**406.

Cumar, F. A., Brady, R. O., Kolodny, E. H., McFarland, V. W., and Mora, P. T., 1970, Enzymatic block in the synthesis of gangliosides in DNA virus transformed tumorgenic mouse cell lines, *Proc. Natl. Acad. Sci. U.S.A.* **67:**757.

Danishefsky, I., Oppenheimer, E. T., Heritier-Watkins, O., and Willhite, M., 1966, Mucopolysaccharides in animal tumors, *Cancer Res.* **26:**229.

David, A. E., 1973, Assembly of the vesicular stomatitis virus envelope: Incorporation of viral polypeptides into the host plasma membrane, *J. Mol. Biol.* **76:**135.

David, A. E., 1977, Assembly of the vesicular stomatitis virus envelope: Transfer of viral polypeptides from polysomes to cellular membranes, *Virology* **76:**98.

Davidson, E. H., 1963, Heritability and control of differentiated function in cultured cells, *J. Gen. Physiol.* **46:**983.

Debray, H., and Montreuil, J., 1977, Isolation and characterization of surface glycopeptides from adult rat hepatocytes in an established line. Abstract in Fourth International Symposium on Glycoconjugates, Woods Hole, Massachusetts.

Defendi, V., Lehman, J., and Kramer, P., 1963, "Morphologically normal" hamster cells with malignant properties, *Virology* **19:**592.

DeLuca, S., Heingard, D., Hascall, V. C., Kimura, J. H., and Caplan, A. I., 1977, Chemical and physical changes in proteoglycans during development of chick limb bud chondiocytes grown *in vitro, J. Biol. Chem.* **252:**6600.

Den, H., Schultz, A. M., Basu, M., and Roseman, S., 1971, Glycosyltransferase activities in normal and polyoma-transformed BHC cells, *J. Biol. Chem.* **246:**2721.

Dorland, L., Haverkamp, J., Schut, B. L., Vliegenthart, J. F. G., Spik, G., Strecker, G., Fournet, B., and Montreuil, J., 1977, The structure of the asialo-carbohydrate units of human serotransferrin as proven by 360 MHz proton magnetic resonance spectroscopy, *FEBS Lett.* **77:**15.

Dorland, L., Haverkamp, J., Vliegenthart, J. F. G, Strecker, G., Michalski, J., Fournet, B., Spik G., and Montreuil, J., 1978, 360 MHz H nuclear-magnetic resonance spectroscopy of sialyl-olyosaccharides from patients with bialidosis (mycolipidosis I and II), *Eur. J. Biochem.* **87:**323.

Dulbecco, R., 1970, Topoinhibition and serum requirements of transformed and untransformed cells, *Nature* **227:**802.

Emmelot, P., 1973, Biochemical properties of normal and neoplastic cell surfaces; a review, *Eur. J. Cancer* **9:**319.

Emmelot, P., and Bos, C. J., 1972, Studies on plasma membranes. XVII. On the chemical composition of plasma membranes prepared from rat and mouse liver and hepatomas, *J. Membr. Biol.* **9:**83.

Endo, Y., and Kobata, A., 1976, Partial purification and characterization of an endo-α-N-acetylgalactosaminidase from the culture medium of Diplococcus pneumonia, *J. Biochem.* **80:**1.

Etchison, J. R., Robertson, J. S., and Summers, D. F., 1977, Partial structural analysis of the oligosaccharide moieties of the vesicular stomatitis virus glycoprotein by sequential chemical and enzymatic degradation, *Virology* **78:**375.

Farber, E., 1973, Hyperplastic liver nodules in: *Methods in Cancer Research* (H. Busch, ed.), Vol. VII, pp. 345–375, Academic Press, New York.

Fox, T. O., Sheppard, J. R., and Burger, M. M., 1971, Cyclic membrane changes in animal cells: Transformed cells permanently display a surface architecture detected in normal cells only during mitosis, *Proc. Natl. Acad. Sci. U.S.A.* **68**:244.

Fratantoni, J. C., Hall, C. W., and Neufeld, E. F., 1968, The defect in Hurlers and Hunters syndromes: Faulty degradation of mucopolysaccharide, *Proc. Natl. Acad. Sci. U.S.A.* **60**:699.

Freedman, V., and Shin, S., 1974, Cellular tumorigenicity in *nude* mice: Correlation with cell growth in semi-solid medium, *Cell* **3**:355.

Fukuda, M., and Matsumura, E., 1975, Endo-β-galactosidase of *Escherichia freundii* hydrolysis of pig colonic mucin and milk oligosaccharides by endo glycosidic action, *Biochem. Biophys. Res. Commun.* **64**:465.

Fukuda, M. N., and Matsumura, G., 1976, Endo-β-galactosidase of *Escherichia freundii* purification and endoglycosidic action on keratin sulfates, oligosaccharides and blood group active glycoprotein, *J. Biol Chem.* **251**:6218.

Gacto, M., and Steiner, S., 1976, Fucolipid metabolism as a function of cell population density in noimal and murine sarcoma virus-transformed rat cells, *Biochim. Biophys. Acta* **444**:11.

Gahmberg, C. G., and Hakomori, S., 1973, External labeling of cell surface galactose and galactosamine in glycolipid and glycoprotein of human erythrocytes, *J. Biol. Chem.* **248**:4311.

Gahmberg, C. G., Utermann, G., and Simons, K., 1972, The membrane proteins of Semliki Forest virus have a hydrophobic part attached to the viral membrane, *FEBS Lett.* **28**:179.

Gahmberg, C. G., Kiehn, O., and Hakamori, S., 1974, Changes in a surface labelled galactoprotein and in glycolipid concentrations in cells transformed by a temperature-sensitive polyoma virus mutant, *Nature* **248**:413.

Gasic, G., and Gasic, T., 1962, Removal and regneration of the cell coating in tumor cells, *Nature* **196**:170.

Glick, M. C., 1974, Chemical components of surface membranes related to biological properties, *Miami Winter Symp.* **7**:213.

Glick, M. C., and Buck, C. A., 1973, Glycoproteins from the surface of metaphase cells, *Biochemistry* **12**:85.

Glick, M. C., Rabinowitz, Z., and Sachs, L., 1973, Surface membrane glycopeptides correlated with tumorigenesis, *Biochemistry* **12**:4864.

Goggins, J. F., Johnson, G. F., and Pastan, I., 1972, The effect of dibutyryl cyclic adenosine monophosphate on synthesis of sulfated acid mucopolysaccharides by transformed fibroblasts, *J. Biol. Chem.* **247**:5759.

Gorin, P. A. J., and Spencer, J. F. T., 1968, Galactomannans of trichosporon fermentous and other yeasts, proton magnetic resonance and chemical studies, *Can. J. Chem.* **46**:2299.

Gorin, P. A. J., Mazurek, M., and Spencer, J. F. T., 1968, Proton magnetic resonance spectra of trichosporon aculeatum mannan and its borate complex and their relationship to chemical structure. *Can. J. Chem.* **46**:2305.

Gorin, P. A. J., Spencer, J. F. T., and Bhattacharjee, S. S., 1969, Structures of yeast mannans containing both α- and β-linked D-manno pyranose units, *Can. J. Chem.* **47**:1499.

Grey, A. and Carver, J. P., 1978, personal communication.

Grimes, W. J., 1970, Sialic acid transferases and sialic acid levels in normal and transformed cells, *Biochemistry* **9**:5083.

Grimes, W. J., 1973, Glycosyltransferase and sialic acid levels of normal and transformed cells, *Biochemistry* **12**:990.

Grimes, W. J., and Greegor, S., 1976, Carbohydrate compositions of normal, spontaneously transformed, and virally transformed cells derived from BALB/c mice, *Cancer Res.* **36**:3905.

Hakomori, S., and Murakami, W. T., 1968, Glycolipids of hamster fibroblasts and derived malignant-transformed cell lines, *Proc. Natl. Acad. Sci. U.S.A.* **59**:254.

Hakamori, S., Saito, T., and Vogt, P. K., 1971, Transformation by Rous sarcoma virus: Effects on cellular glycolipids, *Virology* **44**:609.

Hamerman, D., Todaro, G. J., and Green, H., 1965, The production of hyaluronate by spontaneously established cell lines and viral transformed lines of fibroblastic origin, *Biochim. Biophys. Acta* **101**:343.

Hartmann, J. F., Buck, C. A., Defendi, V., Glick, M. C., and Warren, L., 1972, The carbohydrate content of control and virus-transformed cells, *J. Cell Physiol.* **80**:159.

Hascall, V. C., Oegema, T. R., Brown, M., and Caplan, A. I., 1976, Isolation and characterization from chick limb bud chondrocytes grown in vitro, *J. Biol. Chem.* **251**:3511.

Heinegard, D., and Axelsson, I., 1977, Distribution of keratan sulfate in cartilage proteoglycans, *J. Biol. Chem.* **252**:1971.

Hirano, H., Parkhouse, B., Nicolson, G. L., Lennox, E. S., and Singer, S. J., 1972, Distribution of saccharide residues on membrane fragments from a myeloma-cell homogenate: Its implications for membrane biogenesis, *Proc. Natl. Acad. Sci. U.S.A.* **69**:2945.

Hirano, S., and Meyer, K., 1971, Enzymatic degradation of corneal and cartilaginous keratosulfates, *Biochem. Biophys. Res. Commun.* **44**:1371.

Hopwood, J. J., and Dorfman, A., 1977, Glycosamino glycan synthesis by cultured human skin fibroblasts after transformation with simian virus 40, *J. Biol. Chem.* **252**:4777.

Huang, C.-C., Mayer, H. E., and Montgomery, R., 1970, Microheterogeneity and pancidispersity of glycoproteins. Part I. The carbohydrate of chicken ovalbumin, *Carbohydr. Res.* **13**:127.

Hudgin, Pricer, W. E., Jr., Aswell, G., Stockert, R. J., and Morell, A. G., 1974, The isolation and properties of a rabbit liver binding protein specific for asialoglycoproteins, *J. Biol. Chem.* **249**:5536.

Hughes, R. C., and Clark, J., 1974, Growth of baby hamster cells in media containing neuraminidase, *Exp. Cell Res.* **85**:362.

Hunt, L. A., and Summers, D. F., 1976, Association of vesicular stomatitis virus proteins with HeLa cell membranes and released virus, *J. Virol.* **20**:637.

Hunt, L. A., Etchison, J. R., and Summers, D. F., 1978, Oligosaccharide chains are trimmed during synthesis of the envelope glycoprotein of vesicular stomatitis virus, *Proc. Natl. Acad. Sci. U.S.A.* **75**:754.

Hunt, L. A., Etchison, J. R., and Summers, D. F., 1980, Characterization of mannose-labeled oligosaccharide chains from newly synthesized glycoproteins in HeLa cell membranes, *J. Biol. Chem.*, manuscript in preparation.

Hynes, R. O., 1973, Alteration of cell surface proteins by viral transformation and by protolipids, *Proc. Natl. Acad. Sci. U.S.A.* **70**:3170.

Hynes, R. O. (ed.), 1976, *Surfaces of Normal and Transformed Cells*, Wiley Interscience, New York.

Hynes, R. O., and Bye, J. M., 1974, Density and cell cycle dependence of cell surface proteins in hamster fibroblasts, *Cell* **3**:113.

Inbar, M., and Sachs, L., 1969, Structural difference in sites on the surface membrane of normal and transformed cells, *Nature* **223**:710.

Ishimoto, N., Temin, H., and Strominger, J. L., 1966, Studies of carcinogenesis by avian sarcoma viruses. II. Virus-induced increase in hyaluronic acid synthetase in chicken fibroblasts, *J. Biol. Chem.* **241:**2052.

Itaya, K., Hakamori, S. I., and Klein, G., 1976, Long-chain neutral glycolipids and gangliosides of murine fibroblast lines and their low- and high-tumorigenic hybrids, *Proc. Natl. Acad. Sci. U.S.A.* **73:**1568.

Ito, S., Muramatsu, T., and Kobata, A., 1975, Release of galactosyl oligosaccharides by endo-β-N-acetylglucosaminidase D, *Biochem. Biophys. Res. Commun.* **63:**938.

Jeanloz, R. W., and Codington, J. F., 1974, Glycoproteins at the cell surface of sublines of TA3 tumor, *Miami Winter Symp.* **7:**241.

Kahn, P., and Shin, S., 1979, Cellular tumorigenicity in nude mice, *J. Cell Biol.* **82:**1.

Kalckar, H. M., 1965, Galactose metabolism and cell "sociology," *Science* **150:**305.

Katz, F. N., Rothman, J. E., Lingappa, V. R., Blobel, G., and Lodish, H. F., 1977, *Proc. Natl. Acad. Sci. U.S.A.* **74:**3278.

Kaufman, R. L., and Ginsburg, V., 1968, The metabolism of L-fucose by HeLa cells, *Exp. Cell Res.* **50:**127.

Kawai, Y., and Spiro, R. G., 1977, Fat cell plasma membranes. II. Studies on the glycoprotein components, *J. Biol. Chem.* **252:**6236.

Kent, J. L., and Pogo, B. G. T., 1974, Rapid, periodate-induced stimulation of permeability and macromolecular synthesis in chicken erythrocytes, *Biochem. Biophys. Res. Commun.* **56:**161.

Kim, J. J., and Conrad, H. E., 1976, Kinetics of mucopolysaccharide and glycoprotein synthesis by chick embryo chondrocytes, *J. Biol. Chem.* **251:**6210.

Kim, Y. S., 1977, Glycoprotein alteration in human colonic adenocarcinoma in mucus in health and disease, *Adv. Exp. Med. Biol.* **89:**443.

Kim, Y. S., and Isaacs, R., 1975, Glycoprotein metabolism in inflammatory and neoplastic diseases of human colon, *Cancer Res.* **35:**2092.

Klenk, H. D., and Choppin, P. W., 1969, Lipids of plasma membranes of monkey and hamster kidney cells and of parainfluenza virus grown in these cells, *Virology* **28:**255.

Klenk, H. D., and Choppin, P. W., 1970, Plasma membrane lipids and parainfluenza virus assembly, *Virology* **40:**939.

Knipe, D. M., Baltimore, D., and Lodish, H. F., 1977, Separate pathways of maturation of the major structural proteins of vesicular stomatitis virus, *J. Virol.* **21:**1128.

Koide, N., and Muramatsu, T., 1974, Endo-β-N-acetylglycosaminidase acting on carbohydrate moieties of glycoproteins, *J. Biol. Chem.* **249:**4897.

Koide, N., Nose, M., and Muramatsu, T., 1977, Recognition of IgG by Fc receptor and complement: Effects of glycosidase digestion, *Biochem. Biophys. Res. Commun.* **75:**838.

Kornfeld, R., 1978, Structure of oligosaccharides of three glycopeptides from calf thymocyte plasma membranes, *Biochemistry* **17:**1415.

Kornfeld, R., and Kornfeld, S., 1970, The structure of a phytohemagglutinin receptor site from human erythrocytes, *J. Biol. Chem.* **245:**2536.

Kornfeld, R., and Kornfeld, S., 1976, Comparative aspects of glycoprotein structure, *Annu. Rev. Biochem.* **45:**217.

Kraemer, P. M., 1967, Sialic acid of mammalian cell lines, *J. Cell Physiol.* **67:**23.

Kraemer, P. M., 1971, Complex carbohydrates of animal cells: Biochemistry and physiology of the cell periphery, in: *Biomembranes* (L. A. Manson, ed.), Vol. I., pp. 67–190, Plenum Press, New York.

Kraemer, P. M., 1979, Mucopolysaccharides: Cell biology and malignancy, in: *Surfaces of Normal and Malignant Cells* (R. O. Hynes, ed.), pp. 149–198, Wiley Interscience, New York.

Lai, M. C., and Duesberg, P. H., 1972, Differences between the envelope glycoproteins and glycopeptides of avian tumor viruses released from transformed and nontransformed cells, *Virology* **50**:359.

Larriba, G., Klinger, M., Sramek, S., and Steiner, S., 1977, Novel fucose-containing components of rat tissues, *Biochem. Biophys. Res. Commun.* **77**:79.

Lee, Y. C., and Scocca, J. R., 1972, A common structural unit in asparagine-oligosaccharides of several glycoproteins from different sources, *J. Biol. Chem.* **247**:5753.

Lembach, K. J., 1976, Enhanced synthesis and extracellular accumulation of hyaluronic acid during stimulation of quiescent human fibroblasts by mouse epidermal growth factor, *J. Cell Physiol.* **89**:277.

Lennarz, W. J., 1975, Lipid linked sugars in glycoprotein synthesis, *Science* **188**:986.

Levine, E. M., Becker, Y., Boone, C. W., and Eagle, H., 1965, Contact inhibition, macromolecular synthesis and polyribosomes in cultured human diploid fibroblasts, *Proc. Natl. Acad. Sci. U.S.A.* **53**:350.

Lie, S. O., McKusick, V. A., and Neufeld, E. F., 1972, Stimulation of genetic mucopolysaccharidosis in normal human fibroblasts by alteration of pH of the medium, *Proc. Natl. Acad. Sci. U.S.A.* **69**:2361.

Lindahl, U., 1972, Enzymes involved in the formation of the carbohydrate structure of heparin, in: *Methods in Enzymology* (V. Ginsburg, ed.), Vol. 28, pp. 676–684, Academic Press, New York.

Makita, A., and Shimojo, H., 1973, Polysaccharides of SV-40 transformed green monkey kidney cells, *Biochim. Biophys. Acta* **304**:571.

Mannino, R. J., Ballmer, K., and Burger, M. M., 1977, Cell surface modulation and growth inhibition of transformed cells. Abstract of Cold Spring Harbor Meeting on "The Transformed Cell," May 18–22, 1977.

Marshal, R. D., 1972, Glycoproteins, *Annu. Rev. Biochem.* **41**:675.

Meager, A., Nairn, R., and Hughes, R. C., 1975, Analysis of transformed cell variants of BHK_{21}/C_{13} isolated as survivors of adenovirus type 5 infections, *Virology* **68**:41.

Meezan, E., Wu, H. C., Black, P. H., and Robbins, P. W., 1969, Comparative studies on the carbohydrate containing membrane components of normal and virus transformed mouse fibroblasts. II. Separation of glycoproteins and glycopeptides by Sephadex chromatography, *Biochemistry* **8**:2518.

Molnar, J., Robinson, G. B., and Winzler, R. J., 1965a, Biosynthesis of glycoproteins. IV. The subcellular sites of incorporation of glucosamine-1-^{14}C into glycoprotein in rat liver, *J. Biol. Chem.* **240**:1882.

Molnar, J., Teegarden, D. W., and Winzler, R. J., 1965b, The biosynthesis of glycoproteins. IV. Production of extracellular radioactive macromolecules by Ehrlich ascites carcinoma cells during incubation with glucosamine-^{14}C, *Cancer Res.* **25**:1860.

Mora, P. T., Brady, R. O., Bradley, R. M., and McFarland, V. W., 1969, Gangliosides in DNA virus-transformed and spontaneously transformed tumorigenic mouse cell lines, *Proc. Natl. Acad. Sci. U.S.A.* **63**:1290.

Morris, C. C., 1960, Quantitative studies on the production of acid munopolysaccharides by replicate cell cultures of rat fibroblasts, *Ann. N.Y. Acad. Sci.* **86**:878.

Morrison, T. G., and McQuain, C., 1977, Assembly of viral membranes. I. Association of vesicular stomatitis virus membrane proteins and membranes in a cell-free system, *J. Virol.* **21**:451.

Moyer, S. A., and Summers, D. F., 1974, Vesicular stomatitis virus envelope glycoprotein alterations induced by host cell transformation, *Cell* **1**:63.

Mudd, J. A., 1974, Glycoprotein fragment associated with vesicular stomatitis virus after proteolytic digestion, *Virology* **62**:573.

Muramatsu, T., 1971, Demonstration of an Endo-glycosidase acting on a glycoprotein, *J. Biol. Chem.* **246**:5535.

Muramatsu, T., Atkinson, P. H., Nathenson, S. G., and Ceccarini, C., 1973, Cell-surface glycopeptides: Growth-dependent changes in the carbohydrate-peptide linkage region, *J. Mol. Biol.* **80**:781.

Muramatsu, T., Koide, N., and Ogata-Arakawa, M., 1975, Analysis of oligomannosyl cores of cellular glycopeptides by digestion with endo-β-N-acetylglucosaminidases, *Biochem. Biophys. Res. Commun.* **66**:881.

Muramatsu, T., Koide, N., Ceccarini, C., and Atkinson, P. H., 1976a, Characterization of mannose-labeled glycopeptides from human diploid cells and their growth-dependent alterations, *J. Biol. Chem.* **251**:4673.

Muramatsu, T., Ogata, M., and Koide, N., 1976b, Characterization of fucosyl glycopeptides from cell surface and cellular material of rat fibroblasts, *Biochim. Biophys. Acta* **444**:53.

Nakazawa, K., and Suzuki, S., 1975, Purification of keratan sulfate-endogalactosidase and its action on keratan sulfates of different origin, *J. Biol. Chem.* **250**:912.

Nameroff, M., and Holtzer, H., 1967, The loss of phenotypic traits by differentiated cells. IV. Changes in polysaccharides produced by dividing chondrocytes, *Dev. Biol.* **16**:250.

Nigam, V. N., and Cantero, A., 1973, Polysaccharides in cancer: Glycoproteins and glycolipids, *Adv. Cancer Res.* **17**:1.

Nishigaki, M., Muramatsu, T., and Kobata, A., 1974, Endo glycosidases acting on carbohydrate moieties of glycoproteins: Demonstration in mammalian tissue, *Biochem. Biophys. Res. Commun.* **59**:638.

Ogata, S.-I., Muramatsu, T., and Kobata, A., 1975, Fractionation of glycopeptides by affinity column chromatography on concanavalin A–Sepharose, *J. Biochem.* **78**:687.

Ogata, S., Muramatsu, T., and Kobata, A., 1976, New structural characteristic of the large glycopeptides from transformed cells, *Nature* **259**:580.

Ohta, N., Pardee, A. B., McAuslan, B. R., and Burger, M. M., 1968, Sialic acid content and controls of normal and malignant cells, *Biochim. Biophys. Acta* **158**:98.

Old, L. J., and Boyse, E. A., 1964, Immunology of experimental tumors, *Annu. Rev. Med.* **15**:167.

Onodera, K., and Sheinin, R., 1970, Macromolecular glucosamine containing component of the surface of cultivated mouse cells, *J. Cell Sci.* **7**:337.

Oseroff, A. R., Robbins, P. W., and Burger, M. M., 1973, The cell surface membrane: Biochemical aspects and biophysical probes, *Annu. Rev. Biochem.* **42**:647.

Ossowski, L., Quigley, J. P., and Reich, E., 1974, Fibrinolysis associated with oncogenic transformation, *J. Biol. Chem.* **249**:4312.

Perdue, J. F., Kletzien, R., and Miller, K., 1971, The isolation and characterization of plasma membrane from cultured cells. I. The chemical composition of membrane isolated from uninfected and oncogenic RNA virus-converted chick embryo fibroblasts, *Biochim. Biophys. Acta* **249**:419.

Perdue, J. F., Kletzien, P., and Wray, V. L., 1972, The isolation and characterization of plasma membrane from cultured cells. IV. The carbohydrate composition of membranes isolated from oncogenic RNA virus-converted chick embryo fibroblasts, *Biochim. Biophys. Acta* **266**:505.

Pollack, R. E., and Burger, M. M., 1969, Surface-specific characteristics of a contact-inhibited cell line containing the SV40 viral genome, *Proc. Natl. Acad. Sci. U.S.A.* **62**:1074.

Reading, C. L., Penhoet, E. E., and Ballou, C. E., 1978, Carbohydrate structure of vesicular stomatitis virus glycoprotein, *J. Biol. Chem.* **253**:5600.

Reissig, J. L., Lai, W., and Glasgow, J. E., 1975, An endogalactosaminidase from *Streptomyces griseus, Can. J. Biochem.* **53**:1237.

Riskin, D. B., and Quigley, J. P., 1974, Virus-induced modification of cellular membranes related to viral structure, *Annu. Rev. Microbiol.* **28**:325.

Robbins, J. C., and Nicolson, G. L., 1975, Surfaces of normal and transformed cells, in:

Cancer: A Comprehensive Treatise, Vol. 4, Biology of Tumors: Surfaces, Immunology, and Comparative Pathology (F. F. Becker, ed.), pp. 3–54, Plenum Press, New York.

Roblin, R., Cou, L.-N., Black, P. H., 1975a, Proteolytic enzymes, cell surface changes, and viral transformation, Adv. Cancer Res. 32:203.

Roblin, R., Albert, S. O., Gelb, N. A., and Black, P. H., 1975b, Cell surface changes correlated with density-dependent growth inhibition, glycosaminoglycan metabolism in 3T3, SV3T3, and conconavalin A selected revertant cells, Biochemistry 14:347.

Rodén, L., Baker, J. R., Heltin, T., Schwartz, N. B., Stoolmiller, A. C., Yamagata, S., and Yamagata, T., 1972, Biosynthesis of chondroitin sulfate, in: Methods in Enzymology (V. Ginsburg, ed.), Vol. 28, Part B, pp. 638–676, Academic Press, New York.

Roseman, S., 1970, The synthesis of complex carbohydrates by multiglycosyltransferase systems and their potential function in intercellular adhesion, Chem. Phys. Lipids 5:270.

Roth, S., 1973, A molecular model for cell interactions, Q. Rev. Biol. 48:541.

Roth, S., and White., D., 1972, Intercellular contact and cell-surface galactosyl transferase activity, Proc. Natl. Acad. Sci. U.S.A. 69:485.

Ruoslahti, E., Vaheri, A., Kuosela, P., and Linder, E., 1973, Fibroblast surface antigen: A new serum protein, Biochim. Biophys. Acta 322:352.

Saito, H., and Uzman, B. G., 1971, Production and secretion of chondroitin sulfates and dermatan sulfate by established mammalian cell lines, Biochem. Biophys. Res. Commun. 43:723.

Saito, H., Yamagata, T., and Suzuki, S., 1968, Enzymatic methods for the determination of small quantities of isomeric chondroitin sulfates, J. Biol. Chem. 243:1536.

Sakakibara, K., Umeda, M., Saito, S., and Nagase, S., 1977, Production of collagen and asialic glycosaminoglycons by an epithelial liver cell clone in culture, Exp. Cell Res. 110:159.

Sakiyama, H., and Burge, B. W., 1972, Comparative studies of the carbohydrate-containing components of 3T3 and simian virus-40 transformed 3T3 mouse fibroblasts, Biochemistry 11:1366.

Satoh, C., Duff, R., Rapp, F., and Davidson, E. A., 1973, Production of mucopolysaccharides by normal and transformed cells, Proc. Natl. Acad. Sci. U.S.A. 70:54.

Satoh, C., Banks, J., Horst, P., Kreider, J. W., and Davidson, E. A., 1974, Polysaccharide production by cultured B-16 mouse melanoma cells, Biochemistry 13:1233.

Schacter, H., 1977, Control of biochemical parameters in glycoprotein production, Adv. Exp. Med. Biol. 89:103.

Segrest, J. P., Kahane, I., Jackson, R. I., and Marchesi, V. T., 1973, Major glycoprotein of the human erythrocyte membrane: Evidence for an amphipathic molecular structure, Arch. Biochem. Biophys. 155:167.

Sharon, N., and Lis, H., 1972, Lectins: Cell-agglutinating and sugar-specific proteins, Science 177:949.

Shen, L., and Ginsberg, V., 1968, in: Biological Properties of the Mammalian Surface Membrane (L. A. Manson, ed.), Monograph 8, pp. 67–71, Wistar Institute Press, Philadelphia.

Shin, S. I., Freedman, V. H., Risser, R., and Pollack, R., 1975, Tumorigenicity of virus-transformed cells in nude mice is correlated specificially with anchorage independent growth in vitro, Proc. Natl. Acad. Sci. U.S.A. 72:4435.

Singer, S. J., 1971, in: Structure and Function of Biological Membranes (L. I. Rothfield, ed.), pp. 146–223, Academic Press, New York.

Singer, S. J., and Nicolson, G. L., 1972, The fluid mosaic model of the structure of cell membranes, Science 175:720.

Smets, L. A., Van Beek, W. P., and Van Rocij, H., 1976, Surface glycoproteins and concanavalin A-mediated agglutinability of clonal variants and tumor cells derived from SV40-virus-transformed mouse 3T3 cells, *Int. J. Cancer* **18**:462.

Smets, L. A., Van Beek, W. P., and Van Nie, R., 1977, Membrane glycoprotein changes in primary mammary tumors associated with autonomous growth, *Cancer Lett.* **3**:133.

Smets, L. A., Van Beek, W. P., Van Rooy, H., and Homburg, Ch., 1978, The relationship between membrane glycoprotein alterations and anchorage independent growth in neoplastic transformation, *Cancer Biochem. Biophys.* **2**:203.

Smith, D. F., Neri, G., and Walborg, E. F., 1973, Isolation and partial chemical characterization of cell-surface glycopeptides from AS-30D rat hepatoma which possesses binding sites for wheat germ agglutinin and concanavalin A, *Biochemistry* **12**:2111.

Spiro, R. G., 1973, Glycoproteins, *Adv. Prot. Chem.* **27**:349.

Spiro, R. G., and Bhoyroo, V. D., 1974, Structure of the O-glycosidically linked carbohydrate units of fetuin, *J. Biol. Chem.* **249**:5704.

Steck, T. L., and Dawson, G., 1974, Topographical distribution of complex carbohydrates in the erythrocyte membrane, *J. Biol. Chem.* **249**:2135.

Steiner, S., and Steiner, M. R., 1976, Fucolipid patterns of cell lines transformed by highly and weakly tumorigenic simian virus 40 and herpes simplex virus, *Intervirology* **6**:32.

Stoker, M. G. P., and Rubin, H., 1967, Density dependent inhibition of cell growth in culture, *Nature* **215**:171.

Stone, K. R., Smith, R. E., and Joklik, W. K., 1974, Changes in membrane polypeptides that occur when chick embryo fibroblasts and NRK cells are transformed with avian sarcoma viruses, *Virology* **58**:86.

Strecker, G., Herlant-Peers, M. C., Fournet, B., Montreuil, J., Dorland, L., Haverkamp, J., Vliegenthart, F. G., and Farriaux, J. P., 1977, Structure of seven oligosaccharides excreted in the urine of a patient with Sandhoff's disease (GM$_2$ gangliosidosis-variant O), *Eur. J. Biochem.* **81**:165.

Tai, T., Yamashita, K., Ogata-Arakawa, M., Koide, N., Muramatsu, T., Iwasita, S., Inove, Y., and Kobata, A., 1975a, Structural studies of two ovalbumin glycopeptide in relation to the endo-β-N-acetylglucosaminidase specificity, *J. Biol. Chem.* **250**:8569.

Tai, T., Ito, S., Yamashita, K., Muramatsu, T., and Kobata, A., 1975b, Asparagine-linked oligosaccharide chains of IgG: A revised structure, *Biochem. Biophys. Res. Commun.* **65**:968.

Tai, T., Yamashita, K., Ito, S., and Kobata, A., 1977a, Structures of the carbohydrate moiety of ovalbumin glycopeptide III and the difference in specificity of endo-β-N-acetylglucosaminidases C$_{II}$ and H, *J. Biol. Chem.* **252**:6687.

Tai, T., Yamashita, K., and Kobata, A., 1977b, The substrate specificities of endo-β-N-acetylglucosaminidases C$_{II}$ and H, *Biochem. Biophys. Res. Commun.* **78**:434.

Takasaki, S., and Kobata, A., 1976, Purification and characterization of an endo-β-galactosidase produced by *Diplococcus pneumoniae*, *J. Biol. Chem.* **251**:3603.

Tarentino, A. L., and Maley, F., 1974, Purification properties of an endo-β-N-acetylglucosaminidase from *Streptomyces griseus*, *J. Biol. Chem.* **249**:811.

Tarentino, A. L., and Maley, F., 1975, A comparison of the substrate specificities of endo-β-N-acetylglucosaminidases from *Streptomyces griseus* and *Diplococcus pneumonae*, *Biochem. Biophys. Res. Commun.* **67**:455.

Tarentino, A. L., Plummer, T. H., Jr., and Maley, F., 1972, A re-evaluation of the oligosaccharide sequence associated with ovalbumin, *J. Biol. Chem.* **247**:2629.

Tarentino, A. L., Plummer, T. H., Jr., and Maley, F., 1973, A β-mannoside linkage in the unit A oligosaccharide of bovine thyroglobulin, *J. Biol. Chem.* **248**:5547.

Tarentino, A. L., Plummer, T. H., Jr., and Maley, F., 1974, The release of intact oligosaccharides from specific glycoproteins by endo-β-N-acetylglucosaminidase H, *J. Biol. Chem.* **249**:818.

Terry, A. H., and Culp, L. A., 1974, Substrate-attached glycoproteins from normal and virus-transformed cells, *Biochemistry* **13**:414.

Thomas, D. B., and Winzler, R. J., 1971, Structure of glycoproteins of human erythrocytes, *Biochem. J.* **124**:55.

Todaro, G. J., and Green, H., 1963, Quantitative studies of the growth of mouse embryo cells in culture and their development into established lines, *J. Cell Biol.* **17**:299.

Tomida, M., Koyama, H., and Ono, T., 1975, Induction of hyaluronic acid synthetase activity in rat fibroblasts by medium change of confluent culture, *J. Cell Physiol.* **86**:121.

Tomida., M., Koyama, H., and Ono, T., 1977, A serum factor capable of stimulating hyaluronic acid synthesis in cultured rat fibroblasts, *J. Cell Physiol.* **91**:323.

Tomita, M., and Marchesi, V. T., 1975, Amino-acid sequence and oligosaccharide attachment sites of human erythrocyte glycophorin, *Proc. Natl. Acad. Sci. U.S.A.* **72**:2964.

Tuszynski, G. P., Baker, S. R., Fuhrer, J. P., Buck, C. A., and Warren, L., 1978, Glycopeptides derived from individual membrane glycoproteins from control and Rous sarcoma-virus transformed hamster fibroblasts, *J. Biol. Chem.* **253**:6092.

Umemoto, J., Bhavanandan, V. P., and Davidson, E. A., 1977, Purification and properties of an endo-α-N-acetyl-D-galactosaminidase from *Diplococcus pneumoniae*, *J. Biol. Chem.* **252**:8609.

Underhill, C. G., and Keller, J. M., 1975, A transformation-dependent different in the heparan sulfate associated with the cell surface, *Biochem. Biophys. Res. Commun.* **63**:448.

Unkeless, J. C., Tobia, A., Ossowski, L., Quigley, J. P., Rifkin, D. B., and Reich, E., 1973, An enzymatic function associated with transformation of fibroblasts by oncogenic viruses. I. Chick embryo fibroblast cultures transformed by avian RNA tumor viruses, *J. Exp. Med.* **137**:85.

Vaheri, A., Ruoslahti, E., and Nordling, S., 1972, Neuraminidase stimulates division and sugar uptake in density-inhibited cell cultures, *Nature (London) New Biol.* **238**:211.

Van Beek, W. P., Smets, L. A., and Emmelot, P., 1973, Increased sialic acid density in surface glycoprotein of transformed and malignant cells—A general phenomenon? *Cancer Res.* **33**:2913.

Van Beek, W. P., Smets, L. A., and Emmelot, P., 1975, Changed surface glycoprotein as a marker of malignancy in human leukaemic cells, *Nature* **253**:457.

Van Beek, W. P., Emmelot, P., and Homburg, C., 1977, Tumour-associated changes in cell surface glycoprotein of rat hepatoma as compared with embryonic rat liver, *Br. J. Cancer* **36**:157.

Van Nest, G. A., and Grimes, W. J., 1977, A comparison of membrane components of normal and transformed Balb/c cells, *Biochemistry* **16**:2902.

Vannucchi, S., and Chiarugi, V. P., 1977, Surface exposure of glycosaminoglycans in resting and virus transformed 3T3 cells, *J. Cell Physiol.* **90**:503.

Wagner, R. R., Snyder, R. M., and Yamazaki, S., 1970, Proteins of vesicular stomatitis virus: Kinetics and cellular sites of synthesis, *J. Virol.* **5**:548.

Wagner, R. R., Kiley, M. P., Snyder, R. M., and Schnaitman, C. A., 1972, Cytosplasmic compartmentalization of the protein and ribonucleic acid species of vesicular stomatitis virus, *J. Virol.* **9**:672.

Walborg, E. F., Lantz, R. S., and Wray, V. P., 1969, Isolation and chemical characterization of a cell surface sialoglycopeptide fraction from Novikoff ascites cells, *Cancer Res.* **29**:2034.

Warren, L., and Glick, M. C., 1968, Membranes of animal cells. II. The metabolism and turnover of the surface membrane, *J. Cell Biol.* **37**:729.

Warren, L., Critchley, D., and Macpherson, I., 1972a, Surface glycoproteins and glycolipids of chick embryo cells transformed by a temeprature-sensitive mutant of Rous sarcoma virus, *Nature* **235**:275.

Warren, L., Fuhrer, J. P., and Buck, C. A., 1972*b*, Surface glycoproteins of normal and transformed cells: A difference determined by sialic acid and a growth-dependent sialyl transferase, *Proc. Natl. Acad. Sci. U.S.A.* **69**:1838.

Warren, L., Zeidman, I., and Buck, C. A., 1975, The surface glycoproteins of a mouse melanoma growing in culture and as a solid tumor, *in vivo, Cancer Res.* **35**:2186.

Weiss, L., and Poste, G., 1976, The tumor cell periphery, in: *Scientific Foundations of Oncology* (T. Symington, and R. Carter, eds.), pp. 25–35, William Heinemann Medical Books, London.

Winzler, R. J., 1970, Carbohydrates in cell surfaces, *Int. Rev. Cytol.* **29**:77.

Wolfe, L. S., Senior, R. G., and Ng Ying Kin, N. M. K., 1974, The structure of oligosaccharides accumulating in the liver of GM_1-gangliosidosis, type-I, *J. Biol. Chem.* **249**:1828.

Yamada, K. M., and Pastan, I., 1976, Cell surface protein and neoplastic transformation, *Trends Biochem. Sci.* **1**:222.

Yamada, K. M., and Weston, J. A., 1974, Isolation of a major cell surface glycoprotein from fibroblasts, *Proc. Natl. Acad. Sci. U.S.A.* **71**:3492.

Yamada, K. M., Yamaoa, S. S., and Pastan, I., 1976, Cell surface protein partially restores morphology, adhesiveness, and contact inhibition of movement to transformed fibroblasts, *Proc. Natl. Acad. Sci. U.S.A.* **73**:1217.

Yamashita, K., Tachibana, Y., and Kobata, A., 1978, The structures of the galactose-containing sugar chains of ovalbumin, *J. Biol. Chem.* **253**:3862.

Yurchenco, P. D., and Atkinson, P. H., 1975, Fucosyl-glycoprotein and precursor pools in HeLa cells, *Biochemistry* **14**:3107.

Yurchenco, P. D., and Atkinson, P. H., 1977, Equilibration of fucosyl glycoprotein pools in HeLa cells, *Biochemistry* **16**:944.

Yurchenco, P. D., Ceccarini, C., and Atkinson, P. H., 1978, Labeling complex carbohydrates of animal cells with monosaccharides, in: *Methods in Enzymology*, Vol. 50, pp. 175–204, Academic Press, New York.

Zatz, M. M., Goldstein, A. L., Blumenfeld, O., and White, A., 1972, Regulation of normal and leukaemic lymphocyte transformation and recirculation by sodium periodate oxidation and sodium borohydride reduction, *Nature (London) New Biol.* **240**:252.

Carbohydrate Recognition Systems for Receptor-Mediated Pinocytosis

Elizabeth F. Neufeld and Gilbert Ashwell

1. INTRODUCTION

Pinocytosis is a mechanism for the transport of molecules from the exterior to the interior of the cell without transfer through the plasma membrane. Transport of molecules occurs when the membrane wraps around the external material to form vesicles which are pinched off and internalized. The vesicles fuse with lysosomes, the contents of which become enriched with the material brought in from the outside. If the area of the plasma membrane which is internalized has receptors to which a particular substance is tightly bound, that substance is delivered to lysosomes far more efficiently and selectively than if it were simply trapped within the pinocytotic vesicle. In recent years, receptor-mediated pinocytosis systems have been described for substances as diverse as cholesterol esters, vitamin B_{12}, hormones and growth factors, circulating glycoproteins, and hydrolytic enzymes. It is the last two groups, which involve the recognition of specific carbohydrate residues, that are the subject of this chapter.

Elizabeth F. Neufeld and Gilbert Ashwell • National Institute of Arthritis, Metabolism and Digestive Diseases, National Institutes of Health, Bethesda, Maryland 20205.

2. GALACTOSE-BINDING RECEPTOR OF MAMMALIAN HEPATOCYTES

2.1. Development of the Problem

The origin and development of a novel concept for the role of the carbohydrate moiety of serum glycoproteins was described in an earlier, detailed review (Ashwell and Morell, 1974). Briefly, the hypothesis was advanced that sialic acid was essential for serum glycoprotein viability in the circulation. This conclusion was prompted by the observation that, upon injection into rabbits of a preparation of ceruloplasmin from which the terminal, nonreducing sialic acid had been removed enzymatically, the resulting asialoglycoprotein was cleared from the circulation within minutes. This value was in marked contrast to that of the fully sialylated protein which exhibited a normal half-life of several days. The penultimate galactosyl residues, exposed by the removal of sialic acid, were shown to be critical determinants of clearance with the demonstration that survival time was significantly prolonged upon modification of the galactose moiety by galactose oxidase or treatment with β-galactosidase (Morell et al., 1968). The rapid clearance from the circulation of the desialylated glycoprotein was shown to result from sequestration by the liver; there was no significant accumulation of the injected material in any of the other organs tested. A more detailed examination of the subcellular distribution in the liver revealed the lysosomes to be the principal site of catabolism (Gregoriadis et al., 1970).

The possibility that the above phenomena were artifacts resulting from enzymatically induced denaturation of the intact molecule was considered. To meet this objection, a partially purified preparation of rat liver sialyltransferase was isolated and incubated with desialylated ceruloplasmin in the presence of CMP-sialic acid. The reconstituted protein, containing 85% of its normal complement of sialic acid was shown to be homogeneous by polyacrylamide gel electrophoresis and to cocrystallize with authentic ceruloplasmin. Upon injection of this material into rabbits, two distinct classes of molecules were recognized with markedly different half-lives. Those molecules in which all, or essentially all, of the missing sialic acid residues had been restored exhibited a normal serum survival time of 56 hr; those in which less than a critical threshold level had been restored disappeared from the circulation within a few minutes after injection (Hickman et al., 1970). With the demonstration that a survival time of 30 min was sufficient to distinguish between these two classes of molecules, it became possible to determine experimentally the relationship between the extent of desialylation and that fraction of the total dose rendered susceptible to hepatic recognition and catabolism. Utilizing the Poisson distribution function for the statistical analysis, it was shown that the exposure of only two of the ten penultimate galactose residues of ceruloplasmin was sufficient to mark any given molecule for hepatic destruction (Van Den Hamer et al., 1970).

The growing evidence for the extent of specificity involved in the rapid removal and lysosomal catabolism of desialylated ceruloplasmin suggested a basic biological mechanism of broad generality. This concept was supported by the finding that all of the asialo derivatives of the plasma proteins tested, with the exception of transferrin, disappeared rapidly from the circulation upon intravenous injection into rabbits or rats (Morell et al., 1971). The rate of clearance of asialotransferrin was subsequently shown to be faster than that of the native protein when the period of observation was increased from 30 min to several days (Regoeczi et al., 1974).

2.2. Binding by Plasma Membranes

Subsequent investigation in vitro revealed the primary locus of binding to be present on the plasma membranes of the hepatocytes (Pricer and Ashwell, 1971). Upon incubation of the isolated membranes with [^{125}I]asialoorosomucoid, followed by filtration and counting of the filters, a highly sensitive inhibition assay became available to probe the qualitative and quantitative parameters of the binding mechanism (Van Lenten and Ashwell, 1972). Treatment of the membranes with a small amount of neuraminidase was found to result in the complete loss of their capacity to bind asialoglycoproteins. The destructive action of neuraminidase was shown to be a repairable lesion in that enzymatic replacement of the sialic acid residues restored their binding potency. This initially baffling property was subsequently clarified by the demonstration that the apparent loss of binding activity reflected merely the ability of the binding protein (receptor) to form a stable complex with its own galactose residues and thereby prevent subsequent binding of the test substance, [^{125}I]asialoorosomucoid. By masking the terminal galactose residues through resialylation or removing the galactose with β-galactosidase, binding activity was restored (Stockert et al., 1977; Paulson et al., 1977).

2.3. Isolation of Binding Protein

The hepatic receptor for galactose-terminated glycoproteins was isolated and purified by affinity chromatography on columns of Sepharose to which asialoorosomucoid has been covalently bound (Hudgin et al., 1974). The purified rabbit liver receptor was characterized as a water-soluble glycoprotein in which 10% of the dry weight consisted of sialic acid, galactose, mannose, and glucosamine. In aqueous solution, the purified receptor exhibited a high degree of aggregation resulting from the self-associating properties of a single oligomeric protein with a minimal molecular weight of 5×10^5. Two subunits were present with estimated molecular weights of 48×10^3 and 40×10^3 (Kawasaki and Ashwell, 1976a). In the presence of Triton X-100, aggregation was reversed with the formation of a single component with an estimated molecular weight of 2.5×10^5.

Pronase digestion of the intact protein resulted in quantitative recovery of the carbohydrate moiety, which was shown to consist of two distinctly different glycopeptides. The carbohydrate sequence of both was determined, and their relative distribution between the two protein subunits was established (Kawasaki and Ashwell, 1976b). The major glycopeptide was shown to consist of a triantennary structure wherein one of the terminal oligosaccharides (sialic acid, galactose, N-acetylglucosamine) was linked to a single α-mannosyl residue. The remaining two outer chains were joined to the β-mannosyl group adjacent to the diacetylchitobiose asparagine fragment (see Chapter 1). The minor glycopeptide contained only mannose and N-acetylglucosamine with an ovalbuminlike structure.

2.4. Nature of the Binding Mechanism

In accordance with the mechanism of binding proposed by Roseman (1970) whereby cell surface-bound glycosyltransferases would interact with adjacent incomplete carbohydrate chains, the purified receptor was examined for glycosyltransferase activity. Under optimally determined conditions, no transferase activity for sialic acid, galactose, N-acetylglucosamine, or fucose was detectable (Hudgin and Ashwell, 1974). An alternative proposal for the mechanism of binding emerged from the unexpected finding that the purified binding protein possessed the property of agglutinating human and rabbit erythrocytes (Stockert et al., 1974). Identification of this protein as the first recognized lectin of mammalian origin was supported by the subsequent observation that this receptor was also capable of inducing mitogenesis in peripheral lymphocytes (Novogrodsky and Ashwell, 1977). These findings prompted the suggestion that the concept of a lectin be broadened to embrace all biological systems and that the function of lectins be envisaged as residing in their ability to transmit postribosomal information essential for the orderly growth and development of cellular structures (Ashwell, 1977).

2.5. Specificity of Binding

A second finding of significance for the specificity of the binding protein emerged from the studies of Stockert *et al.* (1974) in which it was shown that group A erythrocytes were more sensitive to agglutination by the rabbit lectin than were group B erythrocytes. From this, it could be inferred that *N*-acetylgalactosamine residues (determinants of blood group A) as well as galactose residues (determinants of blood group B) were recognized by the binding protein. Indeed, the α-*N*-acetylgalactosamine-terminated macromolecule, asialoovine submaxillary mucin, proved to be a better inhibitor of agglutination than was the β-galactose-terminated protein, asialoorosomucoid. Subsequent studies, employing an agarose-immobilized rabbit binding protein, have since established the order of glycoside inhibitory capacity as α-methyl-*N*-acetylgalactosamine > β-methly-*N*-acetylgalactosamine > β-methylgalactose > α-methylgalactose (Sarkar *et al.*, 1979).

Yet another degree of latitude in the specificity of binding has been reported recently by Stowell and Lee (1978). Synthetic "neoglycoproteins" were synthesized in which specific thio sugars were attached to the bovine serum albumin via a stable amidine linkage. As was expected, when these derivatives were assayed, the thiogalactosyl neoglycoprotein showed a strong affinity for the purified rabbit binding protein whereas the *N*-acetylglucosaminyl and the mannosyl analogues did not. Surprisingly, the glucosyl neoglycoprotein proved to be as good, or better, than the galactosyl derivative. All of the data obtained were consistent with the thesis that the hepatic binding protein could not discriminate between the D-galacto and the D-gluco configurations.

2.6. Circulating Asialoglycoproteins

Utilizing the plasma membrane inhibition assay described above, Marshall *et al.* (1974) reported the presence of small amounts of inhibitory substances in the serum of normal patients which were presumed to represent a low level of circulating asialoglycoproteins. These values were significantly increased in sera obtained from patients with clinically diagnosed cirrhosis or hepatitis but were unchanged in a variety of non-hepatic pathologies. A more definitive approach to this problem became possible with the availability of the purified hepatic receptor. This was accomplished by covalent linkage of the binding protein to Sepharose. Upon addition of serum to a column prepared with this material, the galactose-terminated proteins were retained and quantitated by virtue of their ability to inhibit the subsequent binding of [^{125}I]asialoorosomucoid

(Lunney and Ashwell, 1976). Subsequently, they were eluted under specific conditions, and characterized and identified as asialoglycoproteins (Lunney, 1976). Both qualitatively and quantitatively, the results confirmed the increased titer in the sera of patients with hepatic disease as reported by Marshall *et al.* (1974). These data were interpreted as supporting the concept that the hepatic binding of desialylated glycoproteins reflects a normal physiological process regulating plasma protein metabolism *in vivo*.

A more convincing correlation between the level of circulating asialoglycoproteins and the status of the hepatic binding protein was obtained by the demonstration that vastly increased serum levels of asialoglycoproteins were present in those species deficient in the galactose-specific receptor. Examination of sera from both avian and reptilian species, by the affinity column technique described above, revealed the existence of an asialoglycoprotein titer in these species which was more than an order of magnitude greater than that seen in mammalian species. The significance of this finding became apparent only after the realization that the asialoglycoprotein binding receptor was absent in avian liver (Lunney and Ashwell, 1976). These correlations constitute the best current evidence in support of participation of the hepatic binding protein in the normal physiological regulation of serum glycoprotein catabolism; direct, hard evidence is not, as yet, available.

2.7. Subcellular Loci of Binding Protein

The initial identification of the rat liver plasma membrane as the major locus of binding activity for galactose-terminated glycoproteins was subsequently expanded to include membranes of the Golgi complex, the smooth microsomes, and the lysosomes (Pricer and Ashwell, 1976). From each of these organelles, as well as from the plasma membranes, the specific binding protein was isolated by affinity chromatography in good yield and with high specific activity. All of the preparations were identical in their specificity towards galactose, their absolute requirement for calcium, their sensitivity to neuraminidase, and their response to carrier dilution with added, nonradioactive asialoglycoprotein. In addition, all of the preparations revealed a similar pattern of band formation on polyacrylamide gel electrophoresis and gave rise to a single, fused precipitin line on double immunodiffusion.

Significantly, it was noted that the optimal conditions for the assay of binding activity by the various subcellular fractions differed in their response to the presence or absence of Triton X-100. In the presence of this detergent, the binding capacities of the Golgi complex and the smooth microsomes were greatly augmented whereas that of the lysosomes was diminished and that of the plasma membrane was unaffected. This initially puzzling observation prompted further examination of the membrane isolated from the organelles as the presumed locus of binding activity. In each case the binding activity was recovered quantitatively in the membranous pellet with complete loss of the differential response to detergent; binding activity was invariant in the presence or absence of Triton X-100. These observations were rationalized by the proposal that the stimulatory effect of detergent on the Golgi and smooth microsomes was to expose, or make available, binding sites within the organelle which were inaccessible to the ligand in aqueous solution. Conversely, the inhibitory effect on lysosomes was ascribed to detergent-induced lysis resulting in the release of hydrolytic enzymes which effectively degraded either the binding protein, the ligand, or both.

The multiple intracellular loci of this unique protein have suggested the possibility of a biosynthetic cycle involving a progression from the smooth microsomes to the Golgi and then to the plasma membranes. The simultaneous presence on the lysosomes, however, has raised a provocative question as to the relationship of the binding protein to the mechanism of pinocytosis whereby the receptor–ligand complex moves from the external surface of the hepatocyte to the interior of the cell and eventually to engulfment by the lysosome.

2.8. Membrane Topology

In an attempt to resolve this problem, Tanabe *et al.* (1979) investigated the topological distribution of the binding protein on the membranes of the various subcellular organelles. This study was facilitated by the availability of a purified antibody to the hepatic receptor which effectively blocked binding of the test ligand, [^{125}I]asialoorosomucoid. Incubation of lysosomes with increasing amounts of this antibody, prior to challenge with the radioactive ligand, progressively reduced the lysosomal binding activity to zero. From this, it could be inferred that all of the measurable activity exhibited by the intact lysosomes arose from receptors oriented on the external, or cytosolic, surface of the membrane. However, the presence of additional cryptic receptors on the inner surface of the vesicular membrane could not be excluded.

To test this possibility, the lysosomal membranes were recovered from each of the antibody blocked lysosomes and examined for the emergence of new binding sites. It was anticipated that exposure of additional receptors on the inner surface of the lysosomal membrane would be reflected in an increased amount of antibody required for neutralization. This was not seen; the exact coincidence of the two antibody neutralization curves substantiated the premise that viable binding receptor was absent on the inner surface of the lysosomal membrane.

In contrast to the results obtained with lysosomes, preparation of membranes prepared from Golgi complex or smooth microsomes gave rise to a marked increase in asialoglycoprotein binding capacity over the intact vesicles as manifested by the incremental amount of antibody required for neutralization. From this and other data (Tanabe *et al.*, 1979) it was concluded that at least 85%, and possibly all, of the binding sites were located on the inner, luminal surface of these organelles.

2.9. Regeneration of Binding Protein

In view of the apparently unique localization of the hepatic binding protein on the cytosolic surface of the lysosomes, the possibility was considered that a specific membrane translocation might accompany the uptake and catabolism of desialylated glycoproteins. Thus it was proposed that the formation of a receptor–asialoglycoprotein complex on the outer surface of the hepatic plasma membrane might be followed by endocytosis with subsequent resolution of the internalized complex whereby the receptor was inserted into the outer lysosomal surface and the ligand (asialoglycoprotein) was delivered to the inner surface as a prelude to catabolism.

Previous studies had demonstrated a time- and dose-dependent hepatic clearance of desialylated glycoproteins from the rat circulation, whereby injection of 5–10 mg of asialoorosomucoid resulted in a delayed clearance time in excess of 2 hr. During this period, hepatic intake and catabolism of the ligand was continuous. Under these conditions, it might be anticipated that the receptor protein responsible for binding would be most susceptible to turnover if, indeed, the receptor–ligand complex was internalized as a unit and subsequently subjected to lysosomal catabolism.

To test this hypothesis, rats were injected intraperitoneally with [^{3}H]leucine. After 16 hr, one group of 10 rats was given an intravenous injection of asialoorosomucoid (5.0 mg) and a control group was treated similarly with intact orosomucoid. Two hours later, the rats were killed, the appropriate livers pooled, the plasma membranes purified, and the receptor protein was isolated by affinity chromatography. The yields of plasma membrane and binding protein were 1.4 mg/g and 1.1 μg/g liver, respectively.

The specific radioactivity of the binding protein recovered from the hepatic plasma membrane was unaffected by the administration of either the rapidly cleared ligand, asialoorosomucoid, or the uncleared control, fully sialylated orosomucoid. Neither of the values was significantly different from that obtained for the average half-life of the rat hepatic binding protein. These results were interpreted as indicating a divergent metabolic pathway for the receptor–ligand complex wherein the receptor was spared and the ligand destroyed.

On the basis of the above results, it has been suggested that the marked difference in the survival rates of the receptor and the ligand may be correlated with the localization of the former on the external surface of the lysosome. Conceivably, endocytosis of the intact complex is followed by a spatial reorientation whereby the receptor is preserved in the lysosomal membranes and the ligand catabolized within the body of this organelle. This postulation, if valid, would provide at least the initial steps for a membrane recycling mechanism such as that described by Tulkens *et al.* (1977). Clearly, however, the possibility has not been excluded that the plasma membrane binding protein provides a stable shuttle mechanism that functions exclusively in the transport of ligand across the cell membrane. If this were the case, the unique distribution of this receptor on the external surface of the lysosome would be unexplained and possibly irrelevant to the catabolic process.

3. N-ACETYLGLUCOSAMINE-BINDING RECEPTOR OF AVIAN HEPATOCYTES

3.1. Historical

Studies designed to identify and quantitate the amount of desialylated (galactose-terminated) glycoproteins in the circulation revealed the asialoglycoprotein titer in the serum of avian and reptilian species to be vastly greater than that found in mammalian serum. The reason for the markedly divergent titers became apparent upon the demonstration that the avian liver was totally deficient in the hepatic binding protein specific for galactose-terminated proteins which had been previously isolated from mammalian liver. Of even greater interest, however, was the identification of an avian hepatic binding activity specific for *N*-acetylglucosamine-terminated glycoproteins, that is, for those proteins from which both sialic acid and galactose had been removed to expose the underlying hexosamine moiety (Lunney and Ashwell, 1976).

3.2. Isolation of Binding Protein

Utilizing [^{125}I]agalactoorosomucoid (*N*-acetylglucosamine-terminated) as the specific ligand, a survey of the various organs of the chicken revealed the binding activity to be restricted to the liver and subsequent studies were confined to that organ (Kawasaki and Ashwell, 1977).

Isolation and purification of the N-acetylglucosamine binding protein was readily accomplished by means of an affinity column prepared by the covalent linkage of agalactoorosomucoid to Sepharose 4B. From 100 g of frozen chicken liver, 6–9 mg of purified binding protein were recovered with a specific binding activity of 150–200 ng ligand/μg protein in an overall yield of 40–50%. In aqueous solution, the binding protein was stable to freezing and was stored at $-20°C$ for periods in excess of one month. Calcium was required for binding; in its absence, or in the presence of 10 mM EDTA, no binding was observed. The K_m for calcium was determined to be 1.3 mM.

3.3. Physical and Chemical Properties

Gel filtration of the purified material on Sepharose 4B in the presence of 0.5% Triton X-100 and 0.5 mM EDTA resulted in the recovery of a single symmetrical peak. Polyacrylamide gel electrophoresis in sodium dodecyl sulfate gave rise to a single subunit with an estimated molecular weight of 26×10^3.

Initial evidence indicating the presence of carbohydrate in the purified binding protein was obtained by a positive periodic acid–Schiff stain on polyacrylamide gel electrophoresis which coincided with the single protein band. The total carbohydrate content of the protein was estimated to be 8% as indicated by recovery of the individual monosaccharides after acid hydrolysis. Sialic acid, galactose, mannose, and glucosamine were shown to be present in approximately equimolar amounts; no detectable amounts of galactosamine, fucose, or glucose were observed. Exposure of the purified protein to neuraminidase was without effect, whereas subsequent treatment with β-galactosidase resulted in complete loss of binding activity, presumably as a result of self-recognition of the exposed N-acetylglucosamine residues.

3.4. Kinetics of Binding

Kinetic studies revealed the binding reaction to be complete within 5 min at room temperature and to be essentially independent of the amount of ligand added. However, the association rate constant for the reaction (k_1) could not be assessed with certainty because of the lack of a reliable method for stopping the reaction instantaneously without dissociating the complex already formed. In contrast, dissociation was a relatively slow process. In the presence of excess ligand, the reverse reaction was shown to be of first order and the dissociation rate constant (k_{-1}) was calculated to be 1.3×10^{-3}/sec.

The binding of [^{125}I]agalactoorosomucoid to the purified avian membrane protein was a saturable process and, at appropriate ligand concentrations, binding was proportional to the protein concentration. A Scatchard plot of the binding data indicated the presence of a single, high-affinity binding site with a dissociation constant of 1.4×10^{-9} M. At saturation, the maximum capacity of the binding protein was 3.9 pmole of [^{125}I]agalactoorosomucoid per μg of protein. On the assumption that the estimated molecular weight of the binding protein as 2.1×10^5 is reasonably accurate, it can be calculated that 0.8 mole of ligand can be bound per mole of binding protein.

Of the proteins tested, specificity was restricted to those with terminal N-acetylglucosamine residues; those with nonreducing β-galactosyl or α-mannosyl groups did not bind to this receptor.

3.5. Comparison of Avian N-Acetylglucosamine and Mammalian Galactose-Binding Proteins

The avian binding receptor shares many properties in common with those of the mammalian protein. In both systems, the major locus of binding activity is restricted to the liver, from which it can be isolated, purified, and assayed by closely analogous techniques. In both cases, calcium is required for a binding reaction specifically directed toward the terminal, nonreducing carbohydrate residues of circulating proteins. Upon purification, both preparations proved to be glycoproteins containing the same set of carbohydrate constituents, including sialic acid, galactose, mannose, and glucosamine. In aqueous solution, the two proteins were recovered in an aggregated state which reversibly converted to a single component by the addition of detergent. Finally, the binding of both proteins was abolished by the action of specific glycosidases.

However, within the framework of the above generalizations, specific properties are clearly distinguishable. The avian protein, in contrast to the mammalian, exhibits only minimal binding activity for asialoglycoproteins and interacts strongly with agalactoglycoproteins. A second major point of differentiation is the ready reversibility of the avian protein–ligand complex in which the dissociation rate constant was calculated to be 1.3×10^{-3}/sec; the mammalian complex was not reversible under experimental conditions. Again, in contrast to the kinetic properties of the mammalian protein, the attainment of equilibrium conditions in the avian system permitted the demonstration of a single, high-affinity binding site with a K_{diss} of 1.4×10^{-9} M as determined by a Scatchard plot.

Although both receptors were shown to be glycoproteins of similar composition, their response to specific glycosidases was characteristically different. Whereas the binding activity of the rabbit protein was destroyed by exposure to neuraminidase, the chicken protein was unaffected. The activity of the latter preparation was abolished by β-galactosidase, a treatment which restored activity to the neuraminidase-inactivated rabbit protein. Finally, the rabbit binding protein possessed the ability to agglutinate both human and rabbit erythrocytes; the avian protein was inert.

4. MANNOSE-6-PHOSPHATE RECOGNITION SYSTEM OF HUMAN FIBROBLASTS

4.1. Recognition of Lysosomal Enzymes—Historical

The existence of a receptor-mediated system in cultured fibroblasts for the uptake of lysosomal enzymes was discovered as a result of studies on genetic disorders of mucopolysaccharide catabolism (Neufeld et al., 1975; see Chapter 7 for a discussion of the catabolism of mucopolysaccharides). Cells cultured from patients with such disorders were shown to accumulate excess sulfated mucopolysaccharide, but the level could be reduced to normal by including in the medium urine concentrates or secretions from fibroblasts derived from normal individuals (Neufeld and Cantz, 1971). Several active principles in these concentrates, or "corrective factors," were extensively purified and their function determined. They were found to be hydrolytic enzymes with specificity towards linkages found in dermatan sulfate and heparan sulfate. For example, the substance which had been purified on the basis of its ability to normalize the mucopolysaccharide catabolism of cells from patients with the Hunter syndrome (i.e., the "Hunter corrective factor") was shown to be a sulfatase specific for sulfated iduronic acid residues (Bach et al., 1973). An analogous substance corrective for cells from patients with the genetically distinct Hurler syndrome was found to be associated with α-L-iduronidase activity (Bach et al., 1972). At the same time, it was shown that the Hunter syndrome resulted from a deficiency of iduronate sulfatase and the Hurler syndrome from a deficiency of α-L-iduronidase (Bach et al., 1972; Matalon and Dorfman, 1972; Sjöberg et al., 1973). The corrective factors were thus understood to be the missing enzymes for the corresponding disorders and correction to be a form of enzyme replacement therapy in vitro.

The finding of α-L-iduronidase activity associated with the Hurler corrective factor prompted Bach *et al.* (1972) to ask how much iduronidase had to be introduced into the cells of Hurler patients in order to effect a significant correction. Replacement to the extent of one-tenth the normal level gave 90% of maximal correction—a result interpreted as encouraging for eventual therapeutic prospects. Unexpectedly, that experiment also showed that over a third of the α-L-iduronidase added to the medium had been taken up by the Hurler cells. Since nonselective pinocytosis could not account for such remarkable concentration of the enzyme, a selective recognition mechanism had to be postulated.

It should be remembered that the "Hurler corrective factor" had been purified by a bioassay which measured the composite of enzyme uptake and catalytic activity (Barton and Neufeld, 1971). When the two parameters were followed independently during purification, two forms of α-L-iduronidase could be separated, one taken up well and corrective, the other poorly taken up and noncorrective (Shapiro *et al.*, 1976).

Analogous studies of another disorder of mucopolysaccharide metabolism showed a deficiency of β-glucuronidase in cells with correction by exogenous β-glucuronidase and the existence of corrective, or high uptake, as well as of noncorrective, or low uptake, forms of that enzyme (Sly *et al.*, 1973; Hall *et al.*, 1973; Brot *et al.*, 1974; Nicol *et al.*, 1974; Glaser *et al.*, 1975).

A number of other lysosomal enzymes, not necessarily related to mucopolysaccharide catabolism, were shown to exist in forms that differed in the extent of uptake into fibroblasts (tabulated in Neufeld *et al.*, 1977). These studies showed that the signal for uptake was independent of catalytic activity. They also showed that the high uptake forms were present in minute amounts, primarily in urine, cell secretions, and platelets, and that most tissue hydrolases were of the low uptake form. Biochemical explanations for this distribution became apparent only later. However, the relative scarcity of the high uptake forms made direct analysis of the recognition signal difficult and generated other experimental approaches. The major and eventually successful approach was the use of competitive inhibitors of the uptake.

4.2. Recognition Signal

The first evidence for a carbohydrate recognition marker was the conversion of a high uptake form of β-hexosaminidase to a low uptake form by oxidation with 0.01 M sodium periodate under conditions believed to be selective for carbohydrates (Hickman *et al.*, 1974). The analogy to the galactose recognition system of hepatocytes (the only receptor-mediated pinocytosis system known at the time) was clear, but the fibroblast system appeared different because uptake of the β-hexo-saminidase was not inhibited by asialofetuin.

Attention was drawn to mannose residues by Hieber *et al.* (1976), who measured the uptake of a highly purified bovine β-galactosidase which contained N-acetylglucosamine and mannose as the only sugar residues. Uptake was inhibited by mannose but not by N-acetylglucos-amine and was lost after treatment of the enzyme with a partially purified α-mannosidase.

However, mannose did not appear to be a satisfactory candidate for the recognition signal of high uptake enzymes because low uptake lyso-somal enzymes also have a high mannose content and because mannose-containing oligosaccharides and glycoproteins were very poor inhibitors of uptake of β-galactosidase and α-L-iduronidase.

Nevertheless, the focus on mannose proved heuristic in that it led to the startling discovery by Kaplan *et al.* (1977*a*) of the potent compet-itive inhibition of β-glucuronidase uptake by mannose-6-phosphate and of the loss of uptake if the β-glucuronidase were pretreated with a phos-phatase. Similar results were obtained with other enzymes, including α-L-iduronidase (Sando and Neufeld, 1977), β-hexosaminidase (Kaplan *et al.*, 1977*b*), and α-N-acetylglucosaminidase (Ullrich *et al.*, 1977). Of all the other phosphorylated compounds tested in the various systems, sim-ilar inhibition was obtained only with fructose-1-phosphate, which in its pyranose form has a strong structural resemblance to mannose-6-phos-phate. No other phosphorylated sugars, phosphorylated amino acids, or nucleotides (including sugar nucleotides and cyclic AMP) had such an effect. Sulfated analogues such as mannose-6-sulfate inhibited α-L-iduron idase (Sando *et al.*, 1978) and β-galactosidase (Jourdian *et al.*, 1978) although less effectively than the phosphate.

The uptake of α-L-iduronidase was also markedly inhibited by a macromolecular urinary glycoprotein fraction composed of other hydro-lytic enzymes recognized by the same receptor. The inhibitor lost potency upon oxidation with 0.01 M $NaIO_4$ and upon preincubation with alkaline phosphatase (Sando and Neufeld, 1977), although it was resistant to treatment with exoglycosidases. Phosphorylated mannans are excellent inhibitors of β-glucuronidase uptake, particularly after hydrolysis of phosphodiester bonds (Kaplan *et al.*, 1977*a*, 1978).

Synthetic ligands such as bovine serum albumin to which mannose-6-phosphate had been coupled on the lysine groups (either as *p*-amino-phenyl mannose-6-phosphate or as a pentamannoside in which the ter-minal mannose is phosphorylated) also inhibited the uptake of α-L-idu-ronidase but no more so than an equivalent concentration of the coupled mannose-6-phosphate (Sando, 1978; Karson *et al.*, 1980). The inhibitory power of model proteins with 30 mannose-6-phosphate prosthetic groups was two orders of magnitude less than that of the urinary glycoprotein fraction, indicating that mannose-6-phosphate alone did not suffice for recognition.

Although chemical proof is still incomplete (Jourdian et al., 1978), the indirect evidence for mannose-6-phosphate as the recognition determinant for hydrolytic enzymes is very strong. One might ask why, if phosphate is present in mammalian glycoproteins, it had not been previously detected. [Phosphorylated mannose has been reported in a partially purified brain glycoprotein fraction; however, the phosphate ester was assigned to a secondary alcohol by NMR, and the proposed structure is therefore different from that suggested for the fibroblast recognition signal (Davis et al., 1976a,b)]. At least part of the reason must lie in the ubiquity of contaminating phosphatases, which would act on phosphorylated glycoproteins during purification unless inhibited by the presence of inorganic phosphate. Another factor may be the transient nature of the high uptake enzyme and its restricted location as discussed below.

4.3. Function

The presence of surface receptors for lysosomal enzymes appeared paradoxical, as did the presence of these enzymes extracellularly. It was suggested by Hickman and Neufeld (1972) that the transport of hydrolytic enzymes from the site of synthesis to their eventual location within lysosomes was by way of secretion and recapture, the latter being highly efficient because of the recognition mechanism. Such a pathway would explain the presence of corrective factors (i.e., high uptake enzymes) in the medium surrounding normal cells. However, the hypothesis was developed primarily to explain the pleiotropic effect of I-cell disease. In this very rare condition (McKusick, 1972), a group of hydrolytic enzymes is deficient intracellularly and present extracellularly; the extracellular enzymes, in contrast to their normal counterparts, are of the low uptake form.

However, various attempts to test the secretion–recapture hypothesis have led to the conclusion that the pathway is not obligatory and occurs to a different extent for different enzymes. For example, cocultivation of cells deficient in β-hexosaminidase with normal cells resulted in a loss of enzyme from the normal cells equivalent to the gain in the deficient ones, thereby indicating that the enzyme pool was shared by all cells in the dish (Reuser et al., 1976; Halley et al., 1978). However, no such intercellular transfer occurred in the case of β-galactosidase, even though that enzyme, like β-hexosaminidase, is affected in I-cell disease and would be expected to have the same recognition and packaging mechanism. Growing cells in the presence of antibody to α-L-iduronidase

(Neufeld *et al.*, 1977) or mannose-6-phosphate (Vladutiu and Rattazzi, 1979; Sly and Stahl, 1978) to trap enzymes has resulted, at best, in a modest reduction in intracellular enzyme.

Perhaps more important, recently formed concepts in cell biology (e.g., Blobel and Dobberstein, 1975) would make secretion and recapture unnecessary to explain the simultaneous occurrence of enzymes in the low uptake form and their inappropriate localization in I-cell disease. As glycoproteins, lysosomal enzymes must presumably have traveled through the cisternae of the endoplasmic reticulum and Golgi, since that is where glycosylating enzymes are located (see Chapters 2 and 3). However, secretory proteins and plasma membrane proteins also pass through these cisternae. Sorting out of proteins for different locations must occur by signals, and it is reasonable to postulate that the mannose-6-phosphate signal for lysosomes might be recognized intracellularly. The absence of such a signal would direct the proteins to be secreted without subsequent reuptake. A defect in the synthesis of the signal in I-cell disease would result in extracellular accumulation of low uptake enzyme. Thus, the surface receptor and presence of high uptake enzyme in surrounding medium might represent a salvage pathway for enzyme that had missed the intracellular route to lysosomes or escaped from the lysosome after incorporation. These alternatives are testable experimentally.

4.4. Generality

The generality of the mannose-6-phosphate system is not yet known. Cultured fibroblasts from rats and some other mammalian species probably have the same recognition system as human fibroblasts (Frankel *et al.*, 1977). Because the pathological manifestation of I-cell disease (i.e., massive lysosomal storage detected by electron microscopy) is limited to certain types of cells [primarily connective tissue, but also Schwann cells and glomerular epithelial cells, (Tondeur and Neufeld, 1975)], the mannose-6-phosphate recognition system may be limited to these. Furthermore, not all hydrolytic enzymes are involved in fibroblasts of I-cell patients; the level of acid phosphatase and of β-glucosidase is normal intracellularly (Neufeld *et al.*, 1975). However, the need for a signal to direct hydrolytic enzymes from cisternae of the endoplasmic reticulum to lysosomes might represent a general principle, even if the specific signal varies among cells or even among groups of enzymes. Some secretion and recapture may also occur *in vivo* in cells other than fibroblasts, as shown by intercellular transfer of β-glucuronidase in chimeric mice that are mosaic with respect to β-glucuronidase content (Feder, 1976; Herrup *et al.*, 1976).

4.5. Binding Protein

The evidence for binding protein is, at the time of this writing, entirely kinetic and based on the finding that the rate of pinocytosis of high uptake enzymes follows Michaelis–Menten saturation kinetics. The concentration of enzyme giving half-maximal uptake is between 10^{-8} and 10^{-9} M; these estimates may be somewhat high, since the preparations tested may have been contaminated with variable amounts of low uptake enzyme (Sando and Neufeld, 1977; Kaplan et al., 1977a). The kinetic model assumes that the receptor concentration remains constant during uptake experiments; this implies rapid resynthesis, or alternatively, a nondestructive shuttle of the binding protein between plasma membrane and lysosomes, as was postulated for the galactose-binding protein of hepatocytes.

Direct measurement of receptor binding and isolation of receptor has been delayed by difficulties in obtaining sufficiently pure high uptake enzyme for radiolabeling. However, direct binding of α-L-iduronidase to fibroblasts has recently been demonstrated by using a very sensitive assay for bound enzyme (Rome et al., 1979). The dissociation constant, K_d, and the K_i for mannose-6-phosphate were found to be 1×10^{-9} M and 1×10^{-4} M, respectively.

5. MANNOSE/N-ACETYLGLUCOSAMINE RECOGNITION SYSTEM OF RETICULOENDOTHELIAL CELLS

5.1. Clearance of Injected Lysosomal Enzymes

The discovery of lysosomal storage diseases by van Hoof and Hers (1964) was accompanied by optimistic predictions that these disorders could be readily treated by enzyme replacement; this sanguine outlook was later supported by the ease of correcting defective fibroblasts in culture. However, after some disappointing clinical experiments, the need for careful animal studies became apparent.

Initial experiments showed that β-glucuronidase from human placenta or from rat liver and preputial gland as well as several other glycosidases were rapidly cleared from rat plasma and appeared primarily in liver lysosomes (Achord et al., 1977a; Stahl et al., 1976a,c; Schlesinger et al., 1976). Spleen and bone were also recipients of the infused enzyme, the latter if the animal had been eviscerated. The rapid clearance was abolished if the enzymes were pretreated with 0.01–0.02 M sodium periodate. Again, carbohydrates were implicated in the uptake of lysosomal enzymes, but since the infused enzymes were all of the low uptake form with respect to cultured fibroblasts, a different recognition system was postulated. That system had to differ also from the hepatocyte recognition of β-galactosyl termini since asialoglycoproteins did not compete with the hydrolases.

5.2. Recognition of Two Carbohydrate Termini

Agalactoorosomucoid, with terminal N-acetylglucosamine residues, was a potent inhibitor of clearance of rat β-glucuronidase, β-galactosidase, and β-hexosaminidase (Stahl *et al.*, 1976*b*). As shown by Achord *et al.* (1977*b*) the clearance of human β-glucuronidase was likewise inhibited by agalactoorosomucoid but, surprisingly, also by yeast mannan. The apparent recognition of two different sugar residues by one pinocytosis system was confirmed by the cross-inhibition of yeast mannan and agalactoorosomucoid. This novel recognition system was also shown to be present on sinusoidal cells, rather than on hepatocytes (Achord *et al.*, 1978; Schlesinger *et al.*, 1978).

Direct demonstration of the mannose/N-acetylglucosamine recognition system on reticuloendothelial cells came from use of isolated rat alveolar macrophages and synthetic conjugates of bovine serum albumin with the thio analogues of the two sugars (Stahl *et al.*, 1978). These neoglycoproteins, as well as the thioglucose analogue, and the enzymes β-glucuronidase and ribonuclease B, were readily taken up by the macrophages with saturation kinetics that indicated receptor binding. The uptake of the galactose analogue was, by contrast, very slight at similar concentrations. Uptake of all the above macromolecules was inhibited by the presence of yeast mannan, and conversely the uptake of [^{125}I]glucose-BSA was inhibited by ovalbumin, ribonuclease B, and agalactoproteins.

5.3. Isolation of a Binding Protein with Dual Specificity

Although cross-inhibition experiments with whole cells strongly suggested the existence of a single receptor recognizing terminal α-mannose or β-N-acetylglucosamine residues, as well as the unnatural β-glucose analogue, they could also be explained by separate but interacting receptors for each sugar. This ambiguity has been resolved by the isolation of a liver membrane protein which recognizes both sugars (Kawasaki *et al.*, 1978). This protein was purified on a mannan–Sepharose 4B affinity column and assayed by its ability to bind to mannan. Inhibitors of binding include, in order of increasing effectiveness, N-acetylglucosamine and mannose, mannose oligosaccharides, ovalbumin, ahexosaminoorosomucoid, agalactoorosomucoid, α-mannosidase, and β-glucuronidase. This binding protein is present in rabbit liver to the extent of 1 mg/150 g. It appears to be a good candidate for the reticuloendothelial receptor, although proof would require demonstration that it originated from the plasma membranes of Kupffer or endothelial cells.

Clearance in rats of glycoproteins with terminal N-acetylglucosamine residues (Stockert et al., 1976) or with terminal mannose residues (Stockert et al., 1976; Winkelhake and Nicolson, 1976; Baynes and Wold, 1976) had been reported previously; it is likely that these studies also dealt with the mannose/N-acetylglucosamine recognition system. There is at present no evidence for mammalian receptor systems specific for mannose, nor, in contrast to the situation in birds (see above) for N-acetylglucosamine.

5.4. Function

Although the mannose/N-acetylglucosamine system was discovered in the course of studying clearance of tissue-derived lysosomal enzymes, this has not been established as its usual, or major function. Presumably any glycoprotein with the appropriate structure could be cleared by this system, including some released in inflammatory processes or derived from infectious microorganisms.

6. FUCOSE RECOGNITION SYSTEM OF MAMMALIAN HEPATOCYTES

Evidence that mammalian hepatocytes contain a receptor that binds glycoproteins specifically through fucose linked $\alpha 1,3$ to N-acetylglucosamine has appeared recently (Prieels et al., 1978). This observation was made possible by comparison of the serum survival time of two closely related glycoproteins: lactoferrin from human milk and transferrin from human serum. Both of these proteins contain two iron-binding sites, an homologous amino acid sequence, and two biantennary oligosaccharide chains in which the core structures are identical. However, in addition to the three terminal sugars of transferrin (sialic acid, galactose, and N-acetylglucosamine), lactoferrin contains additional fucose residues in $\alpha 1,3$ linkage with the N-acetylglucosamine adjacent to galactose; those chains containing fucose are devoid of sialic acid (see also Chapter 1).

[125I]Lactoferrin injected intravenously into mice and rats was rapidly cleared from the circulation; less than 10% of the injected radioactivity remained in the serum 10 min after injection and 85% was recovered in the liver. In order to determine the cellular distribution of [125I]lactoferrin, the liver was perfused with collagenase shortly after injection and the hepatocytes separated from the sinusoidal cells (Kupffer cells plus endothelial cells). At 25 min after the injection, 98% of the radioactivity was recovered in the isolated hepatocytes; 1% was found in the sinusoidal cells.

Evidence was obtained to indicate that the mechanism of hepatic uptake involved the recognition of one or more of the carbohydrate residues on the oligosaccharide chain. Thus, partial destruction of the carbohydrate moiety by exposure to periodic acid, or extensive hydrolysis with mixed glycosidases, markedly reduced the rate of clearance from the serum. More specifically, glycopeptides isolated from lactoferrin inhibited uptake of the intact molecule; the inhibitory potency of these glycopeptides was significantly diminished by periodic acid treatment. Additional inhibition studies revealed that, whereas a variety of macromolecules terminating in galactose, mannose, or N-acetylglucosamine residues were without effect, fucoidin, which contains some $\alpha 1,3$-linked fucose, significantly prolonged the clearance of injected [^{125}I]lactoferrin.

Since an unusual structural feature of the oligosaccharide groups of lactoferrin is fucose in $\alpha 1,3$ linkage to N-acetylglucosamine, the potential role of this sugar in mediating uptake was examined. The availability of the close homologue of lactoferrin, serum transferrin and its desialylated derivative, neither of which are cleared from the circulation rapidly, provided a unique opportunity to test the specificity of this sugar. Asialotransferrin was fucosylated with GDP-[^{14}C]fucose and GDP-fucose-$\alpha 1,3$-N-acetylglucosaminylfucosyltransferase. Upon injection into mice, the fucosylated preparation was cleared much more rapidly than asialotransferrin and, of critical importance, its rate of clearance was inhibited by added lactotransferrin. That the fucosylated asialotransferrin was bound by a receptor other than the galactose binding protein of mammalian hepatocytes was shown by the failure of this derivative to bind to the purified galactose receptor. At the present writing, insufficient data are available to characterize the nature of the fucose receptor since the requisite binding studies *in vitro* have not, as yet, been reported.

7. CONCLUDING REMARKS

Five systems are now known for receptor-mediated pinocytosis of glycoproteins. Each system depends on the recognition of a key terminal carbohydrate residue and is specific to certain cells: galactose (mammalian hepatocytes), L-fucose (mammalian hepatocytes), N-acetylglucosamine (avian hepatocytes), N-acetylglucosamine or mannose (mammalian reticuloendothelial cells), and mannose-6-phosphate (human fibroblasts). The function of these systems is usually viewed in the perspective in which they were discovered. Thus the fibroblast system is believed to be necessary for introducing hydrolytic enzymes into lysosomes whereas the mammalian and avian hepatocyte systems are thought to function in the clearance and catabolism of circulating glycoproteins. However, it is possible that all the systems have multiple functions, related to glycoproteins of both endogenous and exogenous origin. It is also possible that these systems are involved in intercellular recognition (Weigel et al., 1978). An interesting example of a specialized adaptation is the capture by hepatocytes of a galactose-terminated cobalamin binding protein secreted by phagocytosing granulocytes (Burger et al., 1975). The protein binds cobalamin and its bacterial analogues; the rapid transfer of these compounds from a site of infection to the liver may restrict the growth of microorganisms as well as protect mammalian cells from the toxic effects of the analogues (Kohlhouse and Allen, 1977).

In addition to the receptor systems described in this chapter, several other carbohydrate binding proteins of animals have been reported. In beef heart and lung, a lectin has been isolated which appears to be specific for β-galactosides (DeWaard et al., 1976). This preparation is similar to, and possibly identical with, binding proteins described in chick embryo muscle (Den and Malinzak, 1977) and the electric organ of the eel (Teichberg et al., 1975). Information is not currently available as to function or possible role of these proteins in pinocytosis.

The five systems show great but not absolute specificity. The mammalian hepatocyte system recognizes galactose, N-acetylgalactosamine, and glucose, whereas the reticuloendothelial system recognizes mannose, N-acetylglucosamine, and glucose. Thus glucose-terminated glycoproteins can be cleared by hepatocytes and by sinusoidal liver cells, and it is perhaps not surprising that such proteins are not present in mammalian plasma.

The fibroblast mannose-6-phosphate system, the reticuloendothelial N-acetylglucosamine/mannose system, and the hepatocyte galactose system are all inhibited by high (0.1 M) concentrations of L-fucose. This is puzzling, since the steric conformation of L-fucose differs significantly from that of the binding sugars.

The specificity of the binding proteins in different types of cells presents a challenge for therapeutic use of pinocytosis. The hope for replacement of missing enzymes in genetic lysosomal storage diseases was based, at least in part, on observations that yeast invertase and horseradish peroxidase were readily taken up by rat tissues (De Duve and Wattiaux, 1966). These observations were made over a decade before the concept of receptor-mediated pinocytosis and were interpreted to mean that any foreign substance could be introduced nonspecifically into lysosomes. It is now clear that the two enzymes are mannose-rich glycoproteins which are recognized by the reticuloendothelial N-acetylglu-cosamine/mannose system (Stahl et al., 1978; Rodman et al., 1978). Not surprisingly, the most promising results for replacing lysosomal enzymes (rich in mannose and N-acetylglucosamine) have been obtained for Fabry and Gaucher diseases, where storage of undegraded metabolites is in reticuloendothelial cells (Brady, 1978). Targeting of enzymes elsewhere may require removal or blocking of the N-acetylglucosamine/mannose signal as well as attachment of sugars recognized by the binding proteins on the surface of the desired recipient cells. The potential usefulness of directing otherwise inert proteins to specific cells has been demonstrated by the covalent linkage of the appropriate carbohydrate signal to serum albumin and lysozyme (Rogers and Kornfeld, 1971) as well as to RNase (Wilson, 1978).

8. REFERENCES

Achord, D., Brot, F., Gonzales-Noriega, A., Sly, W., and Stahl, P., 1977a, Human β-glucuronidase. II. Fate of infused human placental β-glucuronidase in the rat, Pediatr. Res. 11:816.

Achord, D. T., Brot, F. E., and Sly, W. S., 1977b, Inhibition of the rat clearance system for agalacto-orosomucoid by yeast mannans and by mannose, Biochem. Biophys. Res. Commun. 77:410.

Achord, D. T., Brot, F. E., Bell, C. E., and Sly, W. S., 1978, Human β-glucuronidase: In vivo clearance and in vitro uptake by a glycoprotein recognition system on reticuloendothelial cells, Cell 15:269.

Ashwell, G., 1977, A functional role for lectins, Trends Biol. Sci. 2:N-186.

Ashwell, G., and Morell, H. G., 1974, The role of surface carbohydrates in the hepatic recognition and transport of circulating glycoproteins, Adv. Enzymol. 41:99.

Bach, G., Friedman, R., Weissmann, B., and Neufeld, E. F., 1972, The defect in the Hurler and Scheie syndromes—Deficiency of α-L-iduronidase, Proc. Natl. Acad. Sci. U.S.A. 69:2048.

Bach, G., Eisenberg, F., Jr., Cantz, M., and Neufeld, E. F., 1973, The defect in the Hunter syndrome: Deficiency of sulfoiduronate sulfatase, Proc. Natl. Acad. Sci. U.S.A. 70:2134.

Barton, R. W., and Neufeld, E. F., 1971, The Hurler corrective factor—Purification and properties, J. Biol. Chem. 246:7773.

Baynes, J. W., and Wold, F., 1976, Effect of glycosylation in the in vivo circulating half-life of ribonuclease, J. Biol. Chem. 251:6016.

Blobel, G., and Dobberstein, B., 1975, Transfer of proteins across membranes, J. Cell Biol. 67:852.

Brady, R. O., 1978, Sphinglipidoses, Annu. Rev. Biochem. 47:687.

Brot, F. E., Glaser, J. H., Roozen, K. J., Sly, W. S., and Stahl, P. D., 1974, In vitro correction of deficient human fibroblasts by β-glucuronidase from different human sources, Biochem. Biophys. Res. Commun. 57:1.

Burger, R. L., Schreider, R. J., Mehlman, C. S., and Allen, R. H., 1975, Human plasma R-type vitamin B_{12}-binding proteins, J. Biol. Chem. 250:7707.

Davis, L. G., Javaid, J. J., and Brunngraber, E. G., 1976a, Identification of phosphoglycoproteins obtained from rat brain, FEBS Lett. 65:30.

Davis, L. G., Costello, A. J. R., Javaid, J. J., and Brunngraber, E. G., 1976b, ^{31}P-Nuclear magnetic resonance studies on the phosphoglycopeptides obtained from rat brain glycoprotein, *FEBS Lett.* **65**:35.

De Duve, C., and Wattiaux, R., 1966, Function of lysosomes, *Annu. Rev. Physiol.* **28**:435.

Den, H., and Malinzak, D. A., 1977, The isolation and properties of a β-D-galactoside-specific lectin from chick embryo thigh muscle, *J. Biol. Chem.* **252**:5444.

DeWaard, A., Hickman, S., and Kornfeld, S., 1976, Isolation and properties of β-galactoside binding lectins of calf heart and lung, *J. Biol. Chem.* **251**:7581.

Feder, N., 1976, Solitary cell and enzyme exchange in tetraparental mice, *Nature* **263**:67.

Frankel, H. A., Glaser, J. H., and Sly, W. S., 1977, Human β-glucuronidase. I. Recognition and uptake by animal fibroblasts suggest animal models for enzyme replacement studies. *Pediatr. Res.* **11**:811.

Glaser, J. H., Roozen, K. H., Brot, F. E., and Sly, W. S., 1975, Multiple isoelectric and recognition forms of human β-glucuronidase activity, *Arch. Biochem. Biophys.* **166**:536.

Gregoriadis, G., Morell, A. G., Sternlieb, I., and Scheinberg, I. H., 1970, Catabolism of desialylated ceruloplasmin in the liver, *J. Biol. Chem.* **245**:5833.

Hall, C. W., Cantz, M., and Neufeld, E. F., 1973, β-Glucuronidase deficiency mucopolysaccharidosis: Studies in cultured fibroblasts, *Arch. Biochem. Biophys.* **155**:32.

Halley, D. J. J., deWit-Verbeeck, H. A., Reuser, A. J. J., and Galjaard, J., 1978, The distribution of hydrolytic enzyme activities in human fibroblast cultures and their intercellular transfer, *Biochem. Biophys. Res. Commun.* **82**:1176.

Herrup, K., Mullen, R. J., and Feder, N., 1976, Histochemical evidence for intercellular exchange of β-glucuronidase in chimeric mice, *Fed. Proc.* **35**:1371.

Hickman, J., Ashwell, G., Morell, A. G., Van Den Hamer, C. J. A., and Scheinberg, I. M., 1970, Physical and chemical studies on ceruloplasmin. VIII. Preparation of N-acetylneuraminic acid-1-^{14}C-labeled ceruloplasmin, *J. Biol. Chem.* **245**:759.

Hickman, S., and Neufeld, E. F., 1972, A hypothesis for I-cell disease: Defective hydrolases that do not enter lysosomes, *Biochem. Biophys. Res. Commun.* **49**:992.

Hickman, S., Shapiro, L. J., and Neufeld, E. F., 1974, A recognition marker required for uptake of a lysosomal enzyme by cultured fibroblasts, *Biochem. Biophys. Res. Commun.* **57**:55.

Hieber, V., Distler, J., Myerowitz, R., Schmickel, R. D., and Jourdian, G. W., 1976, The role of glycosidically bound mannose in the assimilation of β-galactosidase by generalized gangliosidosis fibroblasts, *Biochem. Biophys. Res. Commun.* **73**:710.

Hudgin, R. L., and Ashwell, G., 1974, Studies on the role of glycosyltransferases in the hepatic binding of asialoglycoproteins, *J. Biol. Chem.* **249**:7369.

Hudgin, R. L., Pricer, W. E., Jr., Ashwell, G., Stockert, R. J., and Morell, A. G., 1974, The isolation and properties of a rabbit liver binding protein specific for asialoglycoproteins, *J. Biol. Chem.* **249**:5536.

Jourdian, G. W., Distler, J., Hieber, V., and Schmickel, R. D., 1978, personal communication.

Kaplan, A., Achord, D. T., and Sly, W. S., 1977a, Phosphohexosyl components of a lysosomal enzyme are recognized by pinocytosis receptors on human fibroblasts, *Proc. Natl. Acad. Sci. U.S.A.* **74**:2026.

Kaplan, A., Fischer, D., Achord, D. T., and Sly, W. S., 1977b, Phosphohexoxyl recognition: A general characteristic of pinocytosis of lysososmal glycosidases by human fibroblasts, *J. Clin. Invest.* **60**:1088.

Kaplan, A., Fischer, D., and Sly, W. S., 1978, Correlation of structural features of phosphomannans with their ability to inhibit pinocytosis of human β-glucuronidase by human fibroblasts, *J. Biol. Chem.* **253**:647.

Karson, E. M., Sando, G. N., and Neufeld, E. F., 1980, manuscript in preparation.

Kawasaki, T., and Ashwell, G., 1976a, Chemical and physical properties of an hepatic membrane protein that specifically binds asialoglycoproteins, J. Biol. Chem. 251:1296.

Kawasaki, T., and Ashwell, G., 1976b, Carbohydrate structure of glycopeptides isolated from an hepatic membrane binding protein specific for asialoglycoproteins, J. Biol. Chem. 251:5292.

Kawasaki, T., and Ashwell, G., 1977, Isolation and characterization of an avian hepatic binding protein specific for N-acetyl-glucosamine-terminated glycoproteins, J. Biol. Chem. 252:6536.

Kawasaki, T., Etoh, R., and Yamashina, I., 1978, Isolation and characterization of a mannan-binding protein from rat liver, Biochem. Biophys. Res. Commun. 81:1018.

Kolhouse, J. F., and Allen, R. H., 1977, Absorption, plasma transport and cellular retention of cobalamin analogues in the rabbit, J. Clin. Invest. 60:1381.

Lunney, J., 1976, Studies on the regulation of serum glycoprotein homeostasis, Ph.D. dissertation, The Johns Hopkins University, Baltimore.

Lunney, J., and Ashwell, G., 1976, A hepatic receptor of avian origin capable of binding specifically modified glycoproteins, Proc. Natl. Acad. Sci. U.S.A. 73:341.

Marshall, J. S., Green, A. M., Pensky, J., Williams, S., Zinn, A., and Carlson, D. M., 1974, Measurement of circulating desialylated glycoproteins and correlation with hepatocellular damage, J. Clin. Invest. 54:555.

Matalon, R., and Dorfman, A., 1972, Hurler syndrome, an α-L-iduronidase deficiency, Biochem. Biophys. Res. Commun. 47:959.

McKusick, V. A., 1972, Heritable Disorders of Connective Tissue, C. V. Mosby, New York.

Morell, A. G., Irvine, R. A., Sternlieb, I., Scheinberg, I. M., and Ashwell, G., 1968, Physical and chemical studies on ceruloplasmin. V. Metabolic studies on sialic acid-free ceruloplasmin in vivo, J. Biol. Chem. 243:155.

Morell, A. G., Gregoriadis, G., Scheinberg, I. M., Hickman, J., and Ashwell, G., 1971, The role of sialic acid in determining the survival of glycoproteins in the circulation, J. Biol. Chem. 246:1461.

Neufeld, E. F., and Cantz, M. J., 1971, Corrective factors for inborn errors of mucopolysaccharide metabolism, Ann. N.Y. Acad. Sci. 179:580.

Neufeld, E. F., Lim, T. W., and Shapiro, L. J., 1975, Inherited disorders of lysosomal metabolism, Annu. Rev. Biochem. 44:357.

Neufeld, E. F., Sando, G. N., Garvin, A. J., and Rome, L. H., 1977, The transport of lysosomal enzymes, J. Supramol. Struct. 6:95.

Nicol, D. M., Langunoff, D., and Pritzl, P., 1974, Differential uptake of human β-glucuronidase isoenzymes from spleen by deficient fibroblasts, Biochem. Biophys. Res. Commun. 59:941.

Novogrodsky, A., and Ashwell, G., 1977, Lymphocyte mitogenesis induced by a mammalian liver protein that specifically binds desialylated glycoproteins, Proc. Natl. Acad. Sci. U.S.A. 74:676.

Paulson, J. C., Hill, R. L., Tanabe, T., and Ashwell, G., 1977, Reactivation of asialo-rabbit binding protein by resialylation with β-D-galactoside $\alpha 2 \to 6$ sialyltransferase, J. Biol. Chem. 252:8624.

Pricer, W. E., Jr., and Ashwell, G., 1971, The binding of desialylated glycoproteins by plasma membranes of rat liver, J. Biol. Chem. 246:4825.

Pricer, W. E., Jr., and Ashwell, G., 1976, Subcellular distribution of a mammalian hepatic binding protein specific for asialoglycoproteins, J. Biol. Chem. 251:7539.

Prieels, J. P., Pizzo, S. V., Glascow, L. R., Paulson, J. C., and Hill, R. L., 1978, Hepatic receptor that specifically binds oligosaccharides containing fucosyl $\alpha 1 \to 3$ N-acetyl-glucosamine linkages, Proc. Natl. Acad. Sci. U.S.A. 75:2215.

Regoeczi, E., Matton, M. W. C., and Woun, K. L., 1974, Studies of the metabolism of asialotransferrins: Potentiation of the catabolism of known asialotransferrin in the rabbit, *Can. J. Biochem.* **52**:155.

Reuser, A., Halley, D., deWit, E., Hoogeveen, A., van der Kamp, M., Mulder, M., and Galjaard, J., 1976, Intercellular exchange of lysosomal enzymes: Enzyme assays in single human fibroblasts after co-cultivation, *Biochem. Biophys. Res. Commun.* **69**:311.

Rodman, J. S., Schlesinger, P., and Stahl, P., 1978, Rat plasma clearance of horseradish peroxidase and yeast invertase is mediated by specific recognition, *FEBS Lett.* **85**:345.

Rogers, J. C., and Kornfeld, S., 1971, Hepatic uptake of proteins coupled to fetuin glycopeptides, *Biochem. Biophys. Res. Commun.* **45**:622.

Rome, L. H., Weissmann, B., and Neufeld, E. F., 1979, Direct demonstration of binding of a lysosomal enzyme, α-L-iduronidase, to receptors on cultured fibroblasts, *Proc. Natl. Acad. Sci. U.S.A.* **76**:2331.

Roseman, S., 1970, The synthesis of complex carbohydrates by multiglycosyltransferase systems and their potential function in intercellular adhesion. *Chem. Phys. Lipids* **5**:270.

Sando, G. N., 1978, Synthetic inhibitors of receptor-mediated endocytosis of a lysosomal enzyme by cultured human fibroblasts, *Fed. Proc.* **37**:1502.

Sando, G. N., and Neufeld, E. F., 1977, Recognition and receptor-mediated uptake of a lysosomal enzyme, α-L-iduronidase, by cultured fibroblasts, *Cell* **12**:619.

Sando, G. N., Karson, E., and Neufeld, E. F., 1978, unpublished results.

Sarkar, M., Liao, J., Kabat, E. A., Tanabe, T., and Ashwell, G., 1979, The binding site of rabbit hepatic lectin, *J. Biol. Chem.* **254**:3170.

Schlesinger, P., Rodman, J. S., Frey, M., Lang, S., and Stahl, P., 1976, Clearance of lysosomal hydrolases following intravenous infusion. The role of liver in the clearance of β-glucuronidase and N-acetyl-β-D-glucosiminidase, *Arch. Biochem. Biophys.* **177**:606.

Schlesinger, P. H., Doebber, T. W., Mandell, B. F., White, R., deSchriyver, C., Rodman, J. S., Miller, M. J., and Stahl, P., 1978, Plasma clearance of glycoprotein with terminal mannose and N-acetylglucosamine by liver non-parenchymal cells, *Biochem. J.* **176**:103.

Shapiro, L. J., Hall, C. W., Leder, I. G., and Neufeld, E. F., 1976, The relationship of α-L-iduronidase and Hurler corrective factor, *Arch. Biochem. Biophys.* **172**:156.

Sjöberg, I., Fransson, L. Å., Matalon, R., and Dorfman, A., 1975, Hunter's syndrome, a deficiency of L-iduronosulfate sulfatase, *Biochem. Biophys. Res. Commun.* **54**:1125.

Sly, W. S., and Stahl, P., 1979, Receptor-mediated uptake of lysosomal enzymes, in: *Transport of Molecules in Cellular Systems* (S. Silverstein, ed.), pp. 229–244, Dahlem Conferenzen, Berlin.

Sly, W. S., Quinton, B. A., McAlister, W. H., and Rimoin, D. L., 1973, β-Glucuronidase deficiency: Report of clinical, radiologic and biochemical features of a new mucopolysaccharidosis, *J. Pediatr.* **82**:249.

Stahl, P., Rodman, J. S., and Schlesinger, P., 1976a, Clearance of lysosomal enzymes following intravenous infusion. Kinetic and competition experiments with β-glucuronidase and N-acetyl-β-D-glucosaminidase, *Arch. Biochem. Biophys.* **177**:594.

Stahl, P., Schlesinger, P. H., Rodman, J. S., and Doebber, T., 1976b, Recognition of lysosomal glycosidases *in vivo* inhibited by modified glycoproteins, *Nature* **264**:86.

Stahl, P., Six, H., Rodman, J. S., Schlesinger, P., Tulsiani, D. R. P., and Touster, O., 1976c, Evidence for specific recognition sites mediating clearance of lysosomal enzymes *in vivo*, *Proc. Natl. Acad. Sci. U.S.A.* **73**:4045.

Stahl, P. D., Rodman, J. S., Miller, M. J., and Schlesinger, P. H., 1978, Evidence for receptor-mediated binding of glycoproteins, glycoconjugates, and lysosomal glycosidases by alvaolar macrophages, *Proc. Natl. Acad. Sci. U.S.A.* **75**:1399.

Stockert, R. J., Morell, A. G., and Scheinberg, I. H., 1974, Mammalian hepatic lectin, *Science* **186**:365.

Stockert, R. J., Morell, A. G., and Scheinberg, I. H., 1976, The existence of a second route for the transfer of certain glycoproteins from the circulation into the liver, *Biochem. Biophys. Res. Commun.* **68**:988.

Stockert, R. J., Morell, A. G., and Scheinberg, I. H., 1977, Hepatic binding protein: The protective role of its sialic acid residues, *Science* **197**:66.

Stowell, C. P., and Lee, Y. C., 1978, The binding of D-glucosyl-neoglycoproteins to the hepatic asialoglycoprotein receptor, *J. Biol. Chem.* **253**:6107.

Tanabe, T., Pricer, W. E., Jr., and Ashwell, G., 1979, Subcellular membrane topology and turnover of a rat hepatic binding protein specific for asialoglycoproteins, *J. Biol. Chem.* **254**:1038.

Teichberg, V. I., Silman, I., Beitsch, D. D., and Resheff, C., 1975, A β-D-galactoside binding protein from electric organ tissue of *Electrophorus electricus*, *Proc. Natl. Acad. Sci. U.S.A.* **72**:1383.

Tondeur, M., and Neufeld, E. F., 1975, The mucopolysaccharidoses—Biochemistry and ultrastructure, in: *Molecular Pathology* (R. A. Good, S. B. Day, and J. J. Yunis, eds.), pp. 600–621, Charles Thomas, Springfield, Illinois.

Tulkens, P., Schneider, E. J., and Trouet, A., 1977, The fate of the plasma membrane during endocytosis, *Biochem. Soc. Trans.* **5**:1809.

Ullrich, R., Mersmann, G., Weber, E., and von Figura, K., 1977, Evidence for lysosomal enzyme recognition by human fibroblasts via a phosphorylated carbohydrate moiety, *Biochem. J.* **170**:643.

Van Den Hamer, C. J. A., Morell, A. G., Scheinberg, I. M., Hickman, J., and Ashwell, G., 1970, Physical and chemical studies on ceruloplasmin. IX. The role of galactosyl residues in the clearance of ceruloplasmin from the circulation, *J. Biol. Chem.* **245**:4397.

van Hoof, F., and Hers, H. G., 1964, L'ultrastructure des cellules hepatiques dans la maladie de Hurler (gargoylisme), *C. R. Acad. Sci. (Paris)* **259**:1281.

Van Lenten, L., and Ashwell, G., 1972, The binding of desialylated glycoproteins by plasma membranes of rat liver: Development of a quantitative inhibition assay, *J. Biol. Chem.* **247**:4633.

Vladutiu, G. D., and Rattazzi, M., 1979, The excretion–reuptake route of β-hexosaminidase in normal and I-cell disease cultured fibroblasts, *J. Clin. Invest.* **63**:595.

Weigel, P. H., Schmell, E., Lee, Y. C., and Roseman, H., 1978, Specific adhesion of rat hepatocytes to β-galactosides linked to polyacrylamide gels, *J. Biol. Chem.* **25**:330.

Wilson, G., 1978, Effect of reductive lactosamination on the hepatic uptake of bovine pancreatic ribonuclease A dimer, *J. Biol. Chem.* **253**:2070.

Winkelhake, J. L., and Nicolson, G. L., 1976, Aglycosyl-antibody; effects of exoglycosidase treatments on autochtonous antibody survival time in the circulation, *J. Biol. Chem.* **251**:1074.

Structure and Metabolism of Connective Tissue Proteoglycans

Lennart Rodén

1. INTRODUCTION

1.1. Structure of Proteoglycans and Their Polysaccharide Components

With the possible exception of hyaluronic acid, the connective tissue polysaccharides are all synthesized by their parent cells as components of proteoglycans. In these substances, a number of polysaccharide chains are covalently linked to a protein core; e.g., in the proteoglycan of bovine nasal cartilage, which is the prototype of molecules of this kind, close to 100 chondroitin sulfate chains, with a molecular weight of approximately 20,000, and slightly fewer keratan sulfate chains are linked to a core protein (mol. wt. 200,000) which constitutes 7–8% of the entire molecule. In many respects, the proteoglycans are similar to other protein-bound complex carbohydrates, and the conspicuous polysaccharide component *per se* does not distinguish the proteoglycans from the class of glycoproteins; e.g., there are members of the glycoprotein class, such as the blood group substances, which have a high relative content of carbohydrate consisting of a substantial number of monosaccharide units. Rather, the segregation of the proteoglycans into a separate category is based on a few specific characteristics: (1) each polysaccharide consists of repeating

Lennart Rodén • University of Alabama in Birmingham, Birmingham, Alabama 35294.

disaccharide units in which a hexosamine, D-glucosamine, or D-galacto-samine is always present; (2) all connective tissue polysaccharides except keratan sulfate contain a uronic acid, either D-glucuronic acid or its 5-epimer, L-iduronic acid, or both; (3) ester sulfate groups are present in all members of the group except in hyaluronic acid; in addition, N-sulfate groups are found in heparin and heparan sulfate. Although certain other bipolymers are known to contain ester sulfate, e.g., some epithelial mucins (Horowitz, 1977), these compounds are clearly distinguishable from the connective tissue polysaccharides by the other criteria indicated above. It may also be mentioned that the D-glucuronic-acid-containing repeating disaccharide of chondroitin, N-acetylchondrosine, has recently been identified as a component of thyroglobulin (Spiro, 1977); however, since the disaccharide is present as a single unit, thyroglobulin may not be considered a proteoglycan.

Three types of carbohydrate–protein linkages have been recognized in the connective tissue proteoglycans: (1) an O-glycosidic linkage between D-xylose and the hydroxyl group of serine, (2) an N-glycosylamine linkage between N-acetylglucosamine and the amide group of asparagine, and (3) an O-glycosidic linkage between N-acetylgalactosamine and the hydroxyl groups of threonine or serine. The xylose–serine linkage is unique to the connective tissue proteoglycan group and has not been found in any other mammalian glycoconjugates. The "linkage fragment," O-β-D-xylopyranosyl-L-serine, was first isolated from heparin (Lindahl and Rodén, 1964), and the same linkage type has since been established for chondroitin 4- and 6-sulfate, dermatan sulfate, and heparan sulfate (for reviews, see Rodén, 1970; Lindahl and Rodén, 1972). One of the connective tissue polysaccharides, corneal keratan sulfate (or keratan sulfate I), is linked to protein via the linkage between N-acetylglucosamine and asparagine which is so common among the glycoproteins. Surprisingly, keratan sulfate may also be linked through N-acetylgalactosamine to threonine or serine hydroxyl groups. This third linkage type is seen in skeletal keratan sulfate (keratan sulfate II), which is always part of a chondroitin sulfate proteoglycan.

The detailed structures of the repeating disaccharides of the various proteoglycans are listed in Table 1 together with the structures of the three carbohydrate–protein linkage fragments.

1.2. Mechanisms of Biosynthesis

It is now well established that all protein-bound complex carbohydrate species of mammalian tissues are synthesized by the same basic processes (Schachter and Rodén, 1973; Schachter, 1978)(see also Chapter 3). For glycoproteins and proteoglycans alike, the first step is the syn-

Table 1. Structures of Linkage Units in Connective Tissue Polysaccharides

Linkage unit	Polysaccharide	Repeating disaccharide units[a]	Carbohydrate–protein linkage region[a]	Substituents[a]
1	Hyaluronic acid	GlcUA $\xrightarrow{\beta1,3}$ GlcNAc		
2	Hyaluronic acid	GlcNAc $\xrightarrow{\beta1,4}$ GlcUA		
3	Chondroitin sulfates, dermatan sulfate, heparin, and heparan sulfate		Xyl $\xrightarrow{\beta}$ Ser	
4			Gal $\xrightarrow{\beta1,4}$ Xyl	
5			Gal $\xrightarrow{\beta1,3}$ Gal	
6			GlcUA $\xrightarrow{\beta1,3}$ Gal	
7	Chondroitin sulfates	GalNAc $\xrightarrow{\beta1,4}$ GlcUA		
8	Chondroitin sulfates	GlcUA $\xrightarrow{\beta1,3}$ GalNAc		
9	Chondroitin 4-sulfate			GalNAc-**4-sulfate**
10	Chondroitin 6-sulfate			GalNAc-**6-sulfate**
11	Keratan sulfate I		GlcNAc $\xrightarrow{\beta}$ Asn	
12	Keratan sulfate II		GalNAc $\xrightarrow{\alpha}$ Thr and GalNAc $\xrightarrow{\alpha}$ Ser	
13	Keratan sulfates	Gal $\xrightarrow{\beta1,4}$ GlcNAc		
14	Keratan sulfates	GlcNAc $\xrightarrow{\beta1,3}$ Gal		
15	Keratan sulfates			GlcNAc-**6-sulfate**

(Continued)

Table 1. (Continued)

Linkage unit	Polysaccharide	Repeating disaccharide units[a]	Carbohydrate–protein linkage region[a]	Substituents[a]
16	Keratan sulfates			Gal-**6-sulfate**
17	Heparin, heparan sulfate	GlcNAc $\xrightarrow{\alpha1,4}$ GlcUA		
18	Heparin, heparan sulfate	GlcUA $\xrightarrow{\beta1,4}$ GlcNAc		
19	Heparin, heparan sulfate			-*N*-**acetyl**-glucosaminyl
20	Heparin, heparan sulfate			GlcN[*N*-**sulfate**]
21	Heparin, heparan sulfate	IdUA $\xrightarrow{\alpha1,4}$ GlcNAc -GlcNAc,[6-sulfate] -GlcN[*N*-sulfate]		
22	Heparin, heparan sulfate	-GlcN[*N*-sulfate;6-sulfate] GlcN[*N*-sulfate] $\xrightarrow{\alpha1,4}$ IdUA — (6-sulfate) — (2-sulfate)		
23	Heparin, heparan sulfate, and dermatan sulfate			-IdUA- | **2-sulfate**
24	Heparin, heparan sulfate			-GlcN[*N*-sulfate or *N*-acetyl] | **6-sulfate**

25 Dermatan   GalNAc $\xrightarrow{\beta 1,4}$ IdUA
    sulfate

26 Dermatan   IdUA $\xrightarrow{\alpha 1,3}$ GalNAc
    sulfate

Summary structures[b]:

Hyaluronic acid:

$\xrightarrow{\beta 1,4}$ GlcUA $\xrightarrow{\beta 1,3}$ GlcNAc $\xrightarrow{\beta 1,4}$ GlcUA $\xrightarrow{\beta 1,3}$ GlcNAc $\xrightarrow{\beta 1,4}$

Chondroitin
sulfates:

$\xrightarrow{\beta 1,4}$ GlcUA $\xrightarrow{\beta 1,3}$ GalNAc $\xrightarrow{\beta 1,4}$ GlcUA $\xrightarrow{\beta 1,3}$ Gal $\xrightarrow{\beta 1,3}$ Gal $\xrightarrow{\beta 1,4}$ Xyl $\xrightarrow{\beta}$ Ser
                                   |
                              4- or 6-sulfate

Keratan sulfates:

$\xrightarrow{\beta 1,4}$ GlcNAc $\xrightarrow{\beta 1,3}$ Gal $\xrightarrow{\beta 1,4}$ GlcNAc $\xrightarrow{\beta 1,3}$ Gal
                                                              | $\overset{\text{(GlcNAc·Man)}}{\diagdown}$ GlcNAc $\xrightarrow{\beta}$ Asn (keratan sulfate I)
                                                         6-sulfate $\diagup$
                                                                              1,6
                                                                      Gal $\diagdown$
                                                                       | GalNAc $\xrightarrow{\alpha}$ Thr(Ser) (keratan sulfate II)
                                                                  6-sulfate |
                                                                             Gal-NANA

Heparin and
heparan sulfate:

                        6-sulfate
                          |
$\xrightarrow{\alpha 1,4}$ IdUA $\xrightarrow{\alpha 1,4}$ GlcN $\xrightarrow{\alpha 1,4}$ GlcUA $\xrightarrow{\beta 1,4}$ GlcNAc $\xrightarrow{\alpha 1,4}$ GlcUA $\xrightarrow{\beta 1,3}$ GalNAc $\xrightarrow{\beta 1,4}$ GlcUA $\xrightarrow{\beta 1,3}$ Gal $\xrightarrow{\beta 1,3}$ Gal $\xrightarrow{\beta 1,4}$ Xyl $\xrightarrow{\beta}$ Ser
              | |
        2-sulfate SO_3^- or Ac

Dermatan sulfate:

$\xrightarrow{\beta 1,4}$ IdUA $\xrightarrow{\alpha 1,3}$ GalNAc $\xrightarrow{\beta 1,4}$ GlcUA $\xrightarrow{\beta 1,3}$ GalNAc $\xrightarrow{\beta 1,4}$ GlcUA $\xrightarrow{\beta 1,3}$ Gal $\xrightarrow{\beta 1,3}$ Gal $\xrightarrow{\beta 1,4}$ Xyl $\xrightarrow{\beta}$ Ser
              | |
        2-sulfate   4-sulfate

---

[a] Abbreviations: (GlcUA) D-glucuronic acid, (IdUA) L-iduronic acid, (GlcN) D-glucosamine, (GalN) D-galactosamine, (Ac) N-acetyl, (Gal) D-galactose, (Xyl) D-xylose, (Ser) L-serine, (Thr) L-threonine, (Asn) L-asparagine, (Man) D-mannose, (NANA) N-acetylneuraminic acid. These abbreviations are also used in the text.
[b] The summary structures are qualitative representations only and do not reflect, for example, the uronic acid composition of hybrid polysaccharides such as heparin and dermatan sulfate, which contain both L-iduronic and D-glucuronic acid. Neither should it be assumed that the indicated substituents are always present; e.g., whereas most iduronic acid residues in heparin carry a 2-sulfate group, a much smaller proportion of these residues are sulfated in dermatan sulfate.

thesis of a core protein to which the carbohydrate chains are subsequently attached. For each monosaccharide component of the glycoconjugates, there is a corresponding nucleotide sugar containing this monosaccharide, and the general pattern of synthesis involves the transfer of the glycosyl group to a suitable acceptor in the growing molecule. One exception to this basic pattern is known, inasmuch as the L-iduronic acid residues of heparin, heparan sulfate, and dermatan sulfate are formed by epimerization of D-glucuronic acid residues already incorporated into the growing polysaccharide (Lindahl, 1976; Lindahl et al., 1977)(see also Section 5.2). Even though in this case the interconversion of structure occurs within a polymer, the rule that all sugars of mammalian complex carbohydrates are formed via nucleotide sugar intermediates is still valid. As has been noted by Leloir (1972), even polysaccharides which are not formed directly from nucleotide sugars are, in all known instances, derived from these compounds in an indirect manner; e.g., the levansucrase of *Leuconostoc mesenteroides* requires sucrose as a substrate (Hestrin et al., 1943), but this disaccharide is itself produced by a nucleotide-sugar-dependent reaction in other organisms. Ultimately, the various monosaccharide components of the mammalian glycoconjugates may be traced back to glucose; the pathways of interconversions have been reviewed elsewhere (see, for example, Schachter and Rodén, 1973)(see also Chapter 3).

1.2.1. Chain Initiation

Details are not yet known concerning the exact temporal relationship between the formation of the core protein of the proteoglycans and the addition of the carbohydrate groups. It is possible that the initiation of the carbohydrate chains begins even before the polypeptide has been completed on the ribosomes, in analogy with the situation observed for the biosynthesis of ovalbumin (Kiely et al., 1976)(see also Chapter 2). However, critical experiments to explore this aspect of proteoglycan formation have not yet been carried out.

The initiation of the polysaccharide chains of the proteoglycans occurs by either of two distinct processes. For those polysaccharides linked by type 1 linkages between xylose and serine, chain initiation takes place by direct transfer of xylose from UDP-xylose to serine hydroxyl groups in the core protein (see Section 3.3). Similarly, we are assuming that the type 3 linkage is formed in basically the same manner by transfer of *N*-acetylgalactosamine from UDP-*N*-acetylgalactosamine to threonine and serine residues of the core proteins of skeletal proteoglycans. Direct evidence for this process has not been obtained, and the suggested mode of chain initiation is only inferred from knowledge of the characteristics of the analogous reaction, which occurs in the biosynthesis of epithelial mucins and other glycoproteins containing type 3 linkages.

Following the discovery of lipid-bound intermediates in the biosynthesis of bacterial complex carbohydrates (Robbins *et al.*, 1967; Lennarz, 1975), it became necessary to consider the possibility that similar mechanisms might be operative in mammalian tissues as well. The dolichols, discovered by Hemming and his collaborators (Hemming, 1974), were recognized by Leloir and his associates as possible mammalian counterparts to the shorter-chain isoprenoid lipids which participate in microbial polysaccharide biosynthesis (Behrens and Leloir, 1970; Behrens *et al.*, 1971; Schachter, 1978). The hypothesis that dolichol-bound sugars may be involved in mammalian glycoprotein biosynthesis has now been amply verified, and it seems that all glycoproteins containing a core region composed of *N*-acetylglucosamine and mannose are formed via such intermediates (see Chapter 2). Since one of the connective tissue polysaccharides (keratan sulfate I) is linked to protein via type 2 linkages between *N*-acetylglucosamine and asparagine, it seemed plausible that its synthesis would occur via the dolichol pathway, and evidence to this effect has recently been presented by Hart and Lennarz (1978). However, details of this process, as it pertains to keratan sulfate I, have not yet been determined.

1.2.2. Chain Elongation and Termination

Once the formation of the polysaccharide chains has been initiated by one of the two mechanisms indicated above, continued growth occurs by stepwise transfer of the monosaccharide units from the corresponding nucleotide sugars. This process is governed largely by the substrate specificities of the glycosyltransferases involved, and this subject has been reviewed in some detail elsewhere (see Schachter and Rodén, 1973; Rodén and Schwartz, 1975) (see also Chapter 3). Briefly, the "one enzyme–one linkage" concept applies here (Hagopian *et al.*, 1968), which implies that a particular glycosyltransferase is specific with regard to the sugar donor (nucleotide sugar), the acceptor, and the anomeric configuration and position of the linkage formed. Together with a topographical and functional compartmentalization of the synthetic processes, these properties of the enzymes contribute to a high-fidelity reproduction of the structures of the complex carbohydrates.

For many glycoproteins, termination of chain growth is a natural consequence of the specificities of the glycosyltransferases involved. Most clearly, this may be exemplified by the "capping" effect of the sialytransferases. The sialyl residues introduced by these enzymes are not capable of serving as acceptors for any other sugars (except in the case of G_{M2} ganglioside synthesis, where a second sialyl residue may be transferred to an acceptor sialyl group), and termination of chain growth thus ensues. The causes of termination of the growth of polysaccharides with repeating disaccharide units are less clear. The introduction of a sulfate group in position 4 on the N-acetylgalactosamine units of chondroitin sulfate may be advocated as a mechanism to this end, since the addition of a glucuronosyl residue is then prevented (Telser *et al.*, 1966). However, a sulfate group in position 6 does not have the same effect, and the observed phenomenon therefore may not be considered a generally valid mechanism for termination of sulfated polysaccharide chains. It is possible that no specific mechanism exists and that the growth ceases when the polysaccharide chains are propelled beyond the sites of contact with the membrane-bound glycosyltransferases by the cellular secretory apparatus.

1.2.3. Polymer Modification

In the course of formation of the oligosaccharide side chains of the glycoproteins which are linked to protein by type 3 linkages, considerable modification of the initial structures occurs. This "processing," which is effected by various glycosidases, is discussed in Chapters 1 and 2 and is applicable only to keratan sulfate I in the group of substances under discussion here. However, specific studies of this process, as it applies to keratan sulfate I, have not yet been undertaken.

On the other hand, a series of specific modifications must be considered for the connective tissue proteoglycans, which will be described largely in the course of discussing the biosynthesis of the individual members of the group. The introduction of sulfate groups into position 4 or 6 of the N-acetylgalactosamine residues of the chondroitin sulfates and dermatan sulfate may be mentioned as an example. A particularly interesting group of five modifications has been observed in heparin biosynthesis (Lindahl *et al.*, 1977). Among these are the epimerization of D-glucuronosyl to L-iduronosyl residues which takes place after the initial biosynthetic product has been deacetylated and N-sulfated. Details of these processes will be discussed in Section 5.2.

1.3. Catabolic Pathways

In large measure, the normal catabolism of the carbohydrate moieties of the connective tissue proteoglycans occurs by a set of reactions which may be regarded as a reversal of the synthetic processes. A battery of exoenzymes remove, in a stepwise manner, the glycosyl groups as well as substituent sulfate groups, beginning from the nonreducing terminus of the molecule. In addition, endoenzymes of different specificities are also involved, the best known being hyaluronidase, which cleaves internal N-acetylhexosaminidic linkages in hyaluronic acid and the chondroitin sulfates. Much of the knowledge which has accumulated concerning the catabolic processes in recent years has emerged from intensive studies in several laboratories of the genetic defects which, as a class, go under the name of "mucopolysaccharidoses" (for reviews, see Dorfman and Matalon, 1972, 1976; Neufeld *et al.*, 1975; Hall *et al.*, 1978; McKusick *et al.*, 1978; see also Chapter 6). As a result of this work, pathways of degradation may now be formulated for all connective tissue polysaccharides, and the existence of enzymatic activities corresponding to the components of these substances, glycosyl groups as well as others, is now well documented. However, it should be pointed out that whereas the individual enzymes are being purified and characterized, there is still insufficient information concerning the integration of their activities in the living cell. It may also be noted that complete degradation of any one polysaccharide to its components has not yet been accomplished with enzymes of mammalian origin [cf. Dietrich *et al.* (1973) describing degradation of heparin by bacterial enzymes]. Even though the properties of the known catabolic enzymes are such that they would be expected to degrade the connective tissue polysaccharide to monosaccharides, confirmation of this assumption by actual experimentation would be desirable.

A further comment is needed concerning the relative contributions of exo- and endogenzymes to the overall degradative processes. Although extensive fragmentation can be achieved by the latter, they are not alone capable of adequate degradation, since in the mucopolysaccharidoses an accumulation of partially degraded material occurs.

A variety of proteases may be assumed to participate in the degradation of the protein cores of the connective tissue proteoglycans. As for the carbohydrate moieties, the available information does not as yet permit the formulation of a sequence of events which has been verified experimentally. (This aspect of proteoglycan metabolism has been reviewed in Barrett and Dingle, 1971; Barrett, 1972.)

2. HYALURONIC ACID

2.1. Structure of Hyaluronic Acid

Hyaluronic acid is widely distributed among the organs and tissues of the mammalian body and is found in such diverse locations as synovial fluid, the vitreous body of the eye, and loose connective tissue (Meyer *et al.*, 1956; Brimacombe and Webber, 1964; Schubert and Hamerman,

1968; Mathews, 1975). Profound physiological effects are exerted even by low concentrations of this polysaccharide, as evidenced by its essential role in the formation of proteoglycan aggregates in cartilage, where its concentration is no more than about 0.5% of that of the chondroitin sulfate proteoglycans. Hyaluronic acid is also unique in not being limited to animal tissues, inasmuch as it is produced by group A streptococcus and many other bacteria. In the former, the polysaccharide is synthesized by the protoplast membrane (Markovitz and Dorfman, 1962) and may be released by relatively gentle means which would not be likely to cleave covalent linkages (Stoolmiller and Dorfman, 1969a). Although the presence of branch points has occasionally been suggested, chemical as well as electron-microscopic data indicate that the molecule is unbranched and is composed of repeating disaccharide units of the structure indicated in Table 1 and Figures 1 and 2. This structure has been established largely by investigations of degradation products obtained after enzymatic digestions (Weissmann et al., 1954; Hirano and Hoffman, 1962; Linker et al., 1955, 1956; Hoffman et al., 1957b) as well as by partial acid hydrolysis (Rapport et al., 1951; Weissman and Meyer, 1952, 1954; Weissmann et al., 1953). Upon acid hydrolysis, the N-acetylglucosaminidic linkages are preferentially cleaved, and the disaccharide unit 1 (Figure 1 and Table 1), which has an N-acetylglucosamine unit at the reducing terminus, linked in β1,3-linkage to a glucuronic acid residue, is therefore the major product. Digestion with testicular hyaluronidase results in cleavage of the N-acetylhexosaminidic linkages exclusively, and the products comprise a series of oligosaccharides with N-acetylglucosamine at the reducing terminus; the mode of action of this enzyme is such that upon exhaustive digestion, the tetrasaccharide is the major product, closely followed by the hexasacharide and with smaller amounts of higher oligosaccharides; only a small proportion of disaccharide is found in a digest of this type. Corroboration of the structure inferred from the indicated

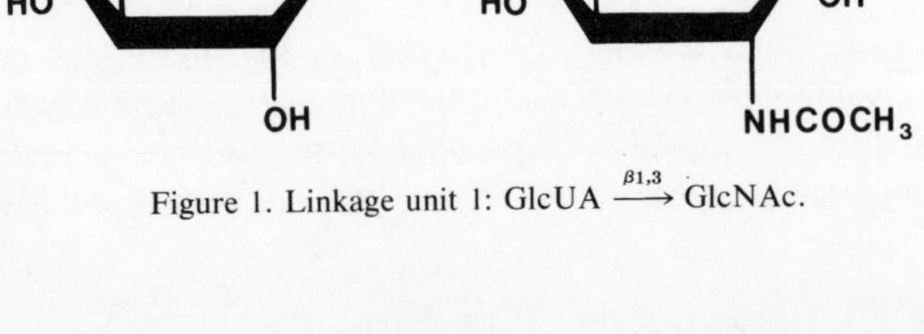

Figure 1. Linkage unit 1: GlcUA $\xrightarrow{\beta 1,3}$ GlcNAc.

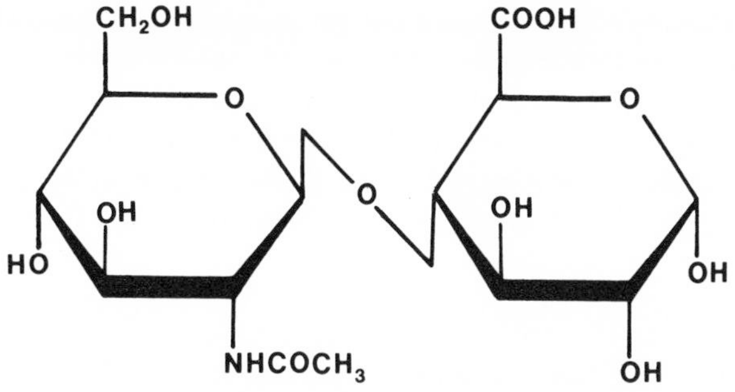

Figure 2. Linkage unit 2: GlcNAc $\xrightarrow{\beta 1,4}$ GlcUA.

chemical and enzymatic approach has come from studies of the degradation products obtained by treatment of hyaluronic acid with leech hyaluronidase. This enzyme is specific for the glucuronidic linkages in hyaluronic acid and does not cleave any other polysaccharide.

It has long been assumed that hyaluronic acid, in analogy with other connective tissue polysaccharides, is covalently linked to protein and exists in the tissues as a proteoglycan. To date, there is no firm evidence that this is indeed the case. Hyaluronic acid from mammalian sources always seems to be associated with a small amount of protein, and some evidence that this protein may be covalently linked to the polysaccharide has been presented by Sandson and Hamerman (1962) and by Hamerman *et al.* (1966). In more recent studies, Scher and Hamerman (1972) isolated hyaluronic acid from synovial fluid by chromatography on ECTEOLA–cellulose and found that the product still contained 0.35% protein. After the protein was labeled with ^{125}I, the radioactivity remained with the hyaluronic acid, even after further fractionation by density-gradient centrifugation in cesium chloride containing 4 M guanidinium chloride. Although this finding is highly indicative of covalent attachment of hyaluronic acid to protein, it does not constitute positive proof, and the isolation of a small oligosaccharide linked to peptide would be more convincing in this respect.

Recent investigations by Toole and collaborators (B. Toole, personal communication) have suggested that hyaluronic acid may initially be synthesized as a proteoglycan, since a portion of newly synthesized hyaluronic acid appears to remain bound to the cell surface, perhaps via covalent linkage to protein. It might then be postulated that release of hyaluronic acid into the surrounding medium would occur by cleavage of linkages in the polysaccharide proper, yielding a product which is devoid of protein. This interesting possibility needs to be explored further.

Evidence from biosynthetic studies relating to this problem is also ambiguous. Whereas formation of dermatan sulfate in cultured fibroblasts is strongly inhibited by puromycin and cycloheximide, only a moderate effect on hyaluronic acid synthesis is observed (Matalon and Dorfman, 1968). This effect cannot be attributed with certainty to inhibition of core protein synthesis, and it seems equally likely that it is a secondary expression of the general decrease in protein synthesis.

2.2. Biosynthesis of Hyaluronic Acid

Some 25 years ago, Glaser and Brown (1955) first achieved enzymatic biosynthesis of hyaluronic acid structures, when they showed that hyaluronic acid oligosaccharides were formed from UDP-glucuronic acid and UDP-N-acetylglucosamine in a cell-free preparation from Rous sarcoma. Subsequently, Markovitz et al. (1959), using a particulate enzyme preparation from group A streptococci, obtained synthesis of macromolecular hyaluronic acid. Since then, several mammalian tissue preparations have been described which catalyze the synthesis of high-molecular-weight hyaluronic acid from the two nucleotide sugars. However, little progress has been made in our understanding of the detailed mechanism of assembly of the hyaluronic acid chain.

2.2.1. Linkage Unit 1

Although our lack of knowledge concerning the mechanism of hyaluronic acid biosynthesis precludes a detailed discussion of the formation of each of the two components of the repeating disaccharide unit, some information pertinent to these individual components has been obtained. Stoolmiller and Dorfman (1969a) observed that the streptococcal system, which synthesizes macromolecular hyaluronic acid (mol. wt. 24,000) in the presence of both nucleotide sugars, is apparently capable of adding single glucuronic acid residues when the particulate enzyme preparation is incubated in the presence of UDP-glucuronic acid alone. Evidence to this effect came from the observation that treatment with β-glucuronidase released a substantial proportion of the label (40%) as free glucuronic acid. Furthermore, treatment with streptococcal hyaluronidase released 25% of the bound radioactivity as N-acetylhyalobiuronic acid. Some of the label was also found in the unsaturated disaccharide, $\Delta4,5$-N-acetylhyalobiuronic acid, indicating that some chain elongation had taken place in addition to the transfer of single uronic acid residues.

2.2.2. Linkage Unit 2

If information concerning the glucuronosyl groups of hyaluronic acid is scanty, the tangible evidence regarding the synthesis of the N-acetylglucosaminyl residues (Figure 2) is even more so. Whereas transfer of single glucuronosyl residues to an undefined endogenous acceptor in the crude streptococcal system has been demonstrated, evidence of the same nature is not available for the N-acetylglucosaminyl residues. It should also be emphasized in this context that the approach taken in investigations of chondroitin sulfate synthesis in Dorfman's laboratory (see Stoolmiller and Dorfman, 1969b; Dorfman, 1974) (see also Section 3.3), which involves the use of well-defined oligosaccharides as acceptors, has not been successful in the study of hyaluronic acid biosynthesis. When hexa- or pentasaccharide from hyaluronic acid was included in incubation mixtures containing streptococcal enzyme and the appropriate nucleotide sugar, no transfer to these oligosaccharides was observed. This is in striking contrast to the ready transfer to chondroitin sulfate oligosaccharides which occurs under similar conditions.

The failure to obtain hyaluronic acid synthesis with small oligosaccharides as primers leads one to the possibility that this polysaccharide may be synthesized via lipid intermediates in much the same fashion as bacterial peptidoglycans and the O-antigens of *Salmonella* and other bacterial polysaccharides. Another possibility would be the formation of the hyaluronic acid repeating disaccharide unit via the dolichol pyrophosphate intermediates which are now known to participate in the formation of the asparagine-N-acetylglucosamine-linked oligosaccharide chains of mammalian glycoproteins. To date, no evidence has been obtained to indicate the existence of a lipid intermediate pathway in hyaluronic acid synthesis. A characteristic feature of such a process is the formation of UMP as a result of transfer of the glycosyl 1-phosphate portion of the nucleotide sugar to the lipid monophosphate intermediate. When this possibility was examined by Ishimoto and Strominger (1967), no significant formation of UMP was observed, whereas production of UDP was easily detected (indicating transfer of a glycosyl group rather than a glycosyl 1-phosphate). Similar results were obtained by Stoolmiller and Dorfman (1969a). Furthermore, bacitracin, which specifically blocks peptidoglycan synthesis by inhibition of dephosphorylation of lipid pyrophosphate, had no effect on hyaluronic acid formation (Siewert and Strominger, 1967; Stoolmiller and Dorfman, 1969a).

In recent studies by Turco and Heath (1977), a disaccharide-lipid, glucuronosyl-N-acetylglucosaminylpyrophosphoryldolichol, was isolated from simian-virus-40-transformed human lung fibroblasts, raising again the possibility that lipid intermediates may be involved in the formation of connective tissue polysaccharides. However, the position of the glucuronidic linkage has been tentatively characterized as 1,4, and such a structure would obviously not be a candidate for participation in the formation of the 1,3-linked units of hyaluronic acid. Rather, the possibility remains that the novel disaccharide-lipid may be an intermediate in heparin formation.

In summary, the biosynthesis of hyaluronic acid by bacterial and animal cells requires the participation of the two nucleotide sugars UDP-glucuronic acid and UDP-N-acetylglucosamine, but it is not yet known what the mechanisms of chain initiation and chain elongation are. The nature of the endogenous acceptors in hyaluronic-acid-synthesizing systems has not been determined. It seems possible that the acceptor might be a protein-bound oligosaccharide and that the chain may be growing on a core protein. However, inhibition of protein synthesis by puromycin and chloramphenicol, which abolishing protein synthesis almost completely (Stoolmiller and Dorfman, 1969a), did not affect the formation of hyaluronic acid.

2.3. Catabolism of Hyaluronic Acid

Theoretically, the two enzymes β-glucuronidase and β-N-acetylglucosaminidase in concerted action are capable of degrading hyaluronic acid to its monosaccharide components. Structurally, this is not true if the residue at the reducing terminus of the polysaccharide is N-acetylglucosamine, since the disaccharide, N-acetylhyalobiuronic acid (GlcUA $\xrightarrow{\beta 1,3}$ GlcNAc), is resistant to digestion with β-glucuronidase. In contrast, the "reverse" disaccharide, GlcNAc $\xrightarrow{\beta 1,4}$ GlcUA, is readily cleaved by β-N-acetylglucosaminidase.

An alternative, or rather complementary, mode of degradation depends on the action of "testicular" hyaluronidase, an endo-β-N-acetylhexosaminidase which is present in many tissues (see, for example, Aronson and Davidson, 1967a,b, 1968; Davidson, 1970; Glaser and Conrad, 1979, and references cited therein). Notably, this enzyme is absent in cultured human skin fibroblasts (Arbogast et al., 1975), which are major hyaluronic-acid-producing cells, but is found in chick embryo skin fibroblasts (Orkin et al., 1977). Upon exhaustive digestion with testicular hyaluronidase, the tetrasaccharide, GlcUA $\xrightarrow{\beta 1,3}$ GlcNAc $\xrightarrow{\beta 1,4}$ GlcUA $\xrightarrow{\beta 1,3}$ GlcNAc, is the major product with small amounts of hexasaccharide and higher oligosaccharides as well as the disaccharide N-acetylhyalobiuronic acid. Complete degradation of the polysaccharide to monosaccharides is apparently not a prerequisite for elimination from the organism, since urine normally contains oligosaccharides and polysaccharide fragments of varying sizes. However, the amounts of bound uronic acid in urine (Kaplan, 1969) are not sufficient to account for the normal metabolic flux of glucuronic acid in man (Neufeld and Fratantoni, 1970). It must therefore be assumed that any oligosaccharides produced by hyaluronidase are degraded further to monosaccharides by the action of the two exoenzymes, β-glucuronidase and β-N-acetylhexosaminidase. In extensive studies of the kinetics of the two enzymes, Weissmann and his collaborators have attempted to define the contribution *in vivo* of the two exoenzymes to the catabolism of hyaluronic acid (Weissman *et al.*, 1975). Based on the observation (Schiller and Dorfman, 1957; Hardingham and Phelps, 1968; Cashman *et al.*, 1969) that in rat skin approximately 0.23 mg hyaluronic acid turns over per day per gram of wet tissue, it was calculated that a minimum rate of N-acetylglucosamine production of 0.05 μmole per hour per unit of enzyme is required to account for the turnover rate solely on the basis of the action of the two exoenzymes. This rate is exceeded at low ionic strength under otherwise physiological conditions. At an ionic strength of 0.18, however, only relatively small hyaluronodextrins are cleaved at a rate sufficient to account for the required monosaccharide production. Since the normal catabolic environment is not likely to be almost salt-free, it must therefore be assumed that the concerted action of β-glucuronidase and β-N-acetylglucosaminidase is not sufficient for physiological hydrolysis of polymeric hyaluronate or larger fragments. Participation of hyaluronidase in the degradative process is therefore likely to occur in all tissues where this enzyme is present.

The levels of the two enzymes in skin and some other tissues are approximately equal as measured with arylglycoside substrates (0.5 unit β-N-acetylglucosaminidase and 0.4 unit β-glucuronidase per gram of moist skin). However, with hyaluronic acid fragments as substrates, the cleavage of the N-acetylglucosamine residues is the rate-limiting step, and the rate of the concerted degradation to monosaccharides decreases significantly only when the ratio of glucuronidase to N-acetylglucosaminidase is substantially reduced.

2.3.1. β-D-Glucuronidase (Linkage Units 1, 6, 8, and 21)

As can be seen from Table 1, several glucuronic-acid-containing disaccharide units are found in the connective tissue polysaccharides, i.e., linkage units 1, 6, 8, and 21. All these glucuronic acid residues may be removed by β-glucuronidase when they are located in nonreducing terminal positions of tetrasaccharides or larger fragments. However, of the disaccharides themselves, only GlcUA $\xrightarrow{\beta 1,3}$ Gal is known with certainty to be susceptible to cleavage by β-glucuronidase, whereas the two disaccharides from hyaluronic acid and the chondroitin sulfates, GlcUA $\xrightarrow{\beta 1,3}$ GlcNAc and GlcUA $\xrightarrow{\beta 1,3}$ GalNAc, are resistant to the enzyme. It is not known whether the disaccharide from heparin and heparan sulfate, GlcUA $\xrightarrow{\beta 1,4}$ GlcNAc (or its N-sulfated analogue), is a substrate for the enzyme.

An extensive literature exists on the biochemistry and physiology of β-glucuronidase (see, for example, Levvy and Conchie, 1966; Wakabayashi, 1970). This enzyme continues to attract attention for several reasons: (1) it possesses interesting genetic features (Ganschow, 1973), including the occurrence of a genetic deficiency in man (Sly *et al.*, 1973); (2) it is found in the microsomal cell fraction as well as in the lysosomes, and a specific protein, egasyn, mediates its binding to the microsomal membranes; and (3) β-glucuronidase exists in high- and low-uptake forms which differ in recognition and uptake by fibroblasts. The latter aspect of the biology of the enzyme is covered in Chapter 6.

It has long been known that the preputial gland of the female rat is extraordinarily rich in β-glucuronidase, with an enzyme content of 5% or more of the total protein. Recently, Himeno *et al.* (1975) described the isolation from this source of pure, crystalline β-glucuronidase, obtained after 20-fold purification (!) in close to 60% yield. The purified enzyme was a tetramer with a molecular weight of 320,000 and contained 5.7% carbohydrate, notably mannose and N-acetylglucosamine. Although the crystalline enzyme showed some charge heterogeneity on isoelectric focusing, it emerged as a single peak on diethylaminoethyl (DEAE)–cellulose chromatography, in some contrast to the behavior of less purified preparations from preputial gland and other sources, which contain multiple forms of the enzyme (see Wakabayashi, 1970). In a recent study of the preputial gland enzyme, Tulsiani *et al.* (1975) have shown that the enzyme may be fractionated into five molecular forms by chromatography on hydroxylapatite. Differences in carbohydrate content explained, in part, the existence of the several forms, and a particularly interesting feature was the presence of glucose in all forms (1.5% in form C and 0.17% in forms A and B) (see also Himeno *et al.*, 1975). Other sugar components in forms A and B were mannose (2.8%), glucosamine (1.9%), fucose (0.2%), and galactose (0.16%). In some of its basic features, bovine liver glucuronidase (Himeno *et al.*, 1974) is similar to the enzyme from rat preputial gland; however, two forms with isoelectric points of 5.1 (major component) and 5.9 were observed, whereas the pI of the latter enzyme is at pH 6.7–6.8.

The lysosomal form (L-form) of β-glucuronidase has been purified by Tomino *et al.* (1975) from mouse liver. The purified L-form is a tetramer with a molecular weight of 280,000–300,000 and is composed of four identical subunits with a molecular weight of 75,000. The total carbohydrate content is approximately 7% (glucosamine, mannose, galactose, and glucose). The paper by Tomino *et al.* (1975) may also be consulted for references concerning the genetics of the lysosomal and the microsomal forms of β-glucuronidase. In a similar study, Himeno *et al.* (1976) isolated and characterized the lysosomal and the microsomal forms of β-glucuronidase from rat liver. Both forms had the same molecular weight, approximately 310,000, but they differed in charge, and three components with molecular weights of 79,000, 74,000, and 70,000 were present in the lysosomal form as compared to a single component of molecular weight 79,000 for the microsomal form.

An important step in the purification described by Himeno et al. (1976) was affinity chromatography on Sepharose-linked immunoglobulin G directed against β-glucuronidase from rat preputial gland. The same approach was used by Brot et al. (1978) in the purification of the enzyme from human placenta. Like β-glucuronidase from other sources, this enzyme had a molecular weight of approximately 300,000 and a carbohydrate content of 6–7%.

A genetic deficiency in β-glucuronidase has been described by Sly et al. (1973). The patients suffering from this condition had clinical features reminiscent of those observed in the Hurler and Hunter syndromes, including an abnormal facies, hepatosplenomegaly, and skeletal deformities. Only moderate elevation of urinary glycosaminoglycans was observed, and the excreted polysaccharide was characterized as chondroitin sulfate of low molecular weight. Chondroitin 4-sulfate and chondroitin 6-sulfate were present in approximately equal amounts. Heparan sulfate and dermatan sulfate, characteristically increased in the Hurler and Hunter syndromes, were within the normal range. Interestingly, no hyaluronic acid was found in the urine. Analysis of lysosomal enzyme activities in cultured skin fibroblasts (Hall et al., 1973) showed complete absence of β-glucuronidase as determined with synthetic substrates. However, the moderate increase in mucopolysaccharide accumulation and turnover time in the cultured fibroblasts and the relatively mild clinical course of the disease would make it likely that significant residual activity toward the natural substrates was present. The β-glucuronidase deficiency was readily detectable in white blood cells, and heterozygous carriers could be distinguished from normal subjects and from the affected patient. The deficiency segregates as an autosomal recessive trait. A concomitant enzymatic abnormality was a decrease of total β-galactosidase activity, which was 40–50% lower than in control cells.

The abnormal mucopolysaccharide metabolism could be corrected by addition of bovine β-glucuronidase, and analysis of intracellular enzyme levels showed that the cells were able to take up the β-glucuronidase in amounts far exceeding the requirement for full correction.

Beaudet et al. (1975) have studied additional patients with β-glucuronidase deficiency and have reported that the phenotypic expressions of the disease may vary considerably. In one patient, the urinary levels of polysaccharides were clearly increased, and the ratio of polymeric to low-molecular-weight material was elevated. The most striking abnormality was a 10- to 20-fold increase in the relative concentration of iduronic-acid-containing glycosaminoglycans. Furthermore, the urine contained detectable amounts of sulfoaminohexose, which is in itself an abnormal finding and indicates defective degradation of heparan sulfate. On reexamination of the patient reported by Sly *et al.* (1973), Beaudet *et al.* (1975) found predominantly dermatan sulfate and heparan sulfate in the urine rather than chondroitin sulfate, as initially suggested. Despite these discrepancies, there seems to be general agreement between the two groups of investigators in that excessive amounts of hyaluronidase-sensitive, glucuronic-acid-containing polysaccharides are not observed. It is thus concluded that the auxiliary hyaluronidase mechanism, perhaps aided by residual glucuronidase activity is capable of degrading the majority of the susceptible polysaccharide molecules to fragments small enough to be eliminated by excretion in the urine.

Recently, Bell *et al.* (1977) have developed an immunoassay for human β-glucuronidase which has enabled them to detect cross-reactive material in patients with β-glucuronidase deficiency.

2.3.2. β-D-N-Acetylhexosaminidase (Linkage Units 2, 7, 13, and 20)

An exoglycosidase activity directed toward β-D-N-acetylhexosaminidic linkages has been known for many years to occur in mammalian tissues. Initial studies with crude enzyme preparations showed that both N-acetylglucosaminidic and N-acetylgalactosaminidic linkages were cleaved, and subsequent work with purified enzymes in many laboratories has indicated that a single enzyme attacks both sugars. β-D-N-acetylhexosaminidase occurs as two major isozymes, A and B, and in addition many minor forms have been observed. Investigations in several laboratories have indicated that the A and B forms share a common subunit, β, which is the only subunit type in form B, whereas the A form contains both an α and a β subunit (see Patel, 1978). There is still uncertainty as to the exact substrate specificities of the various hexosaminidase isozymes, and the reader is referred to other reviews for a description of the complex situation encountered in the study of this enzyme (see, for example, Patel, 1978; Leaback, 1970). In the following discussion, we shall limit ourselves to some aspects of the mode of action of the enzyme which are pertinent to glycosaminoglycan catabolism.

It has been clearly established that the A and B forms are associated with the breakdown of gangliosides. Lack of the A form causes Tay–Sachs disease, while absence of both the A and B forms results in Sandhoff's disease (O'Brien, 1978). The question whether the two major forms of the enzyme are both involved in the degradation of connective tissue polysaccharides was first addressed by Thompson *et al.* (1973).

When a heptasaccharide from chondroitin sulfate, containing a radioactive *N*-acetylgalactosamine residue at the nonreducing terminus, was digested with an extract of normal fibroblasts, cleavage of the hexosaminidic linkage occurred. However, neither Tay–Sachs nor Sandhoff fibroblasts were capable of degrading this substrate, and cleavage of as little as 0.2% of the substrate would have been detected. It was therefore concluded that hexosaminidase A is normally involved in the degradation of the connective tissue polysaccharides and that hexosaminidase B does not participate in this process.

In a subsequent investigation, Cantz and Kresse (1974) obtained further information concerning this problem which was, in part, at variance with the conclusions of Thompson *et al.* (1973). It was found that fibroblasts from patients with Sandhoff's disease accumulate excessive amounts of sulfated polysaccharides, probably chondroitin sulfate and dermatan sulfate, and that the turnover time of the polysaccharides is greater than normal. Addition of β-D-*N*-acetylhexosaminidase A *or* B normalized the catabolism of glycosaminoglycans, provided the enzymes were of a form that could be internalized by these cells. It is difficult to reconcile the corrective activity of the B isozyme with the seemingly clear-cut results of Thompson *et al.* (1973). However, a role for this isozyme in normal catabolism, or at least as an auxiliary agent, is possible in view of the behavior of fibroblasts from patients with Tay–Sachs disease. These cells do not display the abnormal metabolic patterns characteristic of the mucopolysaccharidoses, although some accumulation and a slight lengthening of the turnover time of the polysaccharides have been observed. It is also possible that the glycosaminoglycans are normally degraded by one of the minor forms of the enzyme, for which the natural substrates have not yet been defined. In this context, it should be emphasized that Sandhoff fibroblasts, despite the near-complete absence of both A and B isozymes, do not exhibit changes in glycosaminoglycan metabolism as severe as those in Hurler and Hunter fibroblasts. Furthermore, from a clinical point of view, it is striking that patients with Sandhoff's disease do not have the marked skeletal deformities which are characteristically seen in the mucopolysaccharidoses, nor does accumulation of polysaccharides in the parenchymatous organs occur.

In the experiments of Cantz and Kresse (1974), isozyme A released *N*-acetylglucosamine from radioactively labeled hyaluronic acid oligosaccharides *in vitro*; however, under similar conditions, isozyme B did not attack these substrates. Bach and Geiger (1978) have subsequently shown that degradation of hyaluronic acid may be effected by either isozyme, but the reasons for the divergent results are not clear at present.

In concluding, it should be emphasized that β-D-*N*-acetylhexosaminidase is a key enzyme in the degradation of the connective tissue polysaccharides. No fewer than five of the seven species of polysaccharides, i.e., hyaluronic acid, chondroitin 4-sulfate, chondroitin 6-sulfate, dermatan sulfate, and keratan sulfate (I and II), contain β-hexosaminidic linkages and are consequently degraded by enzymes of this type. It is therefore important that remaining questions in this area be answered through continued experimentation.

3. CHONDROITIN SULFATE PROTEOGLYCANS

The chondroitin sulfates occur in the tissues as components of proteoglycans which usually contain a large number of polysaccharide chains, both chondroitin sulfates and keratan sulfate II, linked to a protein core. In cartilage and at least one other tissue, i.e., aorta, the majority of the proteoglycan molecules are bound noncovalently to hyaluronic acid in large aggregates (mol. wt. up to 200×10^6) which are stabilized by specific "link proteins." Further details of these structures will be discussed in Section 3.2.

3.1. Structure of Chondroitin Sulfates (Linkage Units 3–10)

Six different "linkage units" (3–8) are found in the carbohydrate backbone of the chondroitin sulfates (Table 1). The linkage to protein is of type 1 (xylose–serine), and the linkage region contains the typical tetrasaccharide, glucuronosyl-galactosyl-galactosyl-xylose, of the structure shown in Table 1 (units 3–6). The elucidation of the structure of this region of the chondroitin sulfate proteoglycan molecule has been reviewed in some detail elsewhere (Lindahl and Rodén, 1972).

The repeating disaccharide unit of chondroitin sulfate (linkage units 7 and 8) is formally similar to that of hyaluronic acid, differing only in that the hexosamine is galactosamine rather than glucosamine, while the uronic acid is D-glucuronic acid in both, and the positions and anomeric configurations of the linkages are the same.

The N-acetylgalactosamine units carry an ester sulfate group in either position 4 or 6 (units 9 and 10), and, accordingly, the polysaccharide is designated as chondroitin 4-sulfate or chondroitin 6-sulfate. Only rarely are all the hexosamine residues sulfated in the same position, and a hybrid structure with both 4- and 6-sulfate groups present in the same molecule (but on separate galactosamine residues) seems to be the rule rather than the exception. This situation was first suspected when fractionation of the total chondroitin sulfate pool from certain tissues yielded patterns which deviated markedly from those observed for mixtures of "standard" polysaccharides (Antonopoulos et al., 1965). More compelling evidence for the existence of hybrid structures was obtained by Seno *et al.* (1975), who showed that hybrid tetrasaccharides were formed on digestion of chondroitin sulfate with testicular hyaluronidase. [The possibility of transglycosylation indicated in the classic studies of Weissmann (1955) and Hoffman *et al.* (1956*b*) was ruled out as a major contributing factor in the formation of the hybrid tetrasaccharides on the basis of the yields of the various oligosaccharides.] More recently, Faltynek and Silbert (1978) have provided additional evidence in favor of hybrid structures: Using bacterial chondroitinase ABC, which does not possess transglycosylation activity, these workers isolated a tetrasaccharide containing one 4-sulfated and one 6-sulfated galactosamine residue from a partial digest of chondroitin sulfate.

Apart from the structural heterogeneity indicated above, the chondroitin sulfate chains may vary in average length from one tissue to another and within the same tissue. In the proteoglycan of bovine nasal cartilage, the size of the chains is on the order of 20,000, corresponding to about 40 repeating disaccharide units.

3.2. Aggregates of Chondroitin Sulfate Proteoglycans

After the chondroitin sulfate proteoglycans had been established as distinct molecular entities through the work of Schubert and collaborators (Shatton and Schubert, 1954; Malawista and Schubert, 1958), extensive studies of their physicochemical properties by Mathews (Mathews and Lozaityte, 1958; Mathews, 1975) laid the groundwork for many of our present concepts of proteoglycan structure (Figure 3). Mathews's initial concept was one of a basic unit with a molecular weight of approximately 4×10^6 in which the protein content was about 20%. Mathews also observed the existence of aggregates with molecular weights on the order of 50×10^6 (Mathews and Lozaityte, 1958).

During the past decade, there has been a tremendous upsurge in the study of connective tissue proteoglycans, largely as a result of the development of new techniques for the isolation of these compounds from cartilage in high yield and in undegraded form (Sajdera and Hascall, 1969; Hascall and Sajdera, 1969, 1970; Hascall, 1977). The basic approach of these investigations involved the use of "dissociative" solvents, 4 M guanidine or 3 M $MgCl_2$, which abolish the interactions within the native aggregates and permit the extraction of up to 85% of the proteoglycans of bovine nasal cartilage. If the extract is maintained in a dissociating medium and subjected to density-gradient centrifugation (commonly in cesium chloride), proteoglycan monomers (mol. wt. $\approx 2.5 \times 10^6$) are separated from other components of the extract and are found in the highest-density fraction which sediments to the bottom of the centrifuge

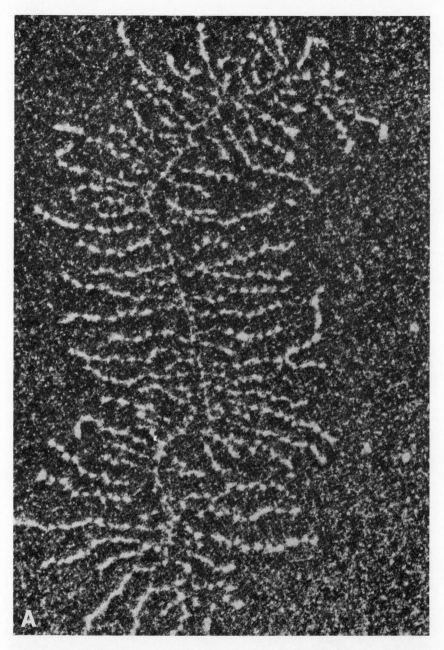

Figure 3. (A) Dark-field electron micrograph of a proteoglycan aggregate of intermediate size in which the proteoglycan subunits and filamentous backbone are particularly well

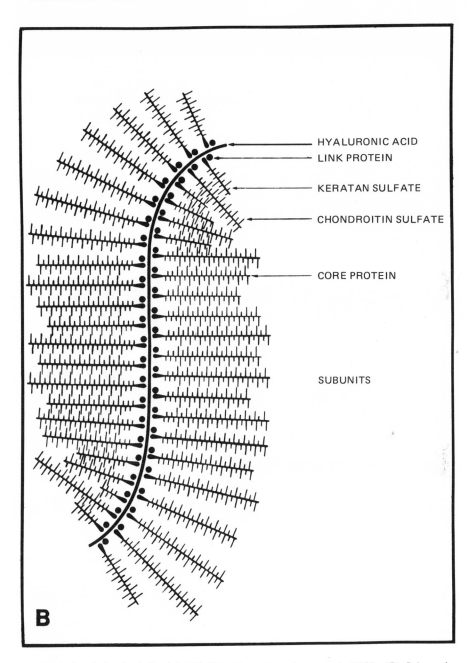

HYALURONIC ACID

LINK PROTEIN

KERATAN SULFATE

CHONDROITIN SULFATE

CORE PROTEIN

SUBUNITS

B

extended and clearly defined (×120,000). From Rosenberg *et al.* (1975). (B) Schematic representation of proteoglycan aggregate (courtesy of Dr. L. Rosenberg).

tube. The dissociation of the native proteoglycan aggregates is apparently a reversible process, and it was observed in the pioneering studies of Hascall and Sajdera (1969) that combination of the isolated monomer fraction with the material at the top of the gradient yielded aggregates detectable by analytical ultracentrifugation. Such aggregate formation naturally required removal or substantial reduction of the concentration of the dissociating agent. In an alternative approach, the initial tissue extract was dialyzed to reduce the guanidine concentration to 0.4 M, and density-gradient centrifugation of this material then yielded the proteoglycan in the form of aggregates. Subsequent centrifugation of the aggregates under dissociative conditions permitted the isolation of proteoglycan monomers separated from other specific components of the aggregates, which were then amenable to further investigation (Figure 4).

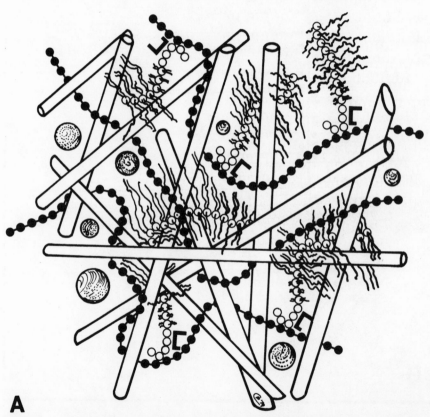

A

Figure 4. (A) Diagrammatic representation of proteoglycan aggregates in a collagen network. Courtesy of Patrick Campbell. (B) Isolation of proteoglycan aggregates and their components (monomers, link proteins, and hyaluronic acid) by extraction with guanidine and density-gradient centrifugation. Modified from Muir and Hardingham (1975).

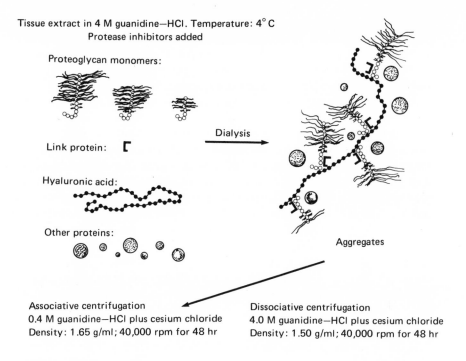

Tissue extract in 4 M guanidine—HCl. Temperature: 4° C
Protease inhibitors added

Proteoglycan monomers:

Link protein:

Hyaluronic acid:

Other proteins:

Dialysis

Aggregates

Associative centrifugation
0.4 M guanidine—HCl plus cesium chloride
Density: 1.65 g/ml; 40,000 rpm for 48 hr

Dissociative centrifugation
4.0 M guanidine—HCl plus cesium chloride
Density: 1.50 g/ml; 40,000 rpm for 48 hr

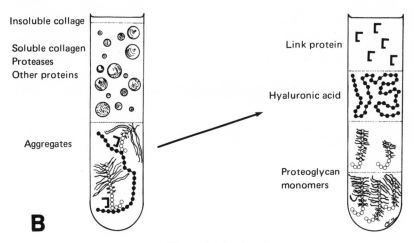

Insoluble collage

Soluble collagen
Proteases
Other proteins

Aggregates

Link protein

Hyaluronic acid

Proteoglycan
monomers

B

Figure 4. (*Continued*)

The initial demonstration (Hascall and Sajdera, 1969) that an aggregating factor (termed "glycoprotein link") was actually present, in the lowest-density region of the gradient, was followed by the important discovery in Muir's laboratory (Hardingham and Muir, 1972; Muir and Hardingham,

1975; Hascall, 1977) that hyaluronic acid interacts specifically with the proteoglycan to form large aggregates (Figure 5). Subsequently, hyaluronic acid was identified by Hardingham and Muir (1974) as a component of the aggregating fraction, and there is now little doubt that this polysaccharide is identical with the major aggregating agent.

Also of major importance in the formation of native aggregates is the participation of "link proteins" which apparently stabilize the interaction between hyaluronic acid and proteoglycans (Hardingham, 1979). In the initial studies of the aggregate components in bovine nasal cartilage proteoglycan, the top fraction [commonly termed A1D4 according to the nomenclature of Heinegård (1972)] obtained in a dissociative density-gradient centrifugation was shown to contain two major components (link proteins 1 and 2) with electrophoretic mobilities corresponding to molecular weights of 51,000 and 47,000, respectively (Keiser *et al.*, 1972; Hascall and Heinegård, 1974; Baker and Caterson, 1977; Caterson and Baker, 1979). Only link protein 2 is found in a chondrosarcoma (Oegama *et al.*, 1975, 1977). Recently, a third, much fainter band, which moves slightly faster than the smallest component, was observed by Baker and Caterson (1977). The link proteins have a strongly hydrophobic character,

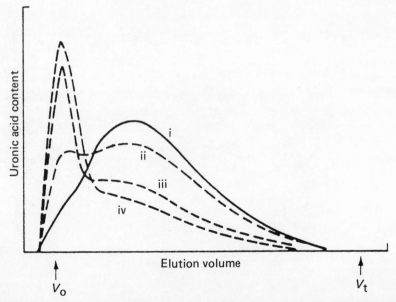

Figure 5. Effect of increasing amounts of hyaluronate on the gel-chromatographic profile of disaggregated proteoglycans. Disaggregated proteoglycans chromatographed on a column of Sepharose 2B with: (i) no hyaluronate, (ii) 0.03% hyaluronate added, (iii) 0.13% hyaluronate added, (iv) 0.63% hyaluronate added. From Muir and Hardingham (1975).

are insoluble in water, and can be handled readily only in the presence of detergent or after modification, e.g., by citraconylation to introduce more hydrophilic residues into the molecule. However, the development of preparative methodology to obtain the link proteins in quantity and in highly purified form by preparative electrophoresis in sodium dodecyl sulfate (SDS) has made possible continued studies of the properties of these interesting compounds (Caterson and Baker, 1978; Baker and Caterson, 1978, 1979). Evidence has been obtained for a generic relationship between the two major link proteins, as indicated by the finding that reduction with mercaptoethanol converts a portion of link protein 1 to a compound which has the same electrophoretic mobility as link protein 2. It appears possible that link protein 1 is exposed *in vivo* to partial cleavage of peptide bonds located in such a position that the resulting fragments are still joined by a disulfide bridge.

Most clearly, the function of the link proteins as stabilizing agents in the proteoglycan–hyaluronic acid interaction has been demonstrated by the finding that oligosaccharides from hyaluronic acid do not dissociate complete aggregates, whereas the hyaluronic acid–proteoglycan complex without the link proteins is readily cleaved (Hardingham, 1979).

Three interactions theoretically are possible within the three-component aggregate: (1) between monomer and hyaluronic acid, (2) between monomer and link proteins, and (3) between hyaluronic acid and link proteins (Figure 6). In addition to the proteoglycan–hyaluronic acid interaction, which is now well documented, binding between proteoglycan monomers and link proteins has recently been demonstrated by Caterson and Baker (1978). Some evidence has also been presented for interaction between hyaluronic acid and link proteins (Hascall and Heinegård, 1974; Oegema *et al.*, 1977). Thus, it appears that the three aggregate components all possess two binding sites, one for each of the other two components. It has also been suspected for some time that the proteoglycan monomers are capable of self-interaction, and evidence for such a phenomenon has been obtained recently in Muir's laboratory (Sheehan *et al.*, 1978).

3.3. Biosynthesis of Chondroitin Sulfate

In analogy with the mechanisms established for glycoprotein biosynthesis, the formation of the core protein of the chondroitin sulfate proteoglycan is the first step in this process. It is not certain, however, that the entire protein molecule (mol. wt. $\approx$ 200,000) needs to be completed before addition of the polysaccharide chains begins. As discussed earlier (Section 1.1.1), if the behavior of glycoproteins is any indication of the mode of synthesis of the proteoglycans, one might assume that the car-

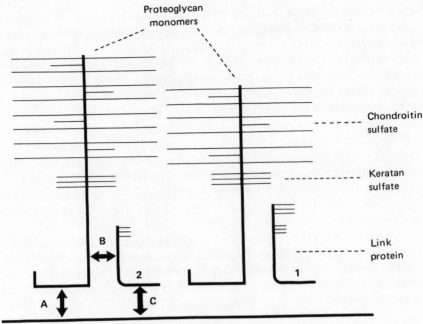

Proteoglycan
monomers

Chondroitin
sulfate

Keratan
sulfate

Link
protein

Hyaluronic acid

Figure 6. Interactions between the components of proteoglycan aggregates: (A) between proteoglycan monomers and hyaluronic acid; (B) between monomers and link proteins 1 and 2; (C) between link protein and hyaluronic acid. From Caterson and Baker (1978).

bohydrate residues are transferred, in part at least, already before completion of the polypeptide on the polysomes.

As may be surmised from the structure of chondroitin sulfate, the polysaccharide chain is assembled by six different glycosyltransferases, responsible for the formation of units 3 through 8.

3.3.1. Linkage Unit 3

The first glycosyl transfer reaction for several of the connective tissue polysaccharides is the transfer of xylose from UDP-xylose to serine hydroxyl groups in their core proteins (Figure 7). This reaction was first demonstrated by Grebner *et al.* (1966*a*), who showed that hen oviduct contains an enzyme which catalyzes transfer to endogenous acceptors present in the crude enzyme preparation. The same type of reaction has subsequently also been found in a mouse mastocytoma (Grebner *et al.*, 1966*b*), embryonic chick cartilage (Robinson *et al.*, 1966), brain, and some other tissues (Stoolmiller *et al.*, 1972). Because all these tissues synthesize more than one type of polysaccharide, it is not possible to state with certainty that a particular cell type contains a xylosyltransferase specific for any one polysaccharide. For example, chondrocytes, while producing a preponderance of chondroitin sulfate, also synthesize heparan sulfate, which is similarly initiated. Indeed, it seems most likely that one and the same xylosyltransferase participates in the formation of the various xylose-linked polysaccharides.

The xylosyltransferase has been purified to homogeneity from embryonic chick cartilage (Schwartz and Rodén, 1974; see also Baker et al., 1972; Stoolmiller et al., 1972) and from a rat chondrosarcoma (Schwartz and Dorfman, 1975). A particularly useful step in the purification procedure was affinity chromatography on a matrix of core protein that had been prepared by Smith degradation of the proteoglycan (this procedure removes the chondroitin sulfate chains and exposes substrate hydroxyl groups in the previously substituted serine residues). The enzyme has a molecular weight of 95,000–100,000 and consists of two pairs of dissimilar subunits with molecular weights of 23,000 and 27,000, respectively, as indicated by gel electrophoresis in SDS after pretreatment with SDS and mercaptoethanol.

The particular importance of the xylosyl-transfer step in the formation of the proteoglycan molecules is not yet clear. However, the observation by Neufeld and Hall (1965) that UDP-D-xylose is a potent inhibitor of UDP-D-glucose dehydrogenase points to a regulatory role for UDP-D-xylose that affects the formation of the entire macromolecule. If an imbalance occurs between the synthesis of the core protein and the nucleotide sugars such that insufficient amounts of core protein are available, an increase in the UDP-D-xylose concentration will be the first response to this situation. This will immediately affect UDP-D-glucose dehydrogenase and decrease the rate of formation of UDP-D-glucuronic acid, thereby decreasing the formation of UDP-D-xylose as well, because this nucleotide sugar is generated from UDP-D-glucuronic acid by decarboxylation. Other than this feedback control, which affects chain initiation and the formation of the glucuronic acid residues of the polysac-

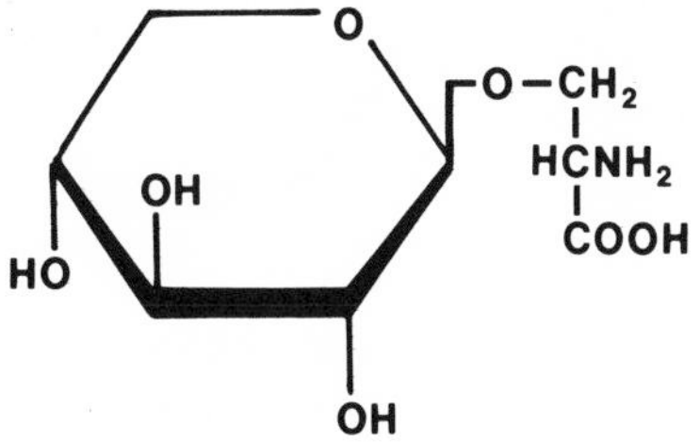

Figure 7. Linkage unit 3: Xyl $\xrightarrow{\beta}$ Ser.

charide, there is also a feedback regulation of hexosamine biosynthesis. UDP-N-acetylglucosamine, which is in ready equilibrium with UDP-N-acetylgalactosamine, is an inhibitor of glutamine:fructose-6-phosphate transamidase, which catalyzes the first specific step in hexosamine synthesis (Kornfeld *et al.*, 1964; Kornfeld, 1967). As a result of these two mechanisms, a finely tuned interplay is possible between the processes involved in core protein synthesis and those leading to the formation of the chondroitin sulfate chains themselves.

The nature of the endogenous acceptors which serve as xylosyl acceptors *in vivo* is not known. We assume, of course, that the acceptor is the core protein of the putative proteoglycan molecule. However, this remains to be established with certainty. We are also still ignorant with respect to several aspects of the process of chain initiation; e.g., we do not know whether xylosyl transfer occurs before the core protein has been completed on the polysomes or whether the protein is released into the cisternae of the endoplasmic reticulum before the xylosylation process begins. Nor do we know whether the chondroitin sulfate chains are initiated in a random fashion along the polypeptide core or whether there is an orderly progression from one end of the molecule to the other.

3.3.2. Linkage Unit 4

The completion of the carbohydrate–protein linkage region is effected by three additional glycosyltransferases, two galactosyltransferases, and one glucuronosyltransferase. The first galactosyl transfer step was investigated by Robinson *et al.* (1966), who showed that a particulate enzyme preparation from embryonic chick cartilage catalyzes transfer of galactose to xylosyl residues of endogenous acceptors (Figure 8); the product of the reaction was characterized by isolation of radioactive galactosyl-xylitol after cleavage of the xylose–serine linkage with alkaline

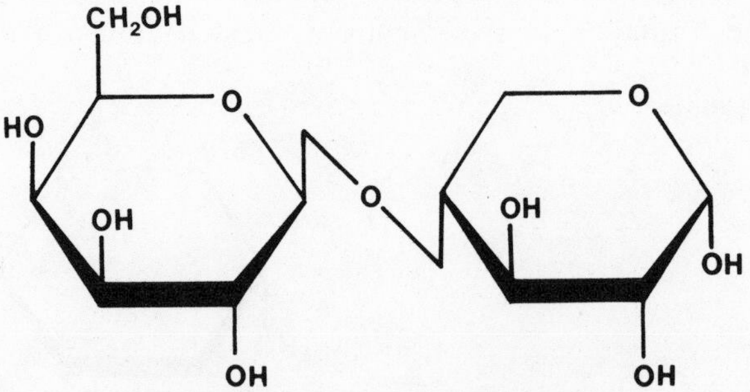

Figure 8. Linkage unit 4: Gal $\xrightarrow{\beta 1,4}$ Xyl.

borohydride. Subsequently, transfer of the second galactose residue was shown by Helting and Rodén (1969a), who isolated labeled 3-O-β-D-galactosyl-D-galactose from a partial acid hydrolysate of a similar reaction mixture.

The drawbacks of relying exclusively on endogenous substrates in the study of glycosyl transfer reactions are discussed in Chapters 2 and 3. Given these drawbacks, a necessary subsequent step in the continued investigation of these reactions was a search for suitable exogenous substrates. As shown in Table 2, several compounds may be used as exogenous acceptors for galactosyltransferase I, including the monosaccharide D-xylose and xylosides such as O-β-D-xylosyl-L-serine, methyl β-D-xylopyranoside, and p-nitrophenyl β-D-xylopyranoside (Helting and Rodén, 1969b; see also Okayama and Lowther, 1973; Okayama et al., 1973; Robinson et al., 1975). It was somewhat surprising to learn that D-xylose is an acceptor, for few glycosyltransferases are capable of transfer to monosaccharide acceptors. However, this situation has been used to advantage in studies of the regulation of proteoglycan biosynthesis, as the addition of a xyloside to cells in culture enables chondroitin sulfate chain growth to occur independently of the formation of core protein. The two parts of the process, core protein formation and polysaccharide synthesis, can therefore be segregated and studied as separate phenomena (see, for example, Okayama and Lowther, 1973; Okayama et al., 1973; Handley and Lowther, 1977; Schwartz et al., 1974a; Galligani et al., 1975; Schwartz, 1977, and references cited therein).

Xylosyltransferase is much more soluble than the other five glycosyltransferases of chondroitin sulfate synthesis. On thorough homogenization and extraction, approximately 80% of this enzyme is obtained in the soluble high-speed supernatant fraction and has been purified from this source. The tight binding to the membranes of the endoplasmic reticulum has been a major obstacle in the studies of the other five glycosyltransferases, but this difficulty has now been overcome by the development of methods for the solubilization of these enzymes. Treatment with the nonionic detergent Nonidet P-40, in the presence of 0.5 M KCl, has yielded soluble preparations amenable to further purification. Galactosyltransferase I has been purified 1100-fold, and the purified material gives two bands on polyacrylamide gel electrophoresis in SDS (Schwartz and Rodén, 1975).

Galactosyltransferase I interacts with a certain degree of specificity with the chain-initiating xylosyltransferase. This interaction has been demonstrated by two different approaches. It was shown by Schwartz (1975) that antibodies to xylosyltransferase also precipitated galactosyltransferase I when the two enzymes were present in the same solution. No precipitation was obtained with galactosyltransferase alone. Evidence

Table 2. Chondroitin Sulfate Glycosyltransferases and Some of Their Exogenous Substrates[a]

Enzyme	Exogenous acceptors
Xylosyltransferase	Smith-degraded cartilage proteoglycan L-Serylglycylglycine D-Xylose
Galactosyltransferase I	O-β-D-Xylosyl-L-serine Methyl, ethyl, butyl, and octyl β-D-xylopyranoside Benzyl β-D-xylopyranoside p-Nitrophenyl β-D-xylopyranoside
Galactosyltransferase II	4-O-β-D-Galactosyl-D-xylose 4-O-β-D-Galactosyl-O-β-D-xylosyl-L-serine
Glucuronosyltransferase I	3-O-β-D-Galactosyl-D-galactose O-β-D-Galactosyl $\xrightarrow{1,3}$ O-β-D-galactosyl $\xrightarrow{1,4}$ D-xylose
N-Acetylgalactosaminyltransferase	GlcUA $\longrightarrow$ GalNAc $\longrightarrow$ GlcUA $\longrightarrow$ GalNAc $\longrightarrow$ GlcUA $\longrightarrow$ GalNAc (chondroitin hexasaccharide) GlcUA $\longrightarrow$ (GalNAc-4S) $\longrightarrow$ GlcUA $\longrightarrow$ (GalNAc-4S) $\longrightarrow$ GlcUA $\longrightarrow$ (GalNAc-4S) (chondroitin 4-sulfate hexasaccharide)
Glucuronosyltransferase II	GalNAc $\longrightarrow$ GlcUA $\longrightarrow$ GalNAc $\longrightarrow$ GlcUA $\longrightarrow$ GalNAc (chondroitin pentasaccharide) (GalNAc-6S) $\longrightarrow$ GlcUA $\longrightarrow$ (GalNAc-6S) $\longrightarrow$ GlcUA $\longrightarrow$ (GalNAc-6S) (chondroitin 6-sulfate pentasaccharide)

[a] From Rodén and Schwartz (1975).

for a specific interaction was also obtained in the course of purification of galactosyltransferase I (Schwartz *et al.*, 1974*b*; Schwartz and Rodén, 1975). When a partially purified preparation of the enzyme was passed through a column containing immobilized xylosyltransferase (either covalently bound to Sepharose or adsorbed to Sepharose-linked core protein), binding occurred, and a considerably purified galactosyltransferase preparation could subsequently be obtained by elution with detergent at elevated ionic strength.

Galactosyltransferase I appears to be dependent on phospholipids for full activity, as indicated by the finding (Schwartz, 1976*a*) that digestion of the microsomal pellet from embryonic chick cartilage with phospholipase C resulted in approximately 75% inactivation of the enzyme. Activity could be restored by the addition of phospholipids, and lysophosphatidylethanolamine and lysophosphatidylcholine were particularly effective in this regard. Dialysis against Nonidet P-40 in the presence of 0.25 M KCl likewise restored activity to almost the initial level.

The obvious differences in the properties of xylosyltransferase and galactosyltransferase I are also reflected in their intracellular metabolism. As shown by Schwartz (1976b), xylosyltransferase turns over with a half-life of 2–3 hr in rapidly growing chondrocytes, whereas the membrane-bound galactosyltransferase I has a half-life of approximately 12 hr.

3.3.3. Linkage Unit 5

The second galactosyl transfer step (Figure 9) in the formation of the linkage region is catalyzed by an enzyme of higher substrate specificity than galactosyltransferase I. This stands to reason, since the acceptor sugar is galactose, which is a common constituent of glycoproteins. Although complete details of the substrate specificity have not yet been established, it seems that this galactosyltransferase requires the disac-

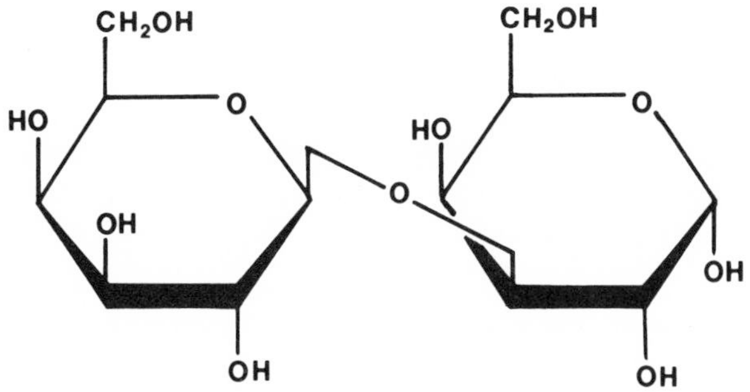

Figure 9. Linkage unit 5: Gal $\xrightarrow{\beta 1,3}$ Gal.

charide structure 4-*O*-β-D-galactosyl-D-xylose for recognition, and transfer to galactose or other galactose-containing disaccharides does not occur (Helting and Rodén, 1969*a*). This enzyme has been purified approximately 40-fold (Schwartz and Rodén, 1975).

3.3.4. Linkage Unit 6

The final step in the formation of the carbohydrate–protein linkage region is the transfer of glucuronic acid to the second galactose residue (Figure 10) by a glucuronosyltransferase which is different from that involved in the formation of the repeating disaccharide units (Helting and Rodén, 1969*b*). The specificity of this enzyme is not as high as that of galactosyltransferase II, and several disaccharides with nonreducing terminal galactose can be utilized as substrates, albeit with different efficiencies. Slight acceptor activity is also observed with free D-galactose.

3.3.5. Linkage Unit 7

Cell-free biosynthesis of chondroitin was first demonstrated with a particulate enzyme preparation from embryonic chick cartilage that synthesized a low-sulfated polymer from UDP-glucuronic acid and UDP-*N*-acetylgalactosamine (Perlman *et al.*, 1964; Silbert, 1964). In a series of elegant studies by Dorfman (1974) and his collaborators, many of the fundamental characteristics of the polymerization process have been determined, and the use of exogenous oligosaccharides as substrates in the various reactions has served as a model for other investigations in this and related fields. It is now clear that the repeating disaccharides (Figures 11 and 12) are formed by alternating transfer of *N*-acetylgalactosamine and glucuronic acid to the nonreducing terminus of the growing chain and that this process requires two distinct enzymes, an *N*-acetylgalac-

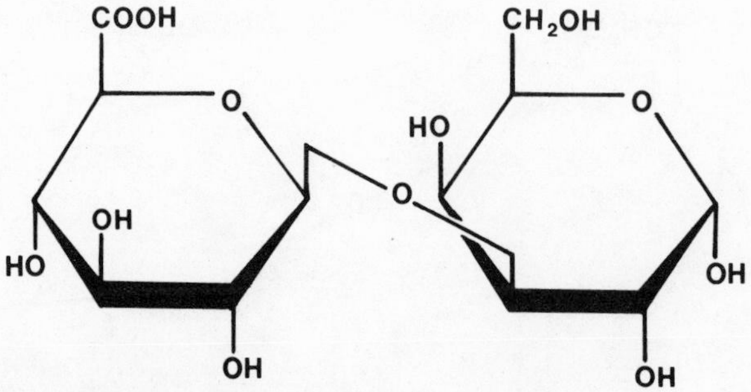

Figure 10. Linkage unit 6: GlcUA $\xrightarrow{\beta 1,3}$ Gal.

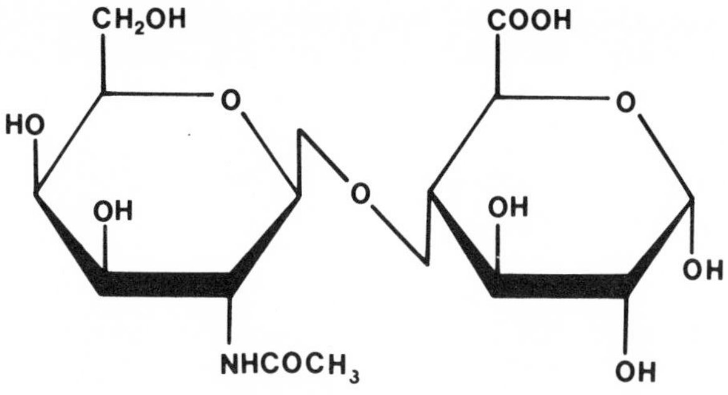

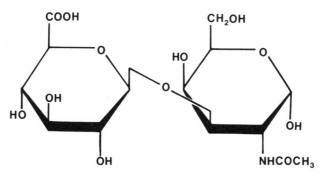

Figure 11. Linkage unit 7: GalNAc $\xrightarrow{\beta1,4}$ GlcUA.

tosaminyltransferase and a glucuronosyltransferase. As can be seen from Table 3, sulfated as well as nonsulfated oligosaccharides with terminal glucuronic acid residues serve as acceptors for N-acetylgalactosaminyl transfer. It should be noted that a hexasaccharide from hyaluronic acid has considerable acceptor activity, indicating that an N-acetylgalactosamine unit in the penultimate position is not absolutely essential.

The constancy in assembly of the polysaccharide chains has been previously commented upon, and it should again be pointed out that the substrate specificities of the enzymes involved would seem to permit certain mistakes; however, this is not known to occur. Although hyaluronic acid is synthesized by chondrocytes, it seems that this process is compartmentalized or regulated in some other fashion so as to avoid the formation of hyaluronic acid–chondroitin sulfate hybrids, a distinct possibility on the basis of the substrate specificities. It may well be that there exists in the membranes a multienzyme complex that helps provide a tightly regulated assembly of any one polysaccharide with little or no chance for interference by irrelevant glycosyltransferases.

Figure 12. Linkage unit 8: GlcUA $\xrightarrow{\beta1,3}$ GalNAc.

Table 3. Acceptor and Donor Specificities of Chondroitin Sulfate "Polymerase"[a]

Oligosaccharide acceptors with nonreducing terminal glucuronic acid[b]

Donor UDP-nucleotide sugar	Chondroitin Hexasaccharide	Chondroitin Tetrasaccharide	Chondroitin 4-sulfate Hexasaccharide	Chondroitin 4-sulfate Tetrasaccharide	Chondroitin 6-sulfate hexasaccharide	Hyaluronic acid hexasaccharide	Desulfated dermatan sulfate tetrasaccharide
UDP-N-acetylgalactosamine	+	–	+	–	+	+	–
UDP-glucuronic acid	–	–	–	–	–	–	
UDP-N-acetylglucosamine	–		–		–	–	

Oligosaccharide acceptors with nonreducing terminal N-acetylhexosamine[b]

		Chondroitin 6-sulfate			
	Chondroitin 4-sulfate pentasaccharide	Heptasaccharide	Pentasaccharide	Trisaccharide	Hyaluronic acid pentasaccharide
UDP-N-acetylgalactosamine	–				–
UDP-glucuronic acid		+	+	+	–

[a] Modified from Dorfman (1969); see also Helting and Rodén (1969b) and Horwitz (1972).

[b] Even-numbered oligosaccharides from chondroitin sulfates, and hyaluronic acid were prepared by digestion with testicular hyaluronidase, which yields homologous oligosaccharides with glucuronic acid at the nonreducing end. Odd-numbered oligosaccharides with N-acetylhexosamine at the nonreducing end were obtained from the even-numbered compounds by digestion with β-glucuronidase. The desulfated dermatan sulfate tetrasaccharide was isolated from an acid hydrolysate of the polysaccharide.

3.3.6. Linkage Unit 8

Glucuronosyltransferase II catalyzes transfer to nonreducing terminal N-acetylgalactosamine residues of oligosaccharides from chondroitin and chondroitin 6-sulfate (Figure 12). However, 4-sulfated pentasaccharide does not serve as acceptor, indicating that sulfation *in vivo* must occur after the addition of a glucuronic acid residue.

Although some properties of this enzyme have been determined, it has not yet been purified extensively, and important questions concerning its characteristics therefore remain to be explored.

3.3.7. Linkage Units 9 and 10

Specific sulfotransferases have been identified which catalyze transfer of sulfate groups from 3'-phosphoadenyl 5'-phosphosulfate (PAPS) to the 4- and 6-positions of the N-acetylgalactosamine residues of chondroitin sulfate (Figures 13 and 14). The 6-sulfotransferase has been extensively purified by Greiling *et al.* (1972). From the early studies of Suzuki and Strominger (1961), it is clear that small oligosaccharides (especially odd-numbered compounds with nonreducing terminal N-acetylgalactosamine) as well as polymeric chondroitin may serve as substrates. Basic kinetic properties of the sulfotransferases have been determined.

3.4. Catabolism of Chondroitin Sulfate Proteoglycans

3.4.1. General Comments

A clear picture of the physiological mode of degradation of the chondroitin sulfate proteoglycans has not yet emerged. It is likely that

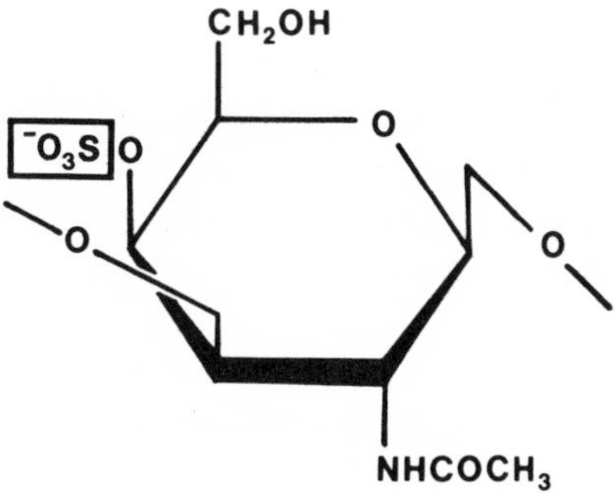

Figure 13. Linkage unit 9: GalNAc-4-sulfate.

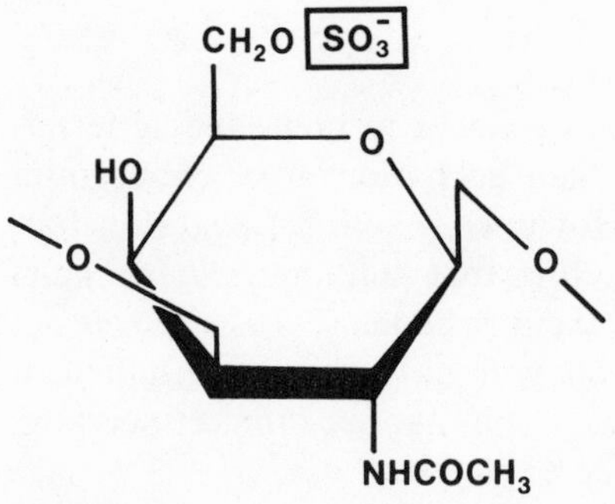

Figure 14. Linkage unit 10: GalNAc-6-sulfate.

this process follows a complex pattern of interwoven pathways determined by such factors as the tissue or cell type in which the degradation occurs and the exact location of the catabolic processes in the tissues. A substantial difference in product patterns is to be expected depending on whether degradation occurs intracellularly in lysosomes or in the extracellular matrix. In the former situation, extensive breakdown to monosaccharides and small oligosaccharides takes place, with some of the monosaccharide fragments being reutilized for synthesis of new proteoglycan molecules. That the catabolic apparatus has the capacity to degrade the proteoglycans to small fragments is shown by the finding that xylosylserine, representing one residue from each of the two major components, is excreted in the urine (Tominaga *et al.*, 1965).

If degradation occurs extracellularly through the action of proteases or carbohydrate-cleaving enzymes, the oligo- and polysaccharide fragments, with attached peptides of varying sizes, are obviously small enough to be excreted in the urine. In a classic experiment, Thomas (1956) (see Revell and Muir, 1972, for additional references) showed that intravenous injection of papain into rabbits caused massive destruction of the extracellular matrix, notably in cartilage, to the point that collapse of the tracheal cartilage rings occurred and loss of the stiff, elastic texture of the ears changed the rabbit's appearance to that of a cocker spaniel. Once released from their linkage to the macromolecular proteoglycan, the chondroitin sulfate chains (mol. wt. $\approx$ 20,000) were apparently able to escape readily from the tissues and were excreted in the urine.

Implicit in these findings is the notion that the proteoglycans themselves cannot be eliminated from the tissues without prior metabolic modification. Indeed, proteoglycans as such are not present in urine, only the fragments from their degradation. Since the proteoglycans in their native aggregate form are of considerable dimensions and are in many locations trapped by a network of collagen and elastin fibrils, it stands to reason that they cannot readily escape into the circulation. Even if this were possible, the size of the aggregates would in all likelihood prevent them from traversing the membranes of the kidney glomeruli, and an experiment by Revell and Muir (1972) is particularly illustrative in this regard. These authors showed that intravenously injected ^{35}S-labeled chondroitin sulfate proteoglycan was not excreted intact; however, over a period of 24 hr, 9–20% of the total radioactivity was found in the urine in the form of single chondroitin sulfate chains. Interestingly, the chain length of the polysaccharide indicated that depolymerization had not occurred, although extensive degradation of the core protein must have taken place. The fate of the major portion of the injected proteoglycan was not determined, but it may be assumed that it was rapidly taken up by various tissues and degraded to small fragments.

In contrast, excretion of individual ^{35}S-labeled chondroitin sulfate chains began immediately following injection, and after 24 hr, between 20 and 58% of the administered radioactivity had been eliminated. Surprisingly, depolymerization of the polysaccharide chains did not occur, although such a change would not have been unexpected in view of the presence of hyaluronidase in serum. However, it was hypothesized that the naturally occurring, nonspecific hyaluronidase inhibitor (Mathews and Dorfman, 1955) might have prevented hyaluronidase action, and it should also be recalled that the pH of blood is far above the pH optimum of the enzyme. (An important methodological point from this study should also be mentioned, i.e., the observation that unless precautions were taken to keep the collected urine sterile, extensive degradation took place over a 24-hr period.)

From the general structure of the chondroitin sulfate proteoglycans, it is evident that proteases as well as carbohydrate-degrading enzymes and sulfatases are required to break down these molecules into their constituent groups. In the following discussion, we shall limit our attention to enzymes acting on the polysaccharide components (for reviews of connective tissue proteases, see Barrett and Dingle, 1971; Barrett, 1972).

Two modes of enzymatic attack on the chondroitin sulfate molecules are possible: (1) hyaluronidase may cleave internal bonds, since, like hyaluronic acid, chondroitin sulfate contains hexosaminidic linkages which are susceptible to the action of this enzyme; and (2) exoenzymes (glycosidases and sulfatases) may degrade the molecule stepwise from the nonreducing end. Evidence available at present indicates that both types of activities are normally operating *in vivo*. However, an accurate assessment of their relative contributions to chondroitin sulfate degradation has not yet been possible, and it has been previously indicated that variation in degradation patterns is to be expected from one tissue or cell type to another (see Davidson, 1970; Dorfman *et al.*, 1972; Glaser and Conrad, 1979, and references cited therein).

In view of the abundance of chondroitin sulfate proteoglycan in cartilage, it is of particular interest to examine the pathways of degradation in this tissue. Although Platt and Dorn (1968) were unable to detect hyaluronidase activity in human cartilage, a hyaluronidaselike endopolysaccharidase and a sulfatase acting on chondroitin sulfate and chondroitin sulfate oligosaccharides have subsequently been found in embryonic chick chondrocytes (Amado *et al.*, 1974; Wasteson *et al.*, 1975). The endoglycosidase had a surprisingly low pH optimum of 2.8, which is in sharp contrast to the pH optimum of 5 observed for testicular hyaluronidase. In a recent study, Glaser and Conrad (1979) showed the presence in cultured chondrocytes from embryonic chick cartilage of a complete enzyme system which depolymerized chondroitin sulfate and released, by exoenzyme action, all components of the repeating disaccharide portion of the molecule including the sulfate groups. When saccharo-1,4-lactone and 2-acetamido-2-deoxy-D-galactonolactone, inhibitors of β-glucuronidase and β-*N*-acetylhexosaminidase, respectively, were added to the incubation mixtures, the polymeric chondroitin sulfate was degraded to oligosaccharides with 10–15 monosaccharides. These findings support the notion that normal catabolism occurs by combined action of hyaluronidase and a group of appropriate exoenzymes acting from the nonreducing terminus. In the system investigated by Glaser and Conrad (1979), removal of the 4-sulfate groups was apparently the rate-limiting step in the overall process, since after a period of digestion, most of the nonreducing termini were occupied by *N*-acetylgalactosamine 4-sulfate residues. The 6-sulfate groups were released more rapidly, as indicated by a progressive decrease in the ratio of 6-sulfate to 4-sulfate groups as the digestion proceeded.

Apart from the actual demonstration of hyaluronidase activity in various tissues, observations on patients with certain mucopolysaccharidoses suggest that hyaluronidase participates extensively in normal mucopolysaccharide catabolism or, at least, that it is readily mobilized to serve this function. Thus, it has been previously mentioned that patients with β-glucuronidase deficiency do not excrete hyaluronic acid or excessive amounts of chondroitin sulfate, whereas dermatan sulfate and heparan sulfate are found in the urine at elevated levels. Were hyaluronidase not at all involved in their catabolism, the two former polysaccharides should also be present. (An alternative possibility is the presence of sufficient residual β-glucuronidase activity; however, this interpretation leaves the elevation of dermatan sulfate and heparan sulfate unexplained.)

Another example: Maroteaux–Lamy patients are deficient in a sulfatase which removes the 4-sulfate groups from nonreducing terminal N-acetylgalactosamine residues of chondroitin 4-sulfate and dermatan sulfate (see Section 3.4.4); in the absence of the enzyme, both polysaccharides ought to accumulate, yet we observe an increase in dermatan sulfate only. This polysaccharide, which is present in the body in much smaller quantity than chondroitin 4-sulfate, can be degraded by hyaluronidase only to a limited degree (see Section 6.1), and the end products of degradation in the absence of 4-sulfatase (or β-glucuronidase) will be of considerable size. (It may be noted that comprehensive investigations into the possible occurrence of oligosaccharides from hyaluronic acid and chondroitin sulfate in urine and tissues have not been carried out and that such fragments may have gone undetected.) Findings in apparent contradiction to our hypothesis are observed in the Morquio syndrome, which results from deficiency in a sulfatase specific for the 6-sulfate groups on the galactose and N-acetylgalactosamine residues of keratan sulfate and chondroitin 6-sulfate, respectively. In patients with Morquio's disease, both keratan sulfate and chondroitin 6-sulfate are excreted, although only the former would have been expected if hyaluronidase plays a substantial role in the *in vivo* degradation of chondroitin 6-sulfate. It should be kept in mind, however, that chondroitin 6-sulfate is a much poorer substrate for hyaluronidase than chondroitin 4-sulfate (which, in turn, is less susceptible to digestion than hyaluronic acid). It would thus appear that hyaluronidase is not available in the tissues in a concentration sufficiently high to cope with all potential substrate molecules, and by its very nature the enzyme does not adequately meet the needs of the organism for extensive degradation of the various substrates. Hyaluronidase should therefore be regarded as an auxiliary enzyme, which complements and amplifies the action of the exoenzymes but is not designed to fully take their place.

An important factor in determining tissue patterns of degradation is the recognition and uptake of the substrates by the cells (see Chapter 6). Little is known of this aspect of proteoglycan metabolism, but a recent study by Truppe *et al.* (1977) illustrates clearly the selectivity in behavior which may be displayed by various cell types. These authors observed that hyaluronate is taken up by cultured rat hepatocytes and human synovial cells but not by human skin fibroblasts and smooth muscle cells. The uptake by hepatocytes was competitively inhibited by hyaluronic acid oligosaccharides, which were not themselves internalized under the conditions tested. In contrast, ^{35}S-labeled chondroitin sulfate proteoglycan was taken up to a much greater extent by skin fibroblasts and smooth muscle cells than by hepatocytes and synovial cells. The pinocytosed hyaluronic acid was rapidly degraded to ethanol-soluble products which remained inside the cells, whereas sulfate liberated from the proteoglycan was found mostly outside the cells.

3.4.2. Exoenzymes Acting on Chondroitin Sulfates

The stepwise degradation of the chondroitin sulfate molecules involves the removal of the residues illustrated by the structure of a fragment from the chondroitin sulfate–protein linkage region:

$$[GlcUA \xrightarrow{\beta 1,3} GalNAc]_n \xrightarrow{\beta 1,4} GlcUA \xrightarrow{\beta 1,3} Gal \xrightarrow{\beta 1,3} Gal \xrightarrow{\beta 1,4} Xyl \xrightarrow{\beta} Ser$$

4- or 6-sulfate

Whereas the glycosyltransferases catalyzing biosynthesis are exacting in their substrate requirements and this structure is assembled by six different glycosyltransferases and two sulfotransferases, the glycosidases are in general not highly specific for the aglycones. Consequently, it may be predicted that degradation of the indicated structure requires the following four glycosidases: (1) β-glucuronidase, (2) β-N-acetylhexosaminidase, (3) β-galactosidase, and (4) β-xylosidase.

In accord with this assumption, liver β-glucuronidase removes glucuronic acid from the nonreducing ends of chondroitin sulfate oligosaccharides larger than disaccharide (Meyer *et al.*, 1956), polymeric dermatan sulfate–chondroitin sulfate hybrids (see, for example, Fransson and Rodén, 1967*a*), and glucuronosyl $\xrightarrow{\beta 1,3}$ galactose (Rodén and Armand, 1966). A not uncommon influence of the linkage position was observed when the rates of cleavage of GlcUA $\xrightarrow{\beta 1,3}$ Gal and GlcUA $\xrightarrow{\beta 1,6}$ Gal were compared: Under conditions yielding quantitative release of glucuronic acid from the 3-linked disaccharide, less than 10% of the 6-linked isomer was cleaved [GlcUA $\xrightarrow{\beta 1,6}$ Gal used in these experiments was a kind gift from Professor Michael Heidelberger, who had first isolated and described it in the same year the author was born (Heidelberger and Kendall, 1929)].

For additional information about β-glucuronidase, the reader is referred to Section 2.3.1.

Once all terminal glucuronic acid residues have been removed from the chondroitin sulfate chains, the structure is open to attack by either of two sulfatases which are specific for the 4- and 6-position, respectively. The properties of these enzymes will be described below. Following the release of terminal sulfate groups, the exposed N-acetylgalactosamine residues are susceptible to cleavage by β-N-acetylhexosaminidase. As has been discussed previously (Section 2.3.2), there is some uncertainty about the isozyme responsible for the removal of the N-acetylgalactosamine residues in chondroitin sulfate, and further clarification of the substrate specificities of the various forms of the enzyme is needed.

The two galactose residues of the linkage region are presumably released by the same acid β-galactosidase which is deficient in G_{M1} gangliosidosis, and some information concerning this process has been presented by Distler and Jourdian (1973) (see also Section 4.3.2 for additional information on this enzyme).

Contrary to expectations, the serine-linked xylose residue does not seem to be a substrate for any mammalian glycosidase discovered so far, as will be discussed in the following section.

3.4.3. β-D-Xylosidase (β-D-Glucosidase)

A specific β-xylosidase has not been found in mammalian tissues. Rather, β-glucosidase also exhibits β-xylosidase activity, as shown with p-nitrophenyl β-D-xyloside as substrate for pig kidney supernatant enzyme by Robinson and Abrahams (1967) and for an enzyme from rat kidney lysosomes by Patel and Tappel (1969a). Similarly, Öckerman (1968) found that a single enzyme contains both β-glucosidase and β-xylosidase activities in homogenates of human liver (see also Patel and Tappel, 1969b; Beck and Tappel, 1968). Quite surprisingly, O-β-D-xylosyl-L-serine was not cleaved by the lysosomal enzyme from rat liver and kidney (Patel and Tappel, 1969b; Fisher and Kent, 1969). In contrast, β-xylosidase from plants and invertebrates, which also exhibited β-glucosidase activity, did indeed hydrolyze the xylose–serine linkage (Fisher *et al.*, 1966; Fukuda *et al.*, 1968; Patel and Tappel, 1969b).

If xylosylserine cannot be cleaved in the mammalian organism, one would expect to find it in the urine. Indeed, Tominaga *et al.* (1965) have isolated xylosylserine in crystalline form from human urine, in amounts corresponding to at least 1 mg per liter of urine. At present, it is difficult to determine whether the excreted amounts are sufficient to account for the normal turnover of the proteoglycans, since reliable information about the total body pools of the proteoglycans and their turnover is not available. In estimating glycosaminoglycan turnover, the excretion of L-xylulose in individuals with essential pentosuria provides some guidance (this sugar is derived from glucuronic acid and cannot be metabolized further in afflicted individuals), but the relative contributions from hyaluronic acid, xylosylserine-linked polysaccharides, and other glucuronides must be known for accurate calculations to be made.

The existence of a genetic deficiency in the enzyme β-aspartylace-tylglucosaminidase, which is associated with severe mental retardation, suggests that the normal functioning of the mammalian organism depends on an enzymatic apparatus capable of degrading the complex carbohydrates completely to their individual components (see Patel, 1978). In the absence of convincing evidence that xylosylserine represents the end point of degradation of chondroitin sulfate and other polysaccharides with the same linkage to protein, a continued search for an appropriate xylosidase would therefore seem justified. Also, the possibility should be considered that the xylose residues may be removed while still attached to a peptide rather than to a serine alone.

3.4.4. *N*-Acetylgalactosamine 4-Sulfate Sulfatase

This enzyme has an interesting history. It has been known for many years under the name of arylsulfatase B, and, like arylsulfatase A, it hydrolyzes *p*-nitrocatechol sulfate. Barium and silver ions inactivate arylsulfatase A, and separate assay of B activity in crude preparations of the two enzymes is therefore possible under appropriately chosen conditions. Credit is due, in particular, to Dodgson and Spencer (Dodgson and Spencer, 1953; Dodgson *et al.*, 1955; Dodgson and Rose, 1970) and Roy (Roy, 1953, 1958, 1976; Roy and Trudinger, 1970) for the early studies on these enzymes. At the time, the natural substrates for the two enzymes were not known; however, it is now clear that arylsulfatase A *in vivo* degrades galactosyl-3-sulfate ceramide (Dulaney and Moser, 1978), whereas the 4-sulfate groups in chondroitin sulfate and dermatan sulfate are the substrates for arylsulfatase B (Dorfman *et al.*, 1976). (A third form, arylsulfatase C, unlike the A and B forms, is membrane-bound and is presumably involved in the cleavage of steroid sulfates.)

The important function of arylsulfatase B in the degradation of connective tissue polysaccharides has been realized through studies of one of the mucopolysaccharidoses, the Maroteaux–Lamy syndrome. In 1973, Stumpf *et al.* (1973) showed that patients with this disease had a deficiency of arylsulfatase B activity in organs (liver, kidney, spleen, and brain) as well as in cultured fibroblasts. The finding that fibroblasts are deficient in the sulfatase, preliminarily reported by Stumpf *et al.* (1973), was subsequently documented more completely by Fluharty *et al.* (1974).

Since the level of dermatan sulfate (but not chondroitin sulfate) is increased in the urine of patients with the Maroteaux–Lamy syndrome, Stumpf *et al.* (1973) suggested that arylsulfatase B may be normally involved in the degradation of the former polysaccharide. Direct evidence for the participation of arylsulfatase B in the catabolism of dermatan sulfate *and* chondroitin 4-sulfate was reported in the following year by

O'Brien *et al.* (1974) and by Matalon *et al.* (1974*a*). A study of the nonreducing termini of ^{35}S-labeled dermatan sulfate accumulating in fibroblasts from patients with the Maroteaux–Lamy disease (O'Brien *et al.*, 1974) showed that these positions were occupied by 4-sulfated *N*-acetylgalactosamine residues. Whereas fibroblasts from normal individuals and patients with Hurler's syndrome and Sandhoff's disease released the sulfate groups from the nonreducing termini, cells from patients with the Maroteaux–Lamy syndrome were deficient in this regard. Similar results were obtained by Matalon *et al.* (1974*a*), who found that extracts of Maroteaux–Lamy fibroblasts released only 15% as much sulfate as controls, when incubated with chondroitin 4-sulfate or a heptasaccharide with an *N*-acetylgalactosamine 4-sulfate residue in the nonreducing terminal position. It was therefore concluded that the enzyme which is deficient in the Maroteaux–Lamy syndrome is a sulfatase for which the substrates are the 4-sulfate groups on the *N*-acetylgalactosamine residues of chondroitin sulfate and dermatan sulfate.

Further support for this conclusion was obtained by Fluharty *et al.* (1975), who showed that UDP-*N*-acetylgalactosamine 4-sulfate was a substrate for arylsulfatase B and that fibroblasts from patients with the Maroteaux–Lamy syndrome were equally deficient in a sulfohydrolase acting on this substrate as in arylsulfatase B.

Fractionation of arylsulfatase B by Gniot-Szulzycka and Donnelly (1976) has indicated that the crude enzyme contains a component, itself without enzymatic activity, which is essential for the activity of the purified enzyme toward poly- and oligosaccharide substrates. The nature of this factor is not yet known. In this investigation, it was also noted, in agreement with a previous observation by Matalon *et al.* (1974*a*), that more than one sulfate group per substrate molecule was released, indicating that internal as well as nonreducing terminal sulfate groups are recognized as substrates by the enzyme. Nevertheless, in view of erroneous conclusions which were drawn regarding the specificity of *N*-acetylgalactosamine 6-sulfate sulfatase in this regard (see Section 3.4.5), definitive studies of the location of susceptible sulfate residues are desirable, in which the participation of β-*N*-acetylhexosaminidase and β-glucuronidase in the degradation of the substrate has been ruled out.

A new form of arylsulfatase B has recently been found by Stevens *et al.* (1977), who showed that a minor, anionic form of the enzyme, B_m, is adsorbed to DEAE–cellulose (together with arylsulfatase A) under conditions where the major arylsulfatase B component is not bound. The two forms appear to be functionally equivalent and generically related; however, the B_m form is more heat-labile and is similar to arylsulfatase A in this regard. Analysis by isoelectric focusing yielded a value of 8.2 for the isoelectric point of placental arylsulfatase B, while the minor form from brain was resolved into three bands, with pI values of 6.8, 7.0, and 7.2.

Arylsulfatase B has also been detected, together with arylsulfatase A, in a rat basophil leukemia tumor (Wasserman and Austen, 1977) and was purified to near homogeneity from this source. Helwig *et al.* (1977) have reported 7700-fold purification of arylsulfatase B from rabbit kidney cortex, yielding a preparation which was homogeneous on polyacrylamide gel electrophoresis at several pH values. In the course of purification, it was observed that the enzyme bound to concanavalin A–Sepharose, indicating that it is a glycoprotein. In an extension of previous studies of the substrate specificity, Helwig *et al.* (1977) found that beside the substrates already mentioned, glucosamine 4,6-disulfate was also cleaved. This finding is somewhat surprising in view of the difference in orientation of the 4-sulfate groups on the two hexosamines.

3.4.5. *N*-Acetylgalactosamine 6-Sulfate Sulfatase

This enzyme was first described by Matalon *et al.* (1974*b*) in the course of their studies of the Morquio syndrome. It was observed that fibroblasts from patients with this disease were deficient in releasing sulfate from chondroitin sulfate with a preponderance of 6-sulfate groups (ratio of 6-sulfate to 4-sulfate, 60:40). In contrast, Morquio fibroblasts liberated sulfate from chondroitin 4-sulfate, although the activity was only half that observed for normal cells. Further studies with a heptasaccharide substrate showed more convincingly that the enzyme deficient in the Morquio fibroblasts is an *N*-acetylgalactosamine 6-sulfate sulfatase. This enzyme was clearly distinct from aryl sulfatases A and B, which were present at normal levels in the Morquio fibroblasts.

Besides chondroitin 6-sulfate, patients with the Morquio syndrome characteristically excrete substantial amounts of keratan sulfate. Since it could be assumed that the disease results from a mutation in a single gene, the question arose whether a substrate structure for the 6-sulfatase was also contained within the keratan sulfate molecule. Initially, Matalon *et al.* (1974*b*) suggested that the 6-sulfated *N*-acetylhexosamine residues constituted the common denominator (*N*-acetylgalactosamine 6-sulfate in chondroitin sulfate and *N*-acetylglucosamine 6-sulfate in keratan sulfate). However, Di Ferrante *et al.* (1978) have subsequently demonstrated that it is the galacto- configuration in *N*-acetylgalactosamine 6-sulfate and galactose 6-sulfate which is important for recognition by the enzyme. Greatly reduced sulfatase activity was observed in extracts of Morquio fibroblasts when the assay was conducted with galactitol 6-sulfate, *N*-acetylgalactosaminitol 6-sulfate, or a chondroitin 6-sulfate tetrasaccharide as substrate. In contrast, normal activity was measured with *N*-

acetylglucosamine 6-sulfate and its alditol. Furthermore, fibroblasts from a patient with a newly discovered mucopolysaccharidosis were deficient in activity toward N-acetylglucosamine 6-sulfate and its alditol but had normal activity for the galactose- and N-acetylgalactosamine-containing substrates. It was therefore concluded that two different hexosamine 6-sulfate sulfatases exist which are specific for either the gluco- or the galacto- configuration of the substrate. The observed accumulation of kearatan sulfate and heparan sulfate in the new mucopolysaccharidosis is in accord with the lack of an enzyme which cleaves the 6-sulfate groups from N-acetylglucosamine residues, and the presence of keratan sulfate and chondroitin 6-sulfate in patients with the Morquio syndrome is to be anticipated from the absence of a galacto-specific enzyme.

Some properties of N-acetylgalactosamine 6-sulfate sulfatase have been determined by Singh *et al.* (1976), and its substrate specificity has been more precisely defined in recent work by Horwitz and Dorfman (1978). Using a turbidimetric assay, Singh *et al.* (1976) measured liberation of sulfate from a nonradioactive tetrasaccharide prepared from chondroitin 6-sulfate by digestion with testicular hyaluronidase. An equally good substrate was obtained by β-glucuronidase digestion of the tetrasaccharide, yielding a trisaccharide with a 6-sulfated N-acetylgalactosamine residue at the nonreducing terminus. In contrast, polymeric chondroitin 6-sulfate and an unsaturated, 6-sulfated disaccharide were not attacked by the enzyme. Since extended incubations resulted in cleavage of both sulfate groups in the tetrasaccharide, it was concluded that internal sulfate groups are accessible to the enzyme; however, this view has been challenged by Horwitz and Dorfman (1978). These investigators demonstrated convincingly that the sulfatase acts exclusively on 6-sulfate groups located on nonreducing terminal N-acetylgalactosamine residues. Previous observations that sulfate was released from oligosaccharides with glucuronic acid in this position could be attributed to the presence of β-glucuronidase in the crude fibroblast extracts, since the reaction was completely inhibited by the addition of saccharo-1,4-lactone. In a control experiment, th β-glucuronidase inhibitor had no effect on sulfate liberation from a pentasaccharide with the proper substrate structure. Horwitz and Dorfman (1978) further noted that only 2% of the pure substrate was cleaved by extracts of fibroblasts from patients with Morquio's syndrome and concluded that previous findings of higher levels of cleavage were likely due to contamination of the substrates with 4-sulfated residues.

A purified sulfatase has not yet been available for study, but some useful information has been obtained by experiments with fibroblast extracts. Singh *et al.* (1976) have reported that the enzyme is inhibited by phosphate and high concentrations of sodium chloride, while it is unaffected by the presence of N-acetylgalactosamine, glucuronic acid, inorganic sulfate, or a tetrasaccharide from chondroitin 4-sulfate. The pH optimum is 4.8.

Very recently an important paper was published by Glössl *et al.* (1979). These investigators purified *N*-acetylgalactosamine 6-sulfate sulfatase approximately 20,000-fold from an extract of human placenta and determined some of its properties. The purified enzyme gave one major and two minor bands on polyacrylamide gel electrophoresis in SDS. The latter components, which comprised about 30% of the total protein, were considered to be impurities. A molecular weight of 78,000 was determined from the mobility of the major band, and a value of about 100,000 was obtained by gel chromatography. The purified enzyme was active towards *N*-acetylgalactosamine 6-sulfate but did not cleave galactose 6-sulfate, leaving unanswered the question whether it is capable of cleaving linkages of the latter type in keratan sulfate. A number of anions inhibited the activity, including sulfate, sulfite, thiosulfate, phosphate, cyanide, and chloride at high concentrations. Stimulation was observed in the presence of reagents containing SH-groups, such as 2-mercaptoethanol and dithiothreitol. Only one sulfate group could be released from the trisaccharide substrate used, *N*-acetylgalactosamine 6-sulfate–glucuronic acid–*N*-acetyl[1-^{3}H]galactosaminitol 6-sulfate, in accordance with the conclusions of Horwitz and Dorfman (1978).

4. KERATAN SULFATE

4.1. Structure of Keratan Sulfate

N-Acetyllactosamine, a commonly occurring disaccharide component in glycoproteins, is the repeating disaccharide unit of keratan sulfate (Table 1 and Figures 15 and 16). Sulfate groups are present in the C-6

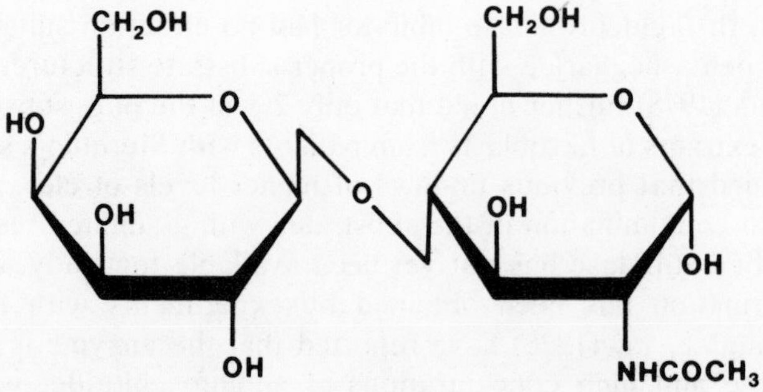

Figure 15. Linkage unit 13: Gal $\xrightarrow{\beta 1,4}$ GlcNAc.

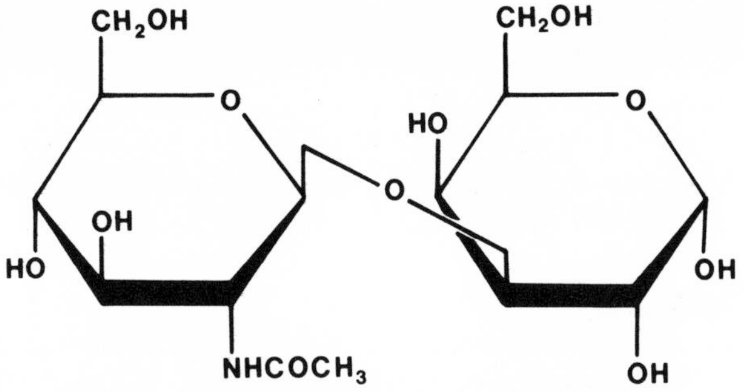

Figure 16. Linkage unit 14: GlcNAc $\xrightarrow{\beta1,3}$ Gal.

position of the glucosamine units (Figure 17) and also on some of the galactose residues (Figure 18). Two keratan sulfate types have been distinguished (Meyer, 1970), one located exclusively in the cornea and the other type being present in several skeletal tissues (nucleus pulposus, cartilage, bone). A major difference between these two types resides in their linkage to protein (Seno *et al.*, 1965; Baker *et al.*, 1969, 1975; Bray *et al.*, 1967). Whereas corneal keratan sulfate (keratan sulfate I) is bound to protein via type 2 linkages, i.e., the *N*-glycosylamine linkage between *N*-acetylglucosamine and the amide group of asparagine (Figure 19), skeletal keratan sulfate (keratan sulfate II) contains type 3 linkages between *N*-acetylgalactosamine and threonine or serine residues (Figure 20). Keratan sulfate I is the sole polysaccharide component of its parent proteoglycan in the cornea (Berman, 1970), while keratan sulfate II is always found together with chondroitin sulfate in the skeletal proteoglycans. In some of their properties, the keratan sulfates are similar to other

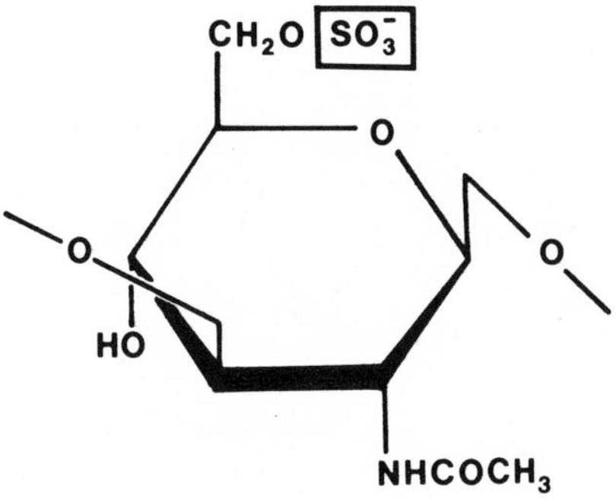

Figure 17. Linkage unit 15: GlcNAc-6-sulfate.

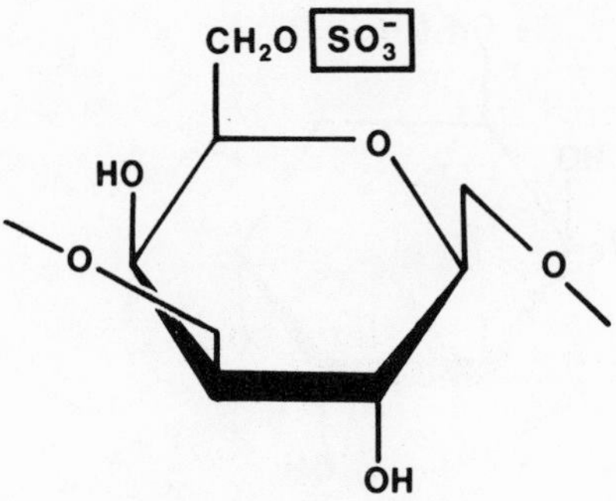

Figure 18. Linkage unit 16: Gal-6-sulfate.

connective tissue polysaccharides, inasmuch as they are composed of a number of repeating disaccharide units and carry ester sulfate groups. On the other hand, the keratan sulfates are in many respects more akin to the glycoproteins; e.g., uronic acid is absent, and the repeating disaccharide is similar to that found in many glycoproteins. In addition, keratan sulfate of both types contains sialic acid (Gregory and Rodén, 1961; Seno *et al.*, 1965), fucose (Seno *et al.*, 1965), and mannose (Bhavanandan and Meyer, 1968). The presence of mannose in keratan sulfate I is not unexpected, since this polysaccharide is linked to protein by the type 2 linkage characteristic of glycoproteins which contain a mannose-*N*-acetylglucosamine core. Recent experiments in Hascall's laboratory have raised some doubts as to the presence of mannose in keratan sulfate II (V. C. Hascall, personal communication). Rather, it appears that mannose is part of a glycoprotein which follows keratan sulfate closely in the purification steps commonly used for the isolation of the polysaccharide. The position of sialic acid has been studied by Hopwood and Robinson

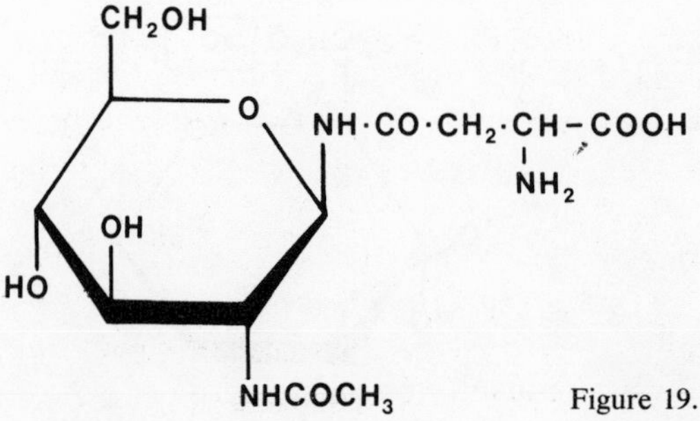

Figure 19. Linkage unit 11: GlcNAc $\xrightarrow{\beta}$ Asn.

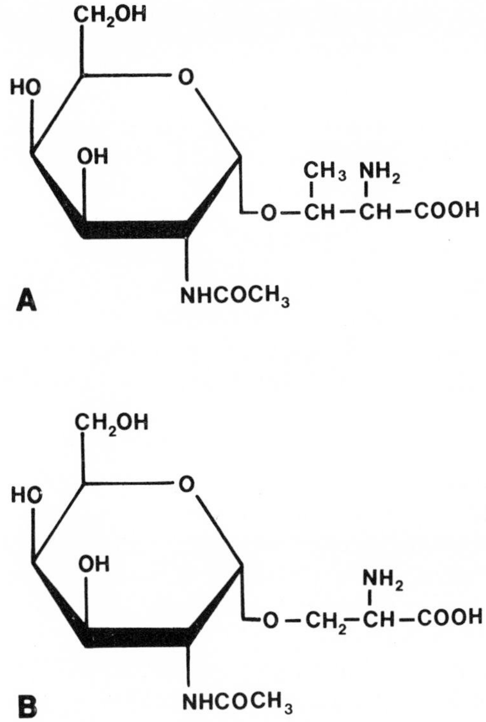

Figure 20. Linkage unit 12: GalNAc $\xrightarrow{\alpha}$ Thr (A) and GalNAc $\xrightarrow{\alpha}$ Ser (B).

(Hopwood, 1972; Hopwood and Robinson, 1974), who have shown that the disaccharide N-acetylneuraminylgalactose is attached to the C-3 position of the N-acetylgalactosaminyl residue which mediates the linkage to protein. The bulk of the keratan sulfate chain is linked to the C-6 position of this same residue. Finally, it should also be noted that the galactose content of keratan sulfate often exceeds that required for the repeating disaccharide structure. Some of the extra galactose residues are presumably linked in side chains, and methylation studies indicated that a majority of the nonreducing terminal residues are galactose (Bhavanandan and Meyer, 1968). However, the exact location of the additional galactose residues remains to be established with greater certainty.

Although molecular weights on the order of 10,000–15,000 have been reported for various keratan sulfate preparations (see, for example, Anseth and Laurent, 1961; Gregory et al., 1964), shorter chains with only a few repeating disaccharide units are also often encountered.

4.2. Biosynthesis of Keratan Sulfate

4.2.1. Linkage Unit 11

As might be expected from the presence of a type 2 carbohydrate–protein linkage in corneal keratan sulfate (Figure 19), the biosynthesis of this polysaccharide shows similarities to that of the analogous glycoproteins. Tunicamycin, an inhibitor of the enzyme catalyzing transfer of *N*-acetylglucosamine 1-phosphate to dolichol monophosphate, drastically decreases the rate of synthesis of keratan sulfate upon incubation of corneas *in vitro* (Hart and Lennarz, 1978). Significant inhibition of the synthesis of other glycosaminoglycans was also observed. However, since a concomitant general decrease in protein synthesis occurred, this effect was considered nonspecific.

Other aspects of the formation of the keratan sulfate–protein linkage region have not yet been investigated, such as the mannosyl-transfer reactions and the introduction of additional *N*-acetylglucosamine in the linkage region. Nor is any specific information available regarding the presumed *N*-acetylgalactosaminyl transfer to threonine and serine residues in the course of initiation of keratan sulfate II chains (linkage unit 12, Figure 20).

4.2.2. Linkage Units 13 and 14

The galactosyl-transfer reaction in keratan sulfate biosynthesis is presumably catalyzed by lactose synthetase, which is involved in the formation of the *N*-acetyllactosamine units (see Figure 15) of many glycoproteins (Chapter 3). However, direct information pertaining to this reaction in keratan sulfate biosynthesis or to the postulated *N*-acetylglucosaminyl-transfer reaction has not been reported so far.

4.2.3. Linkage Units 15 and 16

Little information is at hand concerning the enzymatic aspects of the sulfotransferase reactions involved in keratan sulfate synthesis. In 1961, Wortman (1961) reported transfer of sulfate from PAPS to an unresolved mixture of corneal polysaccharides which probably contained keratan sulfate as a major component.

4.3. Catabolism of Keratan Sulfate

From the known composition of keratan sulfate, it is evident that a large battery of glycosidases is required to completely degrade this polysaccharide to its monosaccharide constituents. In addition, two sulfatases are needed to remove the sulfate groups on galactose and N-acetylglucosamine residues. In the following discussion, we shall limit our attention to the components of the repeating disaccharide units, i.e., N-acetylglucosamine, galactose, and the sulfate groups. The enzymes responsible for the degradation of this portion of the molecule are: (1) β-D-N-acetylhexosaminidase, (2) β-D-galactosidase, (3) N-acetylglucosamine 6-sulfate sulfatase, and (4) galactose 6-sulfate sulfatase (N-acetylgalactosamine 6-sulfate sulfatase). For information regarding glycosidases acting on other components (sialic acid, mannose, fucose), the reader is referred to the review by Patel (1978), which also describes the enzymes which cleave the carbohydrate–protein linkage fragments GlcNAc-Asn and GalNAc-Thr(Ser) (linkage units 11 and 12).

It may be assumed that the β-D-N-acetylglucosamine residues in keratan sulfate are removed, i.e., by β-D-N-acetylhexosaminidase A, which is known to cleave such residues in $\beta1,3$ linkage to galactose; however, specific information concerning this aspect of the degradation of keratan sulfate is not available. It might be postulated that keratan sulfate as well as glycoprotein fragments should accumulate in Tay–Sachs and Sandhoff's diseases, with deficiencies in isozyme A, and in both A and B, respectively. This possibility has not been thoroughly investigated; however, it has been noted earlier that Tay–Sachs cells behave paradoxically in showing little or no signs of mucopolysaccharide accumulation despite their deficiency in a key enzyme and that isozymes other than the A form may participate in the physiological degradation of the complex carbohydrates under discussion.

As has been the case throughout the history of research on the catabolism of polysaccharides and proteoglycans, much of our knowledge about this area of keratan sulfate metabolism has come from studies of patients with genetic defects of degradation. At present, three genetic defects are known which result in defective keratan sulfate catabolism, manifesting itself in keratan sulfaturia and tissue accumulation of incompletely degraded products: (1) the classic Morquio syndrome, caused by deficiency in galactose 6-sulfate sulfatase (N-acetylgalactosamine 6-sulfate sulfatase); (2) deficiency in N-acetylglucosamine 6-sulfate sulfatase, which clinically gives a composite picture with traits of both the Morquio and the Sanfilippo syndromes; and (3) β-galactosidase deficiency, a mild form of the Morquio disease, where the defective keratan sulfate degradation is more prominent than the derangement of ganglioside metabolism. N-acetylgalactosamine 6-sulfate sulfatase has been discussed in Section 3.4.5, and the following presentation will be concerned with β-galactosidase and the second sulfatase.

4.3.1. β-D-Galactosidase

Several forms of β-D-galactosidase exist in mammalian tissues, which differ in their pH optima (acid vs. neutral as well as intermediate forms), molecular weights, substrate specificity, and other properties. The wealth of information concerning this enzyme which has accumulated in recent years cannot be properly covered here, and reference is made to more extensive reviews by Patel (1978) and O'Brien (1978).

Four galactose-containing disaccharide structures have been characterized within the connective tissue polysaccharides: Gal $\xrightarrow{\beta 1,3}$ Gal and Gal $\xrightarrow{\beta 1,4}$ Xyl from the carbohydrate–protein linkage region of chondroitin sulfate and other members of the group; Gal $\xrightarrow{\beta 1,4}$ GlcNAc, which is the characteristic repeating disaccharide unit of keratan sulfate; and Gal $\xrightarrow{\beta 1,3}$ GalNAc, which is located at the linkage of keratan sulfate II to protein. In addition, keratan sulfate contains extra galactose residues of undetermined position. Data pertinent to the cleavage of three of the four disaccharide units will be discussed here.

Two forms of β-D-galactosidase (A and B) with an acid pH optimum have been recognized. The major A form has been purified from human liver, and a number of its properties have been determined (Norden *et al.*, 1974). It has a molecular weight of 72,000 and has antigenic determinants in common with the much larger B form (mol. wt. 600,000–800,000), from which it can be generated under appropriate conditions. Deficiency of the acid β-galactosidase results in G_{M1} gangliosidosis, and, at the same time, keratan-sulfate-like material (Tsay and Dawson, 1973) and glycoprotein fragments accumulate (Wolfe *et al.*, 1974). These observations are in keeping with studies of the substrate specificity of β-galactosidase which have shown that it cleaves not only G_{M1} ganglioside but also lactose, N-acetyllactosamine, and the Gal-GlcNAc linkage in asialofetuin and other similar glycoproteins. It is not known whether glycopeptides containing the Gal ⟶ Gal ⟶ Xyl ⟶ Ser sequence accumulate in G_{M1} gangliosidosis; however, Distler and Jourdian (1973) have purified a β-galactosidase from bovine testis which is similar in its properties to the A enzyme of Norden *et al.* (1974) and is capable of cleaving the two linkage-region galactoses.

Deficiency in acid β-galactosidase activity does not necessarily result in full-blown manifestations of G_{M1} gangliosidosis. Two patients have recently been described (O'Brien *et al.*, 1976; Arbisser *et al.*, 1977; see also Spranger, 1977) who displayed, instead, the clinical symtoms of a mild form of Morquio's disease with only moderate involvement of the nervous system. β-Galactosidase deficiency was the only enzymatic abnormality detectable, and N-acetylgalactosamine 6-sulfate sulfatase activity was normal in the patient of Arbisser *et al.* (1977), although a deficiency in this enzyme would have been expected on the basis of the clinical classification. It is interesting to note that the patient observed by O'Brien *et al.* (1976) had a higher residual activity for G_{M1} ganglioside than for asialofetuin in cultured fibroblasts. In the same vein, Arbisser *et al.* (1977) reported biochemical evidence for defective keratan sulfate degradation. These findings may be explained by the occurrence, within the same heterocatalytic enzyme, of different mutations which affect in a nonuniform manner the catalytic activities toward different substrates. It appears that in the patients under consideration, a mutation in β-galactosidase has occurred which has eliminated most of the catalytic activity toward keratan-sulfate-like substrates, while activity toward G_{M1} ganglioside has been substantially retained.

4.3.2. *N*-Acetylglucosamine 6-Sulfate Sulfatase

This enzyme, which was recently discovered by Di Ferrante and co-workers (Di Ferrante *et al.*, 1978), has been purified from normal human urine by Basner *et al.* (1979). After only 136-fold purification, a homogeneous preparation was obtained which was subjected to further characterization. The purified enzyme is a glycoprotein with a molecular weight of approximately 97,000 and shows considerable charge heterogeneity. Multiple forms with pI values between 5.4 and 8.3 were detected, with a maximum activity at pH 7.7. Studies of the substrate specificity of the enzyme indicate that it acts almost exclusively as an exosulfatase and has a pH optimum of 5.5. It utilizes *N*-acetylglucosamine 6-sulfate and glucose 6-sulfate as substrates, with K_m values of 1.5 and 7.7 mM, respectively. A trisaccharide substrate had a K_m value of 0.15 mM. The enzyme was present in all human tissues tested, with highest activities in kidney and fibroblasts.

5. HEPARIN AND HEPARAN SULFATE PROTEOGLYCANS

5.1. Structure of Heparin

The chemistry of heparin has been the subject of several recent reviews (Lindahl, 1976; Lindahl *et al.*, 1977; Lindahl and Höök, 1978; Rodén and Horowitz, 1978). Through the efforts of many laboratories, a picture of this polysaccharide has emerged which is remarkably different from that held only a few years ago. As can be seen from Table 1, the repeating disaccharide unit of heparin is composed of glucosamine and a uronic acid residue. The latter may be either L-iduronic acid or D-glucuronic acid, and in "full-fledged" heparin, iduronic acid is by far the major uronic acid component and accounts for 70–90% of the total uronic acid. Most of the amino groups carry an N-sulfate group, although a small proportion of the glucosamine residues are N-acetylated (Cifonelli and King, 1972). Most glucosamine residues also carry an ester sulfate group in the C-6 position, and in addition, most of the iduronic acid residues are sulfated at C-2. A heptasaccharide illustrating these structural features is shown in Figure 21.

The existence of a xylose–serine linkage between the core protein and the polysaccharide chains of several connective tissue proteoglycans was first established by a study of heparin preparations which had been subjected to proteolysis in the course of manufacture but had not undergone the final, more drastic steps including bleaching (Lindahl et al., 1965; Lindahl and Rodén, 1964, 1965; Lindahl, 1966a,b). Although these studies represented the first tangible evidence for the occurrence, in vivo, of a heparin proteoglycan, the search for such a native molecule has only recently been successful. Some years back, Horner (1971) reported the isolation of a "macromolecular" form of heparin which was obtained from rat skin following exhaustive pronase digestion and had a molecular weight of 10^6 or more. Since the proteolytic treatment to which this material had been exposed would have degraded a chondroitin sulfate proteoglycan extensively, no serious consideration was given to the possibility that the macromolecular heparin might be a proteoglycan, and Horner suggested that heparin chains with a molecular weight of approximately 36,000 (estimated on the basis of serine content related to hexosamine content) are held together by a link structure which is distinct from a protein core and spans several polysaccharide chains. More recently, it has been shown (Robinson et al., 1978) that the macromolecular heparin does indeed have a "classic" proteoglycan structure, inasmuch as it consists of several heparin chains linked to a common protein core. The protein is unusual in that it is composed of serine and glycine residues only, and conservative estimates indicated that at least two-thirds of the serine residues were polysaccharide-bound. This unusual structure is, not unexpectedly, resistant to pronase and other proteolytic enzymes but is cleaved upon treatment with strong alkali in the presence of borohydride to yield terminal xylitol groups. Characterization of the heparin chains liberated by this procedure showed that they were of unusually high molecular weight, 60,000–100,000, as compared to the molecular-weight range of commercial heparin, which is on the order of 10,000–15,000. The formation of the smaller species seen in commercial heparin apparently occurs by enzymatic cleavage of the larger chains at specific glucuronic acid residues, leaving heparin chains devoid of the typical carbohydrate–protein linkage region and terminating in glucuronic acid (Ögren and Lindahl, 1975). Further details of this process and the

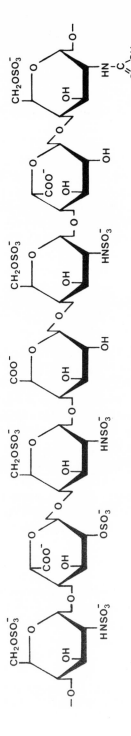

Figure 21. A heptasaccharide fragment of heparin, illustrating some characteristic structural features.

properties of the macromolecular heparin are reviewed elsewhere (Lindahl and Höök, 1978; Rodén and Horowitz, 1978).

In summary, the following features of the structure of heparin now appear reasonably well established:

1. The bulk of the polysaccharide chain consists of alternating units of uronic acid and glucosamine, bound in 1,4 linkages.
2. The majority of the uronic acid residues are L-iduronic acid (70–90%); the remainder are D-glucuronic acid.
3. Most of the amino groups of the glucosamine residues are sulfated, yielding a structure that is unique to heparin and heparan sulfate; a minor proportion of the amino groups is acetylated, and yet another small fraction is unsubstituted.
4. Most, probably all, of the glucuronic acid residues are β-linked; glucosamine and iduronic acid residues are α-linked.
5. Ester sulfate (O-sulfate) groups are present on C-6 of the glucosamine residues and perhaps to some extent on C-3.
6. O-sulfate groups are also present on C-2 of many but not all of the iduronic acid units; the glucuronic acid residues are never sulfated.
7. A variable fraction of the heparin chains are linked to serine or small peptides and contain the specific carbohydrate–protein linkage region ($\rightarrow$ GlcUA $\rightarrow$ Gal $\rightarrow$ Gal $\rightarrow$ Xyl $\rightarrow$ Ser) identical to that found in several other connective tissue polysaccharides. The presence of this linkage region suggests that heparin is synthesized *in vivo* as a proteoglycan.
8. In most tissues studied, heparin occurs primarily as single polysaccharide chains with molecular weights ranging from 5000 to 15,000. Some of these chains have a full complement of carbohydrate–protein linkage region components (serine or serine-containing peptides, xylose, and galactose); others are deficient in this regard and carry a glucuronic acid residue at the reducing terminus.
9. In addition to the single-chain type, a macromolecular form of heparin with a molecular weight of approximately 1×10^6 has been isolated from rat skin after proteolysis, and a group of similar but smaller compounds has been found in a mouse mast cell tumor. The macromolecular heparin has recently been shown to be a proteoglycan containing several large heparin chains (mol. wt. 60,000–100,000) linked to a core composed of equimolar amounts of serine and glycine residues.
10. The macromolecular heparin may be degraded to products with a molecular weight of approximately 14,500 by an endoglucur-

onidase present in certain tissues, including the Furth mouse mastocytoma, normal rat and mouse intestine, and human spleen.

5.2. Biosynthesis of Heparin

Much of the recent interest in heparin biosynthesis has centered around the enzymatic aspects of the formation of the polysaccharide chain and the modifications occurring after polymer formation. From a methodological standpoint, it may be noted that cell-free preparations from transplantable mast cell tumors (Furth and Dunn–Potter types) have been invaluable in providing relatively easily accessible systems in which to investigate these processes. Most of the basic information was furnished by the early studies of Silbert (1963, 1967a–c), who demonstrated that polysaccharide chain formation occurred when a microsomal fraction of the tumor was incubated with UDP-N-acetylglucosamine and UDP-glucuronic acid. The resulting polysaccharide had the same charge density as hyaluronic acid, as shown by chromatography on DEAE–cellulose, but was clearly different from this polysaccharide insofar as it was not degraded by testicular hyaluronidase. Rather, it could be related to heparin, since it was cleaved to small oligosaccharides by a heparinase from *Flavobacterium heparinum*. In the presence of PAPS, a sulfated polymer with the properties of heparin was formed, in which about half the total sulfate appeared to be N-sulfate. In the course of N-sulfation, a corresponding loss of N-acetyl groups was demonstrated (Silbert, 1967c). This N-deacetylation occurred both when PAPS was present during polymerization and when it was added subsequent to the polymerization process.

Since this pioneering effort, there have been considerable advances in our understanding of the structure of heparin as well as the mechanisms by which complex carbohydrate molecules are synthesized. As a result, we have been, for some time, in a position to formulate certain hypotheses regarding the routes by which a molecule like heparin is most likely to be formed at the enzymatic as well as the subcellular level. It was indicated earlier that the "one enzyme–one linkage" hypothesis requires each monosaccharide component of a complex carbohydrate molecule to be transferred from the corresponding nucleotide sugar to an appropriate acceptor of the correct structure to form a specific linkage. In this vein, we have postulated, in the past, that the formation of the iduronic acid residues of dermatan sulfate and heparin would require a nucleotide sugar containing iduronic acid, from which the sugar moiety is transferred to a growing polysaccharide chain. Although such a nucleotide sugar has been described (Jacobson and Davidson, 1962, 1963), its existence has never been confirmed, and it has never been shown to participate in the synthesis of iduronic-acid-containing polysaccharides. In contrast, the process whereby the iduronic acid residues are formed in the course of heparin biosynthesis is an epimerization taking place within the polymer. The discovery of this route of synthesis by Lindahl and his collaborators (Lindahl *et al.*, 1972; Höök *et al.*, 1974) has opened up a new and intriguing line of research in the field of complex carbohydrate formation and especially in the area of polymer modifications. These modifications are particularly extensive in heparin biosynthesis and include deacetylation, *N*-sulfation, and two *O*-sulfation steps (involving iduronic acid and glucosamine) in addition to uronosyl epimerization.

From the available data concerning biosynthesis in combination with the structural information, we may then suggest the following tentative scheme for the major steps in heparin synthesis:

1. Formation of a core protein.
2. Synthesis of the polysaccharide chain by a process involving:
   a. Chain initiation by a xylosyltransferase.
   b. Completion of the carbohydrate–protein linkage region by three additional glycosyltransferases (two galactosyltransferases and a glucuronosyltransferase).
   c. Formation of repeating disaccharide units by the alternating actions of an *N*-acetylglucosaminyltransferase and a glucuronosyltransferase (different from the one involved in the completion of the linkage region).
3. Polymer modifications, including *N*-deacetylation, *N*-sulfation of the amino groups of the glucosamine units, epimerization of D-glucuronic acid residues to L-iduronic acid, and *O*-sulfation of the C-2 position of L-iduronic acid and the C-6 position of glucosamine units, respectively.
4. Proteolytic degradation of the core protein.
5. Cleavage of the macromolecular product by a specific endoglucuronidase.

At present, no information is available concerning the biosynthesis of the protein core of the heparin proteoglycan. For the next step, i.e., formation of the carbohydrate–protein linkage region, it is presumed that the same enzymes that catalyze chondroitin sulfate synthesis are also involved in this process, and they will therefore not be discussed separately here. It should be noted, however, that some of our knowledge of the properties of these enzymes is derived from studies of a heparin-synthesizing mastocytoma (Grebner *et al.*, 1966*b*; Helting, 1971, 1972).

According to the "one enzyme–one linkage" concept, we may postulate the existence of two distinct glycosyltransferases catalyzing (1) transfer of N-acetylglucosamine to nonreducing terminal glucuronic acid residues and (2) transfer of glucuronic acid to terminal N-acetylglucosamine units yielding linkage units 17 and 18, respectively (Figures 22 and 23). Both these enzyme activities have indeed been detected by the use of appropriate substrates, as will be described below.

5.2.1. Linkage Unit 17

In the absence of native acceptors of well-defined structure, certain oligosaccharides from the heparin–protein linkage region have been used to demonstrate the existence of the N-acetylglucosaminyltransferase (Helting and Lindahl, 1972), in analogy with the approach used by Dorfman and collaborators in their studies of chondroitin sulfate biosynthesis (Stoolmiller and Dorfman, 1969b; Rodén, 1970; Dorfman, 1974). The substrates tested so far are shown in Table 4. These compounds were all isolated after degradation of heparin with nitrous acid, and the fragments from the carbohydrate–protein linkage region were separated from the oligosaccharide products by chromatography on Dowex 50. From the mode of preparation, it could be assumed that the compounds listed would all carry a uronic acid residue at the nonreducing terminus, either glucuronic acid or iduronic acid, and they would therefore be potential acceptors for N-acetylglucosaminyl transfer (an exception was Fraction B_1-β, which was obtained by digestion of Fraction B_1 with β-glucuronidase). Of the linkage-region fragments shown in Table 4, only Fraction B_1 had substantial acceptor activity. This material consisted of two repeating disaccharide units linked to the tetrasaccharide of the linkage region and a serine residue. Approximately half the nonreducing termini were glucuronic acid and the other half iduronic acid. Significantly, the product of β-glucuronic digestion, Fraction B_1-β, was a poor substrate,

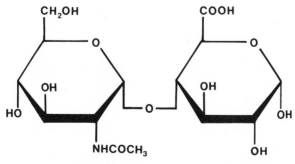

Figure 22. Linkage unit 17: GlcNAc $\xrightarrow{\alpha1,4}$ GlcUA.

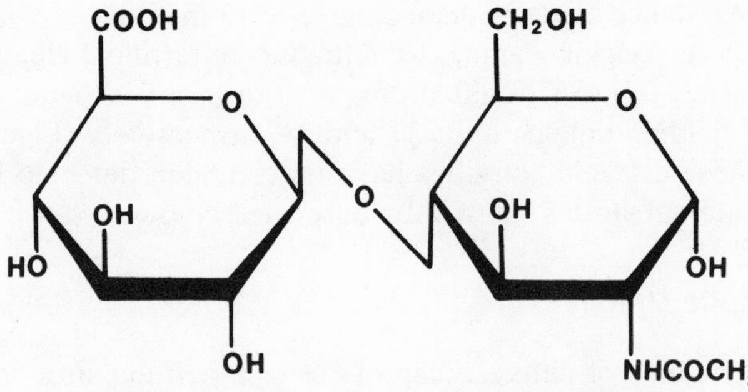

Figure 23. Linkage unit 18: GlcUA $\xrightarrow{\alpha1,4}$ GlcNAc.

indicating that little or no transfer to iduronic acid residues occurred. Some acceptor activity was observed with a hyaluronic acid hexasaccharide with glucuronic acid at the nonreducing terminus. In the specific target area, this oligosaccharide differs from the heparin oligosaccharides only in having a $\beta1,3$ rather than a $\beta1,4$ linkage. It is noteworthy that no transfer occurred to the smaller heparin fragments, such as B_2, which has one disaccharide unit less than B_1. This observation is somewhat disturbing. Although trivial explanations can be envisaged, such as a relatively great difference in K_m values between the two homologues, other alternatives must also be considered. For example, we cannot yet rule out the possibility that lipid intermediates participate in the synthesis of the region close to the carbohydrate–protein linkage.

Since the preparation of the carbohydrate–protein linkage fragments from heparin is a cumbersome process, alternative substrates have been sought, and Forsee *et al.* (1978) have recently reported the use of oligosaccharides from heparan sulfate for the assay of N-acetylglucosaminyltransferase activity (Table 4). In keeping with the observations of Helting and Lindahl (1972), only the larger oligosaccharides were good acceptors.

In contrast to the "polymerizing" glucuronosyltransferase (see Section 5.2.2), the N-acetylglucosaminyltransferase was not appreciably solubilized by treatment with Tween 20 and alkali (Helting and Lindahl, 1972). However, more than half the activity has been solubilized by extraction with Triton X-100, and the enzyme has now been partially purified by conventional procedures in combination with affinity chromatography on heparan sulfate–Sepharose (W.T. Forsee and L. Rodén, unpublished results).

5.2.2. Linkage Unit 18

The glucuronosyltransferase catalyzing the formation of this unit (Figure 23) has been demonstrated in mouse mastocytoma by Helting

Table 4. Potential Acceptors for N-Acetylglucosaminyl Transfer in Heparin Biosynthesis

Hyaluronic acid hexasaccharide   GlcUA ⟶ GlcNAc ⟶ GlcUA ⟶ GlcNAc ⟶ GlcUA ⟶ GlcNAc

Compound A₂   IdUA ⟶ GlcNAc ⟶ GlcUA ⟶ Gal ⟶ Gal ⟶ Xyl ⟶ Ser

Compound B₂   GlcUA ⟶ GlcNAc ⟶ GlcUA ⟶ Gal ⟶ Gal ⟶ Xyl ⟶ Ser

Fraction B₁   { IdUA ⟶ GlcNAc ⟶ UA ⟶ GlcNAc ⟶ GlcUA ⟶ Gal ⟶ Gal ⟶ Xyl ⟶ Ser
                 GlcUA ⟶ GlcNAc ⟶ UA ⟶ GlcNAc ⟶ GlcUA ⟶ Gal ⟶ Gal ⟶ Xyl ⟶ Ser

Fraction B₁-β   { IdUA ⟶ GlcNAc ⟶ UA ⟶ GlcNAc ⟶ GlcUA ⟶ Gal ⟶ Gal ⟶ Xyl ⟶ Ser
                   GlcNAc ⟶ UA ⟶ GlcNAc ⟶ GlcUA ⟶ Gal ⟶ Gal ⟶ Xyl ⟶ Ser

Oigosaccharides prepared from desulfated   (UA-GlcNAc)ₙ-UA-anhydromannose

heparan sulfate or desulfated heparin by

treatment with HNO₂

and Lindahl (Helting and Lindahl, 1971, 1972; Helting, 1972). This enzyme catalyzed transfer to linkage-region fragments containing only one N-acetylglucosamine residue as well as to Fraction B_1-β, with two N-acetylglucosamine residues (Table 5). The need for an N-acetylglucosamine residue linked in α linkage was indicated by the finding that a pentasaccharide from hyaluronic acid (GlcNAc $\rightarrow$ GlcUA $\rightarrow$ GlcNAc $\rightarrow$ GlcUA $\rightarrow$ GlcNAc) was inactive as acceptor. The crucial difference between this oligosaccharide and the substrates listed in Table 5 is the anomeric configuration of the linkage of the nonreducing terminal N-acetylglucosamine residue, which is β rather than α.

Treatment with detergent and alkali solubilized approximately 70% of the glucuronosyltransferase activity in the Furth mouse mastocytoma (Helting, 1972), but subsequent gel chromatography failed to separate the enzyme from the bulk of the solubilized protein. Further, purification by other methods has not yet been attempted.

5.2.3. Modification of the Polymer

Following the synthesis of the polysaccharide backbone, several modifications are required before the molecule has assumed the structure typical of a completed heparin molecule. These modifications include (1) deacetylation, (2) N-sulfation, (3) epimerization of glucuronic acid residues to iduronic acid, and (4) O-sulfation in two positions: on C-2 of the iduronic acid residues and on C-6 of the glucosamine units. It has been apparent for some time that modifications (1), (2), and (4) occur at the polymer level, but only in the last couple of years has it become evident that this is also true for the epimerization. Furthermore, the investigations of Lindahl and collaborators (Lindahl, 1976; Lindahl et al., 1977) have provided strong evidence in favor of the notion that the modification reactions do not occur at random but in a strictly ordered manner, progressing through several distinct intermediate polymer species. Because the overall process is rapid, and formation of fully sulfated heparin chains requires less than 30 sec, the design of these studies has, in general terms, been aimed at experimental segregation of the individual steps by substrate limitation, thus permitting various intermediates to accumulate. Specifically, characterization of the products formed in the presence and absence of PAPS has led to considerable clarification of the sequence of events, and some of the important experiments of this kind will be briefly reviewed (see Lindahl, 1976; Lindahl et al., 1972, 1973, 1976; Höök et al., 1974, 1975a; Jansson et al., 1975).

In a typical experimental protocol, the microsomal fraction of the Furth mastocytoma is incubated with UDP-N-acetylglucosamine and UDP-[^{14}C]glucuronic acid; after a certain period of time, PAPS is added,

Table 5. Acceptors for Glucuronosyl Transfer in Heparin Biosynthesis

GlcNAc $\longrightarrow$ GlcUA $\longrightarrow$ Gal

GlcNAc $\longrightarrow$ GlcUA $\longrightarrow$ Gal $\longrightarrow$ Gal

GlcNAc $\longrightarrow$ GlcUA $\longrightarrow$ Gal $\longrightarrow$ Gal $\longrightarrow$ Xyl

Fraction B_1-β:

GlcNAc $\longrightarrow$ UA $\longrightarrow$ GlcNAc $\longrightarrow$ GlcUA $\longrightarrow$ Gal $\longrightarrow$ Gal $\longrightarrow$ Xyl $\longrightarrow$ Ser

IdUA $\longrightarrow$ GlcNAc $\longrightarrow$ UA $\longrightarrow$ GlcNAc $\longrightarrow$ GlcUA $\longrightarrow$ Gal $\longrightarrow$ Gal $\longrightarrow$ Xyl $\longrightarrow$ Ser and incubation is continued. Alternatively, all three precursors are added simultaneously at the beginning of the incubation. Following digestion with papain, separation of the reaction products in the various incubation mixtures is carried out by high-resolution chromatography on DEAE–cellulose.

Representative product patterns obtained by this approach are shown in Figure 24. In the absence of PAPS, two products are formed; one elutes in the same position as hyaluronic acid, while the other emerges even earlier, indicative of a lower charge density. In the presence of PAPS, three additional peaks are observed; the most retarded of these appears in approximately the same position as standard heparin.

Characterization of the five reaction products with respect to the content of N-acetyl groups, free amino groups, N-sulfate groups, O-sulfate groups, and uronic acid composition gave the results illustrated in Figure 24, which also shows the major disaccharides found in each product. This information easily lends itself to interpretation in terms of a biosynthetic scheme, which may be formulated as follows:

1. The primary product of biosynthesis is the material appearing in peak 2, which consists of a nonsulfated, fully N-acetylated polymer composed of glucuronic acid and N-acetylglucosamine; no iduronic acid is found in this material.

2. Component 2 is deacetylated to yield component I, which has a lower charge density because of its free amino groups and therefore emerges from the DEAE–cellulose column in an earlier position; deacetylation is only partial, and approximately 50% of the glucosamine residues are still acetylated; sulfate groups are absent, and glucuronic acid is the only uronic acid component.

3. The free amino groups become sulfated to yield component 3, in which the content of acetyl groups has also decreased further to about 25%; glucuronic acid is the major or only uronic acid component of this fraction, but some iduronic acid has been detected on occasion.

4. The N-sulfated polymer is exposed to the action of an epimerase, which effects conversion of a large proportion of the glucuronic acid residues to iduronic acid; at the same time, an O-sulfotransferase introduces a sulfate group on C-2 of the newly formed iduronic acid residues; this material appears in peak 4. It should be noted that epimerization and O-sulfation are tightly coupled processes; however, their exact relationship is not yet clear.

5. Finally, the heparin molecule is completed by introduction of a second O-sulfate group on C-6 of the glucosamine units.

The very fact that five distinct components may be isolated from the *in vitro* heparin-synthesizing system is in itself strong evidence for an

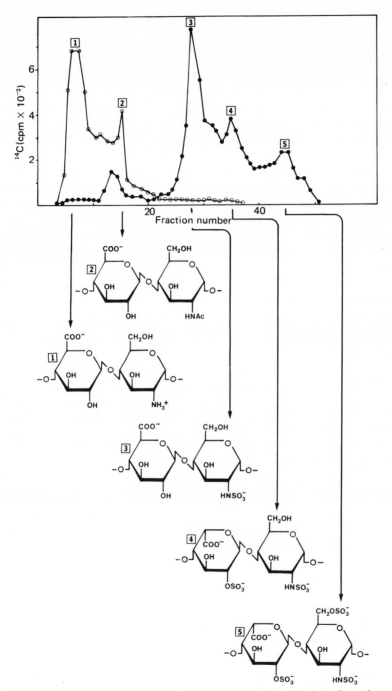

Figure 24. Chromatography on DEAE–cellulose of radioactive products formed upon incubation of mastocytoma microsomes with UDP-glucuronic acid and UDP-N-acetylglucosamine in the presence (●—●) or absence (○—○) of PAPS. Structures of the major disaccharides in each product are also shown. Modified from I. Jacobsson, Ph.D. thesis, Swedish Agricultural University, Uppsala, 1979.

ordered progression of events that passes through several discrete stages. Were the assembly of the polysaccharide a completely random process with simultaneous deacetylation and sulfation of the same molecule, the chromatographic pattern would have appeared totally different and in all likelihood would have consisted of a single broad peak with a gradually changing composition. Since the system described above is an artificial experimental construction, it remains to be seen whether the process *in vivo* follows the same pattern. It would seem, however, that the stricter regulation possible under physiological conditions might tend to sharpen the distinctions among the various phases even further. Nevertheless, it should be kept in mind that the products of *in vivo* synthesis always show some remnants of earlier biosynthetic stages and that, for example, acetyl groups are still present in the most highly sulfated heparin preparations.

In the following discussion, some additional information concerning the individual steps will be presented.

5.2.4. Linkage Unit 19

The acetyl groups of the *N*-acetylglucosamine residues (Figure 25) in component 2 are removed by a deacetylase present in the particulate fraction (100,000*g* pellet) of a homogenate of the Furth mastocytoma (J. Riesenfeld and U. Lindahl, personal communication). The demonstration of this enzyme was made possible by the development of an assay procedure based on the liberation of [^{3}H]acetic acid from chemically *N*-acetylated component I (see Figure 24). The enzyme requires manganous ions for full activity and has a pH optimum around 6.5. It has been solubilized by the detergent–alkali method of Helting (1971) but has not yet been purified further.

The deacetylation reaction is of considerable interest in the context

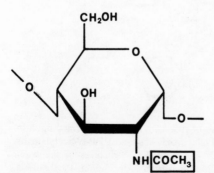

Figure 25. Linkage unit 19: Acetyl group of *N*-acetylglucosamine in heparin and its precursor polysaccharides.

of the overall regulation of the polymer-modification processes. Because deacetylation is a *sine qua non* in terms of the continuation of the modification processes, the deacetylase must be regarded as a key enzyme and is perhaps one of the prime targets for metabolic regulation. Without deacetylation, the intermediate polysaccharide will be unable to continue through the series of reactions involving N-sulfation, uronic acid epimerization, and O-sulfation, a sequence which leads to the fully completed heparin. Indeed, the level of deacetylase activity and its regulation in tissues may conceivably be important factors in determining whether a given precursor polymer will ultimately be converted to heparin or whether it will remain at a lower level of modification and consequently be classified as a heparan sulfate. In a similar fashion, the formation of heparinlike block structures within a heparan sulfate molecule could be regulated via the extent of initial N-deacetylation. Recent studies of the acceptor specificity of mastocytoma O-sulfotransferases lend support to this concept, showing that O-sulfate groups are preferentially incorporated into the N-sulfated (and thus previously N-deacetylated) rather than the N-acetylated regions of the heparan sulfate molecule.

5.2.5. Linkage Unit 20

Enzymatic N-sulfation in cell-free systems, yielding linkage unit 20 (Figure 26), has been studied in several laboratories, and the earlier literature has previously been reviewed (Rodén, 1970; Dodgson and Rose, 1970). Tissue sources used in these studies include mouse mastocytoma (Ringertz, 1963; Balasubramanian *et al.*, 1968), hen uterus (Johnson and Baker, 1973), ox lung (Foley and Baker, 1973), and rat brain (Balasubramanian and Bachhawat, 1964). The N-sulfotransferase from mastocytoma has previously been purified 27-fold by Balasubramanian *et al.* (1968). Enzyme preparations from the various tissue sources catalyze the incorporation of sulfate groups from PAPS into heparinlike polysaccharides.

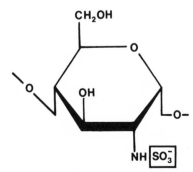

Figure 26. Linkage unit 20: N-Sulfate group of the glucosamine residues in heparin.

Characterization of the products has most often consisted of mild acid hydrolysis under conditions that liberate N-sulfate groups, and transfer to hydroxyl groups has usually been calculated as the difference between total sulfation and N-sulfation. Among the substrates used in the study of the sulfation process are endogenous, microsomal mastocytoma polysaccharide, N-desulfated heparin, and N-desulfated heparan sulfate (Balasubramanian et al., 1968; Johnson and Baker, 1973). It is now well established that the desulfated derivatives are better sulfate acceptors than the corresponding intact polysaccharides. Furthermore, it has been shown that N-acetylation of heparin reduces the acceptor activity by more than 85%. These results are consistent with transfer to free amino groups exposed by previous chemical or enzymatic deacetylation (Lindahl et al., 1973). Characterization of N-sulfate vs. O-sulfate groups can now also conveniently be carried out by degradation of the products of the reaction with nitrous acid, the N-sulfate group being released as free inorganic sulfate, whereas the O-sulfate groups remain with the oligosaccharide products.

Although the N-sulfotransferase has not yet been extensively purified, some of the kinetic parameters for the enzyme have been determined on crude preparations. These and other properties of the enzyme will be discussed in the section on O-sulfotransferases (Section 5.2.7).

5.2.6. Linkage Units 21 and 22

Extensive studies of the problem of iduronic acid formation in Lindahl's laboratory have now established beyond reasonable doubt that this component of the heparin molecule (Figures 27 and 28) is formed by epimerization of glucuronic acid residues in a polymeric heparin precursor. In the early stages of this work, a search for UDP-iduronic acid in mastocytoma tissue was undertaken, with a totally negative outcome (Lindahl, personal communication). Furthermore, it was observed that whereas enzymatic N-acetylglucosaminyl transfer occurred readily to heparin fragments containing nonreducing terminal glucuronic acid residues, analogous fragments with iduronic acid did not serve as acceptors (Helting and Lindahl, 1972) (see Section 5.2.1). Because compounds of the latter type would be obligatory intermediates in a polymerization sequence involving UDP-iduronic acid, these results suggested that the iduronic acid residues were formed by a mechanism different from the direct transfer from a nucleotide sugar. An alternative possibility had been suggested by the unexpected findings of Haug and Larsen (Haug and Larsen, 1971; Larsen and Haug, 1971) that C-5 inversion of uronic acid residues may occur at the polymer level. These authors found that the L-guluronic acid residues of alginic acid were formed by epimerization

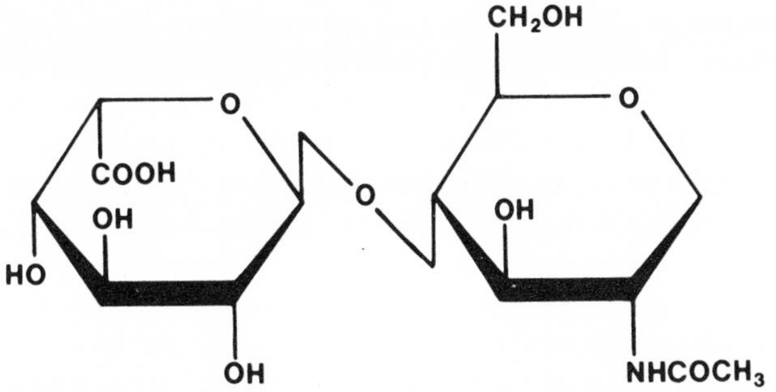

Figure 27. Linkage unit 21: IdUA $\xrightarrow{\alpha 1,4}$ GlcNAc. The glucosamine residues in this linkage unit may be *N*-acetylated, as shown in the figure, but are most often *N*-sulfated and also *O*-sulfated in position C-6. Most of the iduronic acid residues are *O*-sulfated at C-2 (not shown).

of D-mannuronic acid already incorporated into the polymer. In a series of elegant experiments, Lindahl and collaborators provided conclusive evidence for a similar reaction in iduronic acid biosynthesis (Bäckström *et al.*, 1975; Höök *et al.*, 1975a). Microsomal enzyme was first incubated with UDP-[¹⁴C]glucuronic acid and unlabeled UDP-*N*-acetylglucosamine to yield the two nonsulfated polysaccharide species described earlier (components 1 and 2 in Figure 24). At this point, PAPS was added to the reaction mixture together with a large excess of unlabeled UDP-glucuronic acid that would reduce further incorporation of radioactivity to negligible levels. After an additional incubation period, the products were

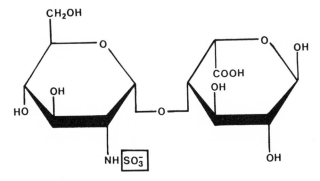

Figure 28. Linkage unit 22: GlcN[*N*-sulfate] $\xrightarrow{\alpha 1,4}$ IdUA. As a consequence of the specificity of the uronosyl 5-epimerase, the glucosamine residues in this linkage unit are never *N*-acetylated. The majority of these residues are also *O*-sulfated, as are the iduronic acid units.

isolated and analyzed for uronic acid composition. While neither of the nonsulfated polysaccharides contained labeled iduronic acid and the radioactivity resided exclusively in glucuronic acid, the O-sulfated species formed during the chase period contained both iduronic acid and glucuronic acid, with approximately one third of the radioactivity in the former component. Since incorporation of radioactivity was limited to the polymerization stage, it was concluded that the labeled iduronic acid must have arisen by epimerization of the glucuronic acid already incorporated into the polysaccharide.

The close relationship between O-sulfation and epimerization is also clearly illustrated by the experiment described above. It has already been indicated that the two sulfated species (components 3 and 4 in Figure 24) differed significantly in uronic acid composition and that epimerization seems to be strongly promoted by concomitant O-sulfation. An attractive explanation of this peculiar phenomenon is possible when we consider that many of the iduronic acid residues are sulfated. Assuming that the epimerization is a relatively freely reversible process, it seems likely that sulfation would lock the iduronic acid residues in position and make them unavailable as substrates for the epimerase. As a consequence of the withdrawal of these residues from the equilibrium, the reaction would be pulled in the direction of epimerization.

Recently, some progress has been made toward an understanding of the mechanism of uronic acid C-5 epimerization (Lindahl et al., 1976). In experiments designed to elucidate the fate of the C-5 hydrogen of the glucuronic acid residues, microsomal enzyme from mastocytoma was incubated with UDP-[5-^{3}H]glucuronic acid, UDP-[^{14}C]glucuronic acid, and unlabeled UDP-N-acetylglucosamine in the presence of PAPS. The resulting labeled polysaccharide was fractionated by ion-exchange chromatography, and the sulfated fraction was degraded to monosaccharides, which were then separated by paper chromatography. Iduronic acid comprised approximately half the total [^{14}C]uronic acid but was devoid of tritium, whereas the glucuronic acid was labeled with both ^{14}C and ^{3}H. Iduronic acid isolated after incubation with 2- or 4-^{3}H-labeled nucleotide sugar retained the tritium label. In both cases, the ratios of ^{3}H to ^{14}C were virtually the same as that for the glucuronic acid component. It can be concluded that C-5 epimerization involves abstraction of the C-5 hydrogen from glucuronic acid units, but further details of the reaction mechanism are not known at this time.

The tritium released from the C-5 position could be quantitatively recovered from the water of the reaction medium by distillation. This measure of enzyme activity was adopted for assay purposes and has been used for determination of the substrate specificity of the epimerase as well as in the purification of the enzyme (Jacobsson et al., 1979;

Malmström *et al.*, 1976, 1980). Polysaccharide intermediates 1–4, ^{3}H-labeled at C-5 of the glucuronic acid units, were tested as substrates for the epimerase by incubation with the microsomal enzyme preparation. With component 3 as substrate (*N*-sulfated but not *O*-sulfated intermediate), up to 60% of the tritium was released into the water, whereas components 1 and 2 (acetylated and deacetylated, nonsulfated polysaccharide) were completely inactive, suggesting that the presence of *N*-sulfate groups is a prerequisite for substrate recognition by the epimerase. During incubation of component 3 with the microsomal enzyme, the iduronic acid content increased somewhat, from 19 to 26% of the total labeled uronic acid. In the presence of PAPS, however, substantial formation of labeled iduronic acid occurred, and a proportionally greater release of tritium took place. Components 4 and 5 also served as substrate, but the tritium release from this compound was less extensive than for intermediate 3.

Extensive purification of the epimerase has been accomplished, although a homogeneous preparation has not yet been obtained. A substantial proportion of the total activity is found in the high-speed supernatant fraction of the mouse mastocytoma, and the enzyme has been purified approximately 8000-fold from this source by a procedure in which affinity chromatography on concanavalin A–Sepharose and heparan sulfate–Sepharose are two important steps (Malmström *et al.*, 1976, 1980).

5.2.7. Linkage Units 23 and 24

The pursuit of studies of the sulfotransferases involved in heparin biosynthesis has been hampered by the lack of specific substrates for the various enzymes. Recent work from Lindahl's laboratory (Jansson *et al.*, 1975) has resulted in the development of assays that enable separate determination of *N*- and *O*-sulfotransferase activities. The substrates used in these assays are *N*-desulfoheparin and *N*-acetylated heparan sulfate, respectively. The former substrate, in accordance with previous results from other laboratories, is an excellent substrate for transfer to amino groups, and only minimal incorporation occurs into other positions. Conversely, the *N*-acetylated polysaccharide serves as a specific substrate for the *O*-sulfotransferases, which are conveniently assayed by precipitation of the radioactive product as the cetylpyridinium complex on filter paper (Wasteson *et al.*, 1973). As yet, it has not been possible to develop separate, convenient assays for the two distinct *O*-sulfotransferases, which catalyze transfer to iduronic acid and glucosamine to yield linkage units 23 and 24, respectively (Figures 29 and 30). Thus, the location of labeled *O*-sulfate groups must at present be determined by a

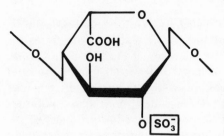

Figure 29. Linkage unit 23: 2-Sulfate group of the iduronic acid residues in heparin (and dermatan sulfate).

laborious degradation procedure which is not suitable for routine analyses.

A major fraction of the N- and O-sulfotransferases in neoplastic mast cells is recovered in the microsomal pellet on subcellular fractionation. Solubilization has been accomplished by treatment with detergent and alkali according to Helting (1971), but further purification has not yet been undertaken. The pH optimum for enzymes of both types is about 7.5. The K_m values for PAPS are 2×10^{-5} M for the N-sulfotransferase and 1×10^{-4} M for the O-sulfotransferases (not specified as to position of incorporation). The enzymes required divalent cation for maximal activity, Mn^{2+} stimulating both the N- and O-sulfotransferases 4- to 5-fold, while Ca^{2+} increased the activity of the N-sulfotransferase, but not that of the O-sulfotransferase. Addition of KCl to the incubation medium in concentrations above 50 mM caused a marked inhibition of both enzymes. In contrast, the N-sulfotransferase was selectively inhibited by NaCl; at an NaCl concentration of 0.125 M, the O-sulfotransferase activity was essentially unaffected, whereas the N-sulfotransferase activity was depressed by 80%. The sulfotransferases also differed with regard to the effects of increased temperature. The O-sulfotransferases were found to be more susceptible to heat inactivation, 60% of the activity being lost after 1 min at 50°C, while 85% of the N-sulfotransferase activity was retained. These results strongly suggest that the N- and O-sulfotransferase activities reside in different enzyme molecules. The existence of two

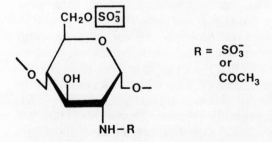

Figure 30. Linkage unit 24: 6-Sulfate group of the glucosamine residues in heparin.

distinct O-sulfotransferases appears highly likely, but a definite conclusion to this effect cannot yet be drawn.

In a narrow sense, linkage units 15 and 24, in keratan sulfate (Section 4.1) and heparin, respectively, may be regarded as identical, since both consist of 6-sulfated glucosamine units. The question may then be asked whether a single enzyme introduces the sulfate groups in this position. However, if we consider, for the two polysaccharides, the anomeric configuration of the glucosaminidic linkages, the identity of the neighboring residues, and the nature of the substituents on the amino groups (sulfate vs. acetyl), it seems more likely that two different sulfotransferases are involved, but specific information in this area is not yet available. On the other hand, the recent discovery by Di Ferrante and collaborators (see Section 4.3) of a genetic deficiency in N-acetylglucosamine 6-sulfate sulfatase, which manifests itself in excretion of both keratan sulfate and heparan sulfate, indicates that only a single enzyme is required for the removal of the 6-sulfate groups of both polysaccharides. This is in keeping with the comparatively low degree of specificity which has been observed for many other catabolic enzymes acting on the connective tissue polysaccharides.

5.3. Heparan Sulfate

From a structural point of view, heparan sulfate is closely related to heparin, and their relationship may best be defined in terms of the modifications at the polymer level which take place during biosynthesis. Focusing on a key step in this process, we may define heparan sulfate as a family of molecules in which deacetylation has occurred to a much lesser degree than in the typical heparins. By and large, the general design of the polymer modifications, including the substrate specificities of the enzymes concerned, is such that the component reactions are tightly coupled and cannot take place out of order. Unless the sequence of events is triggered by the removal of the N-acetyl groups, N-sulfation is impossible for obvious reasons; without N-sulfate groups, the proper substrate structure for the uronic acid 5-epimerase is not at hand, and no iduronic acid residues are made available as acceptors for the 2-sulfotransferase. For optimal activity, even the O-sulfotransferase catalyzing transfer to the C-6 position of the glucosamine residues appears to require a structure containing N-sulfated glucosamine residues. As a consequence of this situation, not only are the heparan sulfates characterized by a higher content of N-acetyl groups than the heparins, but also N-sulfate groups are correspondingly lower, glucuronic acid is the predominant uronic acid component rather than iduronic acid (the proportion of the latter being roughly equal to the N-sulfate content), and the degree of O-sulfation is lower and corresponds to the iduronic acid and N-sulfate content.

From a biological viewpoint, heparin and heparan sulfate should perhaps be regarded as different substances. Heparin is characteristically stored in the granules of the mast cell, from which it can be released in response to certain stimuli and where it may actually be carrying out important intracellular functions. Heparan sulfate, on the other hand, is a ubiquitous component of cell surfaces of many or all cell types and is present in the form of a proteoglycan. Following Kraemer's original identification of heparan sulfate among the cell-surface components (Kraemer, 1971), a number of investigations have confirmed and expanded this finding (Dietrich and DeOca, 1970, 1978; Roblin et al., 1975; Oldberg et al., 1977; Glimelius et al., 1978). Recently, Oldberg et al. (1979) have reported extensive purification of a heparan sulfate proteoglycan from rat liver membranes. The product had a molecular weight of 80,000 and was estimated to contain 4 heparan sulfate chains with an average molecular weight of 14,000. The size of the protein core could not be determined reliably, but different methods yielded a lower-limit value of 17,000 and an upper-limit value of 40,000. Obviously, this proteoglycan is distinct in its properties from the heparin proteoglycan described by Robinson et al. (1978), which had a molecular weight in excess of 900,000 and contained 10–15 heparin chains with molecular weights of about 80,000. Furthermore, the heparan sulfate proteoglycan contained a normal complement of amino acids, whereas the heparin proteoglycan was composed exclusively of serine and glycine. (It should be emphasized, however, that the heparin proteoglycan had been isolated after extensive proteolytic digestion with pronase and that a portion of the native core protein might well have been removed.) In any event, it is possible that the proteoglycans containing heparin and heparan sulfate may represent genetically distinct entities with respect to their core proteins despite the similarities in their polysaccharide components.

The question remains what determines the fate of the common precursor polysaccharide molecules and directs them toward either heparin or heparan sulfate. Since heparin remains intracellularly and heparan sulfate is transported to the cell surface, it is conceivable that they are synthesized in different locations within the cell by multienzyme systems where the relative proportions of the polymer-modifying enzymes vary or where as yet unknown factors regulate the relative activities of the various enzymes. It is even plausible that the heparin molecules are accompanied in the mast cell granules by a small amount of modifying enzymes which continue to act upon the polysaccharide over a long period of time. At present, we are not in a position to answer questions of this nature, and more must be learned about the properties of the synthetic enzymes and the intracellular routes of synthesis before these problems can be fully resolved.

5.4. Catabolism of Heparin and Heparan Sulfate

To properly evaluate the mecchanisms whereby homeostasis is maintained in the metabolism of connective tissue proteoglycans, we must have complete information concerning a number of factors, such as the levels of activity of synthetic as well as catabolic enzymes, substrate concentrations, regulatory mechanisms, existence of alternative pathways, and others. By and large, such complete information is not yet available for any one tissue, and our concepts of the mode of degradation of the connective tissue polysaccharides are mostly based on qualitative data. It has been mentioned previously that the degradation of several polysaccharides is the result of the combined action of an endoglycosidase (hyaluronidase) and exoenzymes which degrade the substrates from the nonreducing end, although a reliable assessment of their relative contributions has not been possible. In the study of the degradation of heparin and heparan sulfate, it has recently become apparent that these polysaccharides may also be degraded by naturally occurring endoglycosidases, which may amplify the action of the exoenzymes. The two groups of enzymes will be briefly reviewed in the following sections.

5.4.1. Endoglycosidases

The presence of a heparinase in liver was first reported by Jaques many years ago (Jaques, 1940; Cho and Jaques, 1956). From structural studies on heparan sulfate isolated from liver of Hurler patients, it has also been evident that endoglycosidase activities exist which are capable of partially degrading the heparinlike polysaccharides *in vivo* (Knecht *et al.*, 1967). More recently, Horner (1972) reported the presence in intestinal mucosa of an enzyme which degrades macromolecular heparin. An enzyme with similar properties has been isolated from a mouse mastocytoma by Ögren and Lindahl (1975) and has been characterized as an endoglucuronidase. It has been indicated previously that this enzyme should perhaps be regarded as part of the biosynthetic apparatus rather than as a catabolic enzyme. Arbogast *et al.* (1977) recently described a heparinase from rat liver which degraded heparin to fragments with a molecular weight of approximately 4000. The enzyme had a broad pH dependence with an optimum at pH 4.4.

From time to time, reports of enzyme activities which degrade heparan sulfate have appeared in the literature. However, only recently have some data as to the nature of these enzymes been appearing. Wasteson *et al.* (1976) have demonstrated an enzyme in platelets which degrades heparan sulfate by cleavage of internal linkages. The smallest product was of approximately the same size as a chondroitin sulfate hexasaccharide. When this enzyme was used to digest polysaccharides on the surfaces of endothelial and glial cells, nearly all the available heparan sulfate was removed, whereas chondroitin sulfate and dermatan sulfate was unaffected. Heparin was also substantially degraded by this enzyme (Wasteson *et al.*, 1977). Höök *et al.* (1975*b*) have shown that the lysosomal fraction from rat liver degrades heparan sulfate to comparatively large oligosaccharides; heparin was a poor substrate for this enzyme, and under the experimental conditions chosen, only a minor shift in size distribution of the polysaccharide occurred after prolonged incubation. The relationship between the enzyme preparations described by Höök *et al.* (1975*b*) and Arbogast *et al.* (1977) is not clear. Höök *et al.* (1977) have also shown that a lysosome preparation from rat spleen degrades both heparin and heparan sulfate rather extensively, yielding some products which are as small as mono- and disaccharides.

Investigations of the exact nature of the susceptible linkages have also been carried out (Klein and von Figura, 1976*a,b*; Klein *et al.*, 1976; Kindler *et al.*, 1977). These workers have presented evidence for the degradation of heparan sulfate by endoglycosidases which cleave glucuronidic as well as hexosaminidic linkages and have partially purified and characterized a heparan-sulfate-specific endoglucuronidase.

5.4.2. Exoenzymes

As shown in Figure 31, a series of enzymes acting from the nonreducing terminus of the heparan sulfate molecule are capable of degrading the polysaccharide to its monosaccharide components. The stepwise degradation requires the action of the following enzymes: (1) iduronate sulfatase, (2) α-L-iduronidase, (3) heparin N-sulfatase, (4) an acetyltransferase catalyzing transfer to free amino groups, (5) α-D-N-acetylglucosaminidase, (6) β-glucuronidase, and (7) N-acetylglucosamine 6-sulfate sulfatase. It should be pointed out that the structure in the figure and the sequence of degradation have been chosen somewhat arbitrarily for illustrative purposes only; an O-sulfate group in the C-6 position on a glucosamine residue is more likely to be found on an N-sulfated than on an N-acetylated residue. It should also be noted that the carbohydrate-protein linkage region of heparan sulfate has not been shown, but it is assumed that the same complement of enzymes is involved in the degradation of this area of the molecule as has been described for the chondroitin sulfates. In the following discussion, some specific information on the various enzymes catalyzing steps 1, 2, 4, 6, and 7 will be

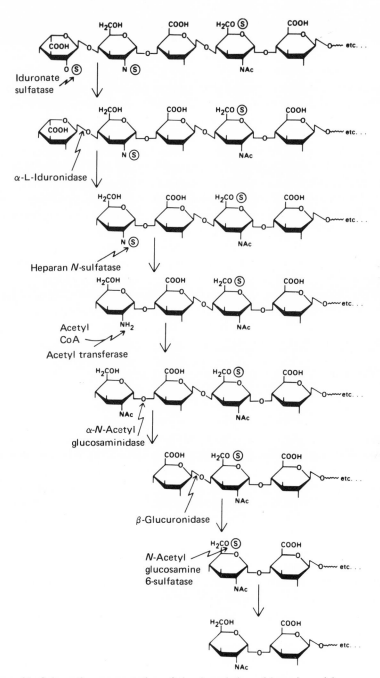

Figure 31. Schematic representation of the degradation of heparin and heparan sulfate. (Courtesy of Dr. Elizabeth F. Neufeld.)

given. The N-acetylglucosamine 6-sulfate sulfatase has been discussed in the context of keratan sulfate catabolism (Section 4.3.2).

5.4.3. Iduronate Sulfatase

The sulfate groups on C-2 of the L-iduronic acid residues of heparin, heparan sulfate, and dermatan sulfate are removed by a specific sulfatase, iduronate sulfatase. This enzyme was discovered by Bach *et al.* (1973) in the course of their studies of the Hunter syndrome. It was shown that the "Hunter corrective factor" (Cantz *et al.*, 1972) released 2% of the sulfate in ^{35}S-labeled mucopolysaccharides prepared from Hunter fibroblasts and that these sulfate groups had been located on iduronic acid units in dermatan sulfate (and presumably also heparan sulfate). Evidence to this effect was obtained by digestion of the labeled polysaccharide with chondroitinase ABC before and after treatment with the Hunter corrective factor. This analysis indicated that the content of disulfated disaccharide decreased from 7.5 to 5% as a result of factor treatment. Since the disulfated disaccharides characteristically contain sulfated iduronic acid residues (Malmström and Fransson, 1971*a*), it was concluded that the sulfate had been released from such residues by the action of a specific iduronate sulfatase. This was also demonstrated with a better-defined substrate, i.e., the disaccharide 4-O-α-L-sulfoiduronosyl-D-sulfoanhydromannose, which had been previously isolated by Lindahl and Axelsson (1971). Upon incubation of this compound with Hunter corrective factor and α-L-iduronidase, iduronic acid was released; however, either enzyme alone was without effect in this regard (liberation of iduronic acid rather than inorganic sulfate was measured, due to the difficulties in detecting the small amounts of sulfate produced).

The conclusion that the basic defect in the Hunter syndrome is a deficiency of iduronate sulfatase was also reached by Coppa *et al.* (1973) and Sjöberg *et al.* (1973). Coppa *et al.* (1973) found that a disulfated, iduronic-acid-containing disaccharide was present in the urine of a patient with the Hunter syndrome and that this disaccharide disappeared when the patient was treated with infusion of a fraction of normal serum. In a similar vein, Sjöberg *et al.* (1973) showed that the saturated, disulfated disaccharide IdUA($-SO_4$)-GalNAc($-SO_4$) was a prominent component of chondroitinase ABC digests of dermatan sulfate from Hunter fibroblasts. This finding indicates that the nonreducing terminus of the polysaccharide is occupied, in part at least by sulfated iduronic acid residues. Some of the chains terminate in N-acetylgalactosamine 4-sulfate, and in dermatan sulfate from normal cells, this is the predominant end group, as judged by the pattern of products obtained on chondroitinase digestion.

Lim *et al.* (1974) have developed an assay for iduronate sulfatase which is based on the use of the tritium-labeled, disulfated disaccharide 4-O-α-L-sulfoiduronosyl-D-[³H]anhydromannitol 6-sulfate. After incubation with the enzyme, the monosulfated radioactive product can be readily separated from the unreacted substrate by paper chromatography or electrophoresis. It has recently been reported by Hopwood (1979) that absence of the sulfate from the anhydromannitol residue substantially decreases the rate of cleavage (similarly, Hopwood observed that iduronosylanhydromannitol 6-sulfate is 25 times more sensitive as a substrate for α-L-iduronidase than the nonsulfated disaccharide). The assay of Lim *et al.* (1974) has been adapted and used for the diagnosis of the Hunter syndrome with serum, lymphocytes, and fibroblasts as enzyme sources (Liebaers and Neufeld, 1976) as well as with amniotic fluid (for prenatal diagnosis) (Liebaers *et al.*, 1977). The serum assay is undoubtedly the method of choice to establish the diagnosis.

Iduronate sulfatase has been purified 10,000-fold from human plasma by Wasteson and Neufeld (unpublished results). The enzyme has not yet been obtained in homogeneous form.

5.4.4. α-L-Iduronidase

Although glycosidases with activity toward nine of the ten monosaccharide components of mammalian complex carbohydrates have been known for many decades, the existence of an enzyme which cleaves nonreducing terminal L-iduronic acid residues was not conclusively documented until the beginning of the 1970s. Some earlier observations had suggested the existence of such an enzyme; e.g., Hoffman *et al.* (1957a) reported that skin extracts degraded a tetrasaccharide derived from dermatan sulfate, and after digestion of dermatan sulfate with a preparation of liver β-glucuronidase, Fransson and Rodén (1967a) observed a compound which was eluted in the same position as L-iduronic acid on Dowex 1 chromatography. Firm evidence for the existence of an α-L-iduronidase was obtained in the beginning of this decade in the laboratories of Dorfman, Neufeld, and Weissmann. Matalon *et al.* (1971) reported that extracts of normal human liver and skin fibroblasts released L-iduronic acid when incubated with desulfated dermatan sulfate. In the same study, it was also observed that the activity of Hurler fibroblasts was much lower than that observed for the normal fibroblasts. The synthesis of phenyl α-L-iduronide by Friedman and Weissmann (1972; Srivastava *et al.*, 1978) has facilitated greatly the detection and quantitative assay of iduronidase activity. Using this substrate, Weissmann and Santiago (1972) showed that rat liver lysosomes contained iduronidase activity and determined some properties of the enzyme. Subsequently, it was conclusively demonstrated that the Hurler syndrome is a deficiency in α-L-iduronidase;

Matalon and Dorfman (1972) found that Hurler fibroblasts contained only 1–2% of the activity observed in normal cells, and Bach et al. (1972) showed that Hurler corrective factor, which had previously been partially purified (Barton and Neufeld, 1971), was associated with α-L-iduronidase activity. Chromatography on hydroxylapatite separated the corrective factor activity into two peaks, and iduronidase activity followed the same pattern. However, the ratio between corrective factor activity and iduronidase activity was considerably higher for the second peak, a finding which was interpreted as a reflection of differences in recognition and uptake of the enzyme by the cells (Chapter 6).

Diagnostic assays of iduronidase (and other lysosomal hydrolases which are deficient in the mucopolysaccharidoses) have been described in detail by Hall and Neufeld (1973) and Hall et al. (1978). In addition to phenyl iduronide, 4-methylumbelliferyl iduronide (Srivastava et al., 1978) and radioactive disaccharides may be used to assay the enzyme and allow greater sensitivity and shorter incubation times. Such disaccharides have been prepared from heparin, i.e., 4-O-α-L-iduronosyl-D-[³H]anhydromannitol (Di Natale et al., 1977; Hall et al., 1978; Thompson, 1978; Hopwood, 1979) and 4-O-α-L-iduronosyl-D-[³H]anhydromannitol 6-sulfate (Hopwood, 1979), and from dermatan sulfate, i.e., 3-O-α-L-iduronosyl-D-[³H]anhydrotalitol (Thompson, 1978). Interestingly, the sulfated disaccharide, iduronosylanhydromannitol 6-sulfate, is a better substrate than the nonsulfated disaccharide (Hopwood, 1979).

α-L-Iduronidase was purified approximately 1000-fold from normal human urine, in its capacity of "Hurler corrective factor" (Barton and Neufeld, 1971). A corrective and a noncorrective form of the enzyme have subsequently been separated by chromatography on heparin–Sepharose (Shapiro et al., 1976) and were found to differ substantially in size (mol. wt. 87,000 and 67,000 for the corrective and noncorrective forms, respectively). More recently, iduronidase has been purified 25,000-fold from the soluble proteins of human kidney by chromatography on heparin–Sepharose, hydroxylapatite, and Bio-Gel P-100 (Rome et al., 1978). The purified enzyme (judged to be approximately 80% pure) was of the low-uptake, noncorrective form and had a molecular weight of 60,000 ± 6500. After reduction of the enzyme with dithiothreitol, polyacrylamide gel electrophoresis in SDS showed one component with an estimated molecular weight of 31,000. Some kinetic properties were also reported for the purified enzyme.

5.4.5. Heparan N-Sulfatase (Heparin Sulfamidase)

The existence of a sulfamidase catalyzing the removal of the N-sulfate groups of heparin and heparan sulfate was first indicated by the finding that inorganic [^{35}S]sulfate appeared in the urine of rats after injection of heparin with ^{35}S-labeled N-sulfate groups (Lloyd et al., 1966; Lemaire et al., 1967). Subsequently, Lloyd et al. (1968) showed that cell-free extracts from rat spleen were able to remove N-sulfate groups from similarly labeled heparin. In confirmation of these results, Dietrich (1970) reported the presence of sulfamidase in lymphoid tissue and purified the enzyme approximately 60-fold from this source. Sulfamidase activity was also present in lung and ileum but was not detected in kidney, heart, or liver. Subsequently, the presence of sulfamidase in liver has been reported by Matalon and Dorfman (1974). Other sources include cultured skin fibroblasts (Kresse and Neufeld, 1972; Kresse, 1973), peripheral leukocytes and lymphocytes (Kresse, 1973; Fabian et al., 1976), and macrophages, in which the activity is particularly high (Fabian et al., 1976).

It is now well documented that heparin sulfamidase is the enzyme deficient in the Sanfilippo A syndrome (Kresse and Neufeld, 1972; Kresse, 1973; Matalon and Dorfman, 1973, 1974), and many of its properties have been established in the course of its purification as the "Sanfilippo A factor." The corrective factor was purified 850-fold from normal human urine (Kresse and Neufeld, 1972) and was judged to be 40% pure as indicated by gel electrophoresis, with β-D-N-acetylhexosaminidase and some arylsulfatase A and β-D-N-acetylgalactosaminidase still present. Further purification has not been possible, since in a more highly purified state, the enzyme rapidly loses activity (Kresse, personal communication). In a recent study, the sulfamidase was purified 12,700-fold from bovine testis by a procedure which included chromatography on concanavalin A–Sepharose and Cibacron Blue–Sepharose (Conary et al., unpublished results); however, it has not yet been obtained in homogeneous form.

The substrate specificity of heparin sulfamidase has not yet been adequately defined. When [^{35}S]-N-sulfated heparin is used as substrate, no more than about 5% of the radioactivity can be liberated by the enzyme, and it would appear that only one sulfate group per substrate molecule is accessible to the enzyme (Dietrich, 1970; Friedman and Arsensis, 1972, 1974; Matalon and Dorfman, 1974). This is borne out more clearly by Dietrich's finding that tetra- and hexasaccharides obtained from heparin by degradation with bacterial heparinase are substrates for the enzyme and that only one N-sulfate group is removed per molecule of substrate.

By analogy with the mode of action of exoglycosidases, it would be reasonable to assume that the N-sulfate groups removed by sulfamidase are located in glucosamine residues at the nonreducing termini of the substrates. In the products of bacterial heparinase action, however, an unsaturated uronic acid residue is located in this position, and its presence obviously does not prevent the action of the sulfamidase. (In this context, it should be recalled that N-acetylgalactosamine 4-sulfate sulfatase appears to cleave not only sulfate groups located in nonreducing terminal position but also some of those in internal positions.) An alternative interpretation of Dietrich's data may be considered, i.e., that the susceptible sulfate group is located at the reducing terminus of the molecule; however, the disaccharide ΔUA-GlcN-2,6-disulfate was not a substrate, nor was the enzyme capable of removing the N-sulfate group from glucosamine 2,6-disulfate. A compelling reason for placing the susceptible residue at the nonreducing terminus, or in the penultimate position, is the established sequence of *in vivo* degradation of the carbohydrate chain, which proceeds from the nonreducing end of the molecule and requires removal of all substituents prior to the action of the appropriate glycosidases. Nevertheless, further studies using well-defined oligosaccharides would clearly be of value in determining the substrate specificity of the sulfamidase in sufficient detail.

The specificity of the mammalian sulfamidase is in sharp contrast to that of a bacterial sulfamidase from *Flavobacterium heparinum* (Dietrich *et al.*, 1973), which preferentially cleaves smaller substrates such as the unsaturated disaccharide and glucosamine 2,6-disulfate.

5.4.6. α-D-N-Acetylglucosaminidase

Forty years ago, Zechmeister *et al.* (1939) first demonstrated the existence, in extracts of snail hepatopancreas, of an enzyme with α-D-N-acetylglucosaminidase activity. An enzyme with α-specificity was subsequently found in mammalian tissues by Roseman and Dorfman (1951), and further characterization of the enzyme has been undertaken by Weissmann *et al.* (1967) (see also von Figura, 1979, for additional references on tissue distribution and purification). Recently, von Figura (1979*a,b*) reported the purification of the enzyme from normal human urine to a state of apparent homogeneity. The purified enzyme has a molecular weight of 307,000 and is a glycoprotein with extensive charge heterogeneity. It is remarkably resistant to proteases, and at an early stage of purification, the enzymatic activity was intact after digestion with pronase or bromelain for 24 hr or trypsin for 4 hr.

In the course of their studies on cultured fibroblasts from patients with various mucopolysaccharidoses, Kresse *et al.* (1971) discovered that the Sanfilippo syndrome consists of two biochemically distinct entities, which are now termed Sanfilippo A and B. Each form required a separate corrective factor for normalization of the mucopolysaccharide metabolism. Subsequently, O'Brien (1972) reported that α-D-N-acetylglucosaminidase was absent from cultured skin fibroblasts and organs of Sanfilippo B homozygotes and also markedly reduced in fibroblasts of heterozygotes. Identity between α-D-N-acetylglucosaminidase and the Sanfilippo B homozygotes and also markedly reduced in fibroblasts of heterozygotes. Identity between α-D-N-acetylglucosaminidase and the Sanfilippo B corrective factor was established in a parallel investigation by von Figura and Kresse (1972), who showed that the factor copurified with the enzyme through a number of purification steps.

The possiblity that patients with the Sanfilippo B syndrome have an altered enzyme without catalytic activity has been investigated by von Figura and Kresse (1976). With the aid of antibodies to purified enzyme from human urine, it was shown that cross-reacting material was present in the urine of three patients with the Sanfilippo B disease. Quantitation by an antibody–Sepharose technique initially suggested that the amount of altered protein was less than 25% of the quantity of enzyme protein in normal urine. However, the cross-reacting material had significantly lower affinity for the antibodies, and taking this into account, it was calculated that it was present at levels exceeding those of the normal enzyme.

Kinetic studies of α-D-N-acetylglucosaminidase have been reported by von Figura (1979a,b) and other investigators, using synthetic substrates (nitrophenyl glycosides) as well as heparin and heparan sulfate. However, continued investigations need to be carried out with special focus on its glycosaminoglycan substrates.

5.4.7. Acetyl-CoA:α-Glucosaminide N-Acetyltransferase

The fate of the terminal glucosamine residues with free amino groups, formed by the action of the N-sulfatase and the 6-sulfatase, has been a problem of long standing. After it was found that α-N-acetylglucosaminidase does not cleave such residues, investigators in several laboratories searched in vain for a specific α-glucosaminidase which would attack the deacetylated sugar. Only recently did Klein et al. (1978) arrive at a solution to the problem through their discovery of an enzyme which acetylates the free amino group of the nonreducing terminal glucosamine residue in a trisaccharide substrate and thereby renders it susceptible to the action of α-N-acetylglucosaminidase. This mode of degradation is essentially a reversal of the biosynthetic pathways, yet the discovery of the key enzyme comes as a surprise, since in contrast to the typical lysosomal enzymes, the acetyltransferase is not a hydrolase but a "synthetic" enzyme.

The elucidation of this step in heparin degradation resulted from investigations of the biochemical lesions in the Sanfilippo syndrome.

Klein *et al.* (1978) observed that several patients who presented a clinical picture typical of the Sanfilippo syndrome had normal levels of both sulfamidase and α-*N*-acetylglucosaminidase. However, they were deficient in the acetyltransferase, and on these grounds Klein and collaborators named this variant of the disease the "Sanfilippo C syndrome."

6. DERMATAN SULFATE

6.1. Structure of Dermatan Sulfate

Dermatan sulfate was first isolated from pig skin by Meyer and Chaffee (1941). Although the skin polysaccharide was similar in composition to chondroitin sulfate from cartilage, certain properties such as its high specific rotation set it apart and led the authors to conclude that it was a distinct entity. Being recognized as a close relative of cartilage chondroitin sulfate, dermatan sulfate was initially named "chondroitin sulfate B" (see Meyer *et al.*, 1956) and has also been called "β-heparin" by Marbet and Winterstein (1951), who isolated it from beef lung heparin by-products. Dermatan sulfate has subsequently been isolated from several other tissues, including hog gastric mucosa (Smith and Gallop, 1953), tendon (Meyer *et al.*, 1956), heart valves (Meyer *et al.*, 1956), and umbilical cord (Danishefsky and Bella, 1966).

Whereas galactosamine was immediately recognized as the hexosamine component of dermatan sulfate and was isolated in crystalline form by Meyer and Chaffee (1941), the identity of the uronic acid remained unknown for many years. When subjected to uronic acid analysis by the carbazole method, pure dermatan sulfate characteristically yielded a value only half that expected on the basis of its hexosamine content. This raised the suspicion that dermatan sulfate might contain a uronic acid different from glucuronic acid, and, indeed, in 1956 Hoffman *et al.* (1956*a*) identified L-iduronic acid as the major uronic acid component. Detailed structural analyses in several laboratories subsequently established that the positions of the glycosidic linkages are analogous to those observed for the chondroitin sulfates (Figures 32 and 33) (it should be noted that the α linkage of L-iduronic acid is analogous to the β-glucuronidic linkages of the chondroitin sulfates); it was also determined that the sulfate groups are located at C-4 of the galactosamine residues (Cifonelli *et al.*, 1958; Mathews, 1958; Stoffyn and Jeanloz, 1960; Hoffman *et al.*, 1960). That a portion of the sulfate groups are located on uronic acid residues was subsequently shown by Suzuki (1960), and continued investigations of this aspect of dermatan sulfate structure by Malmström and Fransson

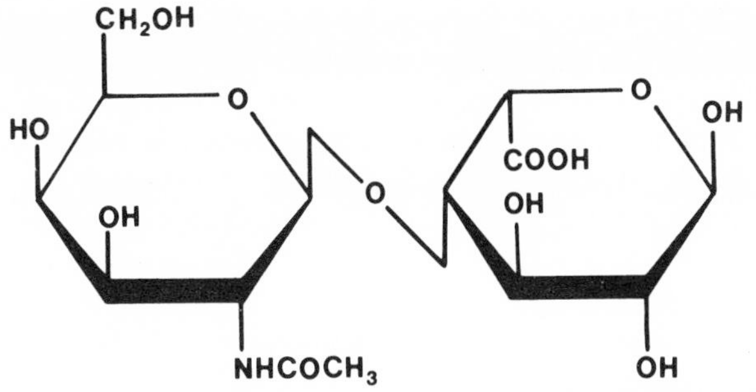

Figure 32. Linkage unit 25: GalNAc $\xrightarrow{\beta 1,4}$ IdUA.

(1971a) and Fransson et al. (1974) have led to the isolation of oligosaccharides containing sulfated iduronic acid residues.

In the early investigations by Meyer and his collaborators, it was
pointed out that a small amount of glucuronic acid was present in dermatan sulfate even after extensive fractionation, and the possibility was
suggested that this uronic acid might also be an integral part of the
dermatan sulfate molecule. This was conclusively demonstrated by
Fransson and Rodén (1967a,b), who found that dermatan sulfate is partially susceptible to cleavage by testicular hyaluronidase and isolated a
hybrid tetrasaccharide containing both iduronic acid and glucuronic acid
from enzymatic digests of the polysaccharide. In this tetrasaccharide,
glucuronic acid occupied the nonreducing terminus, as expected from the
known substrate specificity of hyaluronidase, and iduronic acid was located in the internal position. Dermatan sulfate may thus be viewed as
a copolymer of disaccharides containing both iduronic acid and glucu-

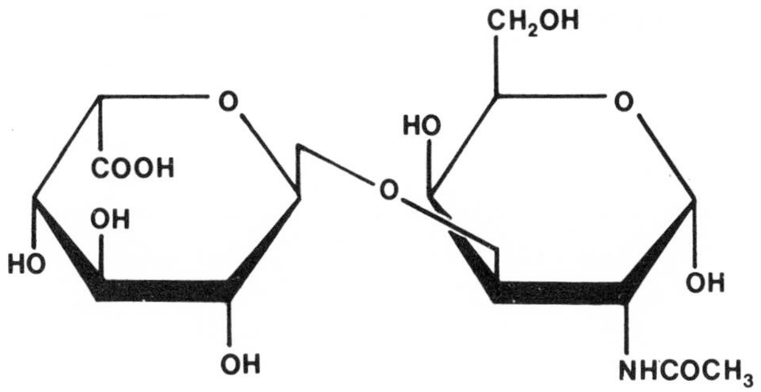

Figure 33. Linkage unit 26: IdUA $\xrightarrow{\alpha 1,3}$ GalNAc.

ronic acid. These observations have been extended to dermatan sulfates from other sources than pig skin, i.e., umbilical cord (Fransson, 1968a), intestinal mucosa (Fransson, 1968b), aorta (Fransson and Havsmark, 1970), sclera (Fransson et al., 1970), joint capsule (Fransson et al., 1970), and fibrous cartilage (Habuchi et al., 1973). In all instances, glucuronic acid is found as an integral part of the polysaccharide and may vary in proportion from a few residues per chain to more than half the total uronic acid. More detailed studies of the distribution (Fransson and Malmström, 1971) have shown that the glucuronic acid usually occurs in one to two clusters per chain.

Like chondroitin sulfate, dermatan sulfate occurs in the tissues as a proteoglycan containing xylose–serine linkages (Bella and Danishefsky, 1968; Stern et al., 1971), and reports of isolation of dermatan sulfate proteoglycans from a number of sources have appeared (Toole and Lowther, 1965, 1968; Preston, 1968; Öbrink, 1972; Cöster and Fransson 1972; Antonopoulos et al., 1974). In general, preprations of dermatan sulfate proteoglycans show signs of containing partially degraded molecules, and it may well be that the tissue location of these proteoglycans is such that they are exposed in vivo to proteases and other catabolic enzymes to a greater extent than the chondroitin sulfate proteoglycans of cartilage. However, ongoing studies of the polysaccharides produced by fibroblasts in culture (see, for example, Malmström et al., 1975a; Sjöberg, 1978) may form the basis for the isolation of less degraded proteoglycan molecules.

6.2. Biosynthesis of Dermatan Sulfate

Insofar as the biosynthesis of dermatan sulfate is known, it follows the same pattern that has been established for the chondroitin sulfates, and, furthermore, the formation of the iduronic acid residues occurs, as in heparin synthesis, by conversion of glucuronic acid residues already incorporated into the growing polymer (Malmström et al., 1975b). The 5-epimerization is tightly coupled to the sulfation process, as shown by the finding that in a cell-free system from fibroblasts, substantially greater conversion occurred if sulfation was allowed to proceed at the same time. Some of the properties of the epimerase catalyzing iduronic acid formation have been preliminarily reported by Malmström et al. (1980), but detailed exploration of this area of connective tissue polysaccharide biosynthesis is only now beginning.

The enzymes catalyzing formation of the repeating disaccharide units have not yet been studied in any detail. The existence of two glycosyl-transferases may be postulated, i.e., a glucuronosyltransferase and an N-acetylgalactosaminyltransferase. Specific information concerning the latter comes from experiments by Malmström and Fransson (1971*b*), who showed that transfer of *N*-acetylgalactosamine from UDP-*N*-acetylgalactosamine occurred to a dermatan sulfate oligosaccharide with iduronic acid at the nonreducing terminus. This interesting finding is in contrast to the results of Helting and Lindahl (1972), who observed that in the course of heparin biosynthesis, the analogous transfer of *N*-acetylglucosamine residues occurred only to glucuronic acid residues and not to iduronic acid.

6.3. Catabolism of Dermatan Sulfate

The presence of a limited number of glucuronic acid residues in dermatan sulfate endows the molecule with certain hyaluronidase-sensitive hexosaminidic linkages. Dermatan sulfate may therefore be degraded, *in vitro* or *in vivo*, to an extent corresponding to the proportion of glucuronic acid residues in the molecule. Continuing stepwise degradation from the nonreducing terminus requires the participation of the following enzymes: (1) β-glucuronidase, (2) *N*-acetylgalactosamine 4-sulfate sulfatase, (3) β-*N*-acetylhexosaminidase, (4) iduronate sulfatase, and (5) α-L-iduronidase. In addition, the action of β-galactosidase is required for complete degradation to the xylosylserine residue of the carbohydrate–protein linkage region. These enzymes have all been discussed in preceding sections, and reference is made to these sections for descriptions of their properties.

7. REFERENCES

Amado, R., Ingmar, B., Lindahl, U., and Wasteson, Å., 1974, Depolymerisation and desulphation of chondroitin sulphate by enzymes from embryonic chick cartilage, *FEBS Lett.* **39**:49.

Anseth, A., and Laurent, T. C., 1961, Studies on corneal polysaccharides. I. Separation, *Exp. Eye Res.* **1**:25.

Antonopoulos, C. A., Engfeldt, B., Gardell, S., Hjertquist, S.-O., and Solheim, K., 1965, Isolation and identification of the glycosaminoglycans from fracture callus, *Biochim. Biophys. Acta* **101**:150.

Antonopoulos, C. A., Axelsson, I., Heinegård, D., and Gardell, S., 1974, Extraction and purification of proteoglycans from various types of connective tissue, *Biochim. Biophys. Acta* **338**:108.

Arbisser, A. I., Donnelly, K. A., Scott, C. I., Di Ferrante, N., Singh, J., Stevenson, R. E., Aylesworth, A. S., and Howell, R. R., 1977, Morquio-like syndrome with beta galactosidase deficiency and normal hexosamine sulfatase activity: Mucopolysaccharidosis IVB, *Am. J. Med. Gen.* **1**:195.

Arbogast, B., Hopwood, J. J., and Dorfman, A., 1975, Absence of hyaluronidase in cultured human skin fibroblasts, *Biochem. Biophys. Res. Commun.* **67**:376.

Arbogast, B., Hopwood, J. J., and Dorfman, A., 1977, Heparinase activity in rat liver, *Biochem. Biophys. Res. Commun.* **75**:610.

Aronson, N. N., Jr., and Davidson, E. A., 1967a, Lysosomal hyaluronidase from rat liver. I. Preparation, *J. Biol. Chem.* **242**:437.

Aronson, N. N., Jr., and Davidson, E. A., 1967b, Lysosomal hyaluronidase from rat liver. II. Properties, *J. Biol. Chem.* **242**:441.

Aronson, N. N., Jr., and Davidson, E. A., 1968, Catabolism of mucopolysaccharides by rat liver lysosomes *in vivo, J. Biol. Chem.* **243**:4494.

Bach, G., and Geiger, B., 1978, Human placental N-acetyl-β-D-hexosaminidase isozymes, *Arch. Biochem. Biophys.* **189**:37.

Bach, G., Friedman, R., Weissmann, B., and Neufeld, E. F., 1972, The defect in the Hurler and Scheie syndromes: Deficiency of α-L-iduronidase, *Proc. Natl. Acad. Sci. U.S.A.* **69**:2048.

Bach, G., Eisenberg, F., Cantz, M., and Neufeld, E. F., 1973, The defect in the Hunter syndrome: Deficiency of sulfoiduronate sulfatase, *Proc. Natl. Acad. Sci. U.S.A.* **70**:2134.

Bäckström, G., Hallén, A., Höök, M., Jansson, L., and Lindahl, U., 1975, Biosynthesis of heparin, *Adv. Exp. Med. Biol.* **52**:61.

Baker, J. R., and Caterson, B., 1977, The purification and cyanogen bromide cleavage of the link proteins from cartilage proteoglycan, *Biochem. Biophys. Res. Commun.* **77**:1.

Baker, J. R., and Caterson, B., 1978, The isolation of "link proteins" from bovine nasal cartilage, *Biochim. Biophys. Acta* **532**:249.

Baker, J. R., and Caterson, B., 1979, The isolation and characterization of the link proteins from proteoglycan aggregates of bovine nasal cartilage, *J. Biol. Chem.* **254**:2387.

Baker, J. R., Cifonelli, J. A., and Rodén, L., 1969, The linkage of corneal keratosulphate to protein, *Biochem. J.* **115**:11P.

Baker, J. R., Rodén, L., and Stoolmiller, A. C., 1972, Biosynthesis of chondroitin sulfate proteoglycan: Xylosyl transfer to Smith-degraded cartilage proteoglycan and other exogenous acceptors, *J. Biol. Chem.* **247**:3838.

Baker, J. R., Cifonelli, J. A., and Rodén, L., 1975, The linkage of corneal keratan sulfate to protein, *Connect. Tissue Res.* **3**:149.

Balasubramanian, A. S., and Bachhawat, B. K., 1964, Enzymic transfer of sulphate from 3′-phosphoadenosine 5′-phosphosulphate to mucopolysaccharides in rat brain, *J. Neurochem.* **11**:877.

Balasubramanian, A. S., Joun, N. S., and Marx, W., 1968, Sulfation of N-desulfoheparin and heparan sulfate by a purified enzyme from mastocytoma, *Arch. Biochem. Biophys.* **128**:623.

Barrett, A., 1972, in: *Lysosomes* (J. T. Dingle, ed.), p. 46, North-Holland, Amsterdam.

Barrett, A., and Dingle, J. (eds.), 1971, *Tissue Proteinases,* North-Holland, Amsterdam.

Barton, R. W., and Neufeld, E. F., 1971. The Hurler corrective factor: Purification and some properties, *J. Biol. Chem.* **246**:7773.

Basner, R., Kresse, H., and von Figura, K., 1979, N-acetylglucosamine-6-sulfate sulfatase from human urine, *J. Biol. Chem.* **254**:1151.

Beaudet, A. L., Di Ferrante, N. M., Ferry, G. D., Nichols, B. L., and Mullins, C. E., 1975, Variation in the phenotypic expression of β-glucuronidase deficiency, *J. Pediatr.* **86**:388.

Beck, C., and Tappel, A. L., 1968, Rat-liver lysosomal β-glucosidase: A membrane enzyme, *Biochim. Biophys. Acta* **151**:159.

Behrens, N. H., and Leloir, L. F., 1970, Dolichol monophosphate glucose: An intermediate in glucose transfer in liver, *Proc. Natl. Acad. Sci. U.S.A.* **66**:153.

Behrens, N. H., Parodi, A. J., Leloir, L. F., and Krisman, C. R., 1971, The role of dolichol monophosphate in sugar transfer, *Arch. Biochem. Biophys.* **143**:375.

Bell, C. E., Sly, W. S., and Brot, F. E., 1977, Human β-b-glucuronidase deficiency mucopolysaccharidoses: Identification of cross-reactive antigen in cultured fibroblasts of deficient patients by enzyme immunoassay, *J. Clin. Invest.* **59**:97.

Bella, A., Jr., and Danishefsky, I., 1968, The dermatan sulfate–protein linkage region, *J. Biol. Chem.* **243**:2660.

Berman, E. R., 1970, Proteoglycans of bovine corneal stroma, in: *Chemistry and Molecular Biology of the Intercellular Matrix* (E. A. Balazs, ed.), p. 879, Academic Press, New York.

Bhavanandan, V. P., and Meyer, K., 1968, Studies on keratosulfates: Methylation, desulfation, and acid hydrolysis studies on old human rib cartilage keratosulfate, *J. Biol. Chem.* **243**:1052.

Bray, B. A., Lieberman, R., and Meyer, K., 1967, Structure of human skeletal keratosulfate, *J. Biol. Chem.* **242**:3373.

Brimacombe, J. S., and Webber, J. M., 1964, Mucopolysaccharides, *Biochim. Biophys. Acta Library*, Vol. 6, Elsevier, Amsterdam.

Brot, F. E., Bell, C. E., and Sly, W. S., 1978, Purification and properties of β-glucuronidase from human placenta, *Biochemistry* **17**:385.

Cantz, M., and Kresse, H., 1974, Sandhoff disease: Defective glycosaminoglycan catabolism in cultured fibroblasts and its correction by β-N-acetylhexosaminidase, *Eur. J. Biochem.* **47**:581.

Cantz, M., Kresse, H., Barton, R. W., and Neufeld, E. F., 1972, Corrective factors for inborn errors of mucopolysaccharide metabolism, *Methods Enzymol.* **28**:884.

Cashman, D. C., Laryea, J. U., and Weissmann, B., 1969, The hyaluronidase of rat skin, *Arch. Biochem. Biophys.* **135**:387.

Caterson, B., and Baker, J. R., 1978, The interaction of link proteins with proteoglycan monomers in the absence of hyaluronic acid, *Biochem. Biophys. Res. Commun.* **80**:496.

Caterson, B., and Baker, J. R., 1979, The link proteins as specific components of cartilage proteoglycan aggregates *in vivo*, *J. Biol. Chem.* **254**:2394.

Cho, M. H., and Jaques, L. B., 1956, Heparinase. III. Preparation and properties of the enzyme, *Can. J. Biochem. Physiol.* **34**:799.

Cifonelli, J. A., and King, J., 1972, The distribution of 2-acetamido-2-deoxy-D-glucose residues in mammalian heparins, *Carbohydr. Res.* **21**:173.

Cifonelli, J. A., Ludowieg, J., and Dorfman, A., 1958, Chemistry of β-heparin (chondroitinsulfuric acid-B), *J. Biol. Chem.* **233**:541.

Coppa, G. V., Singh, J., Nichols, B. L., and Di Ferrante, N., 1973, Urinary excretion of disulfated disaccharides in Hunter syndrome: Correction by infusion of a serum fraction, *Anal. Lett.* **6**:225.

Cöster, L., and Fransson, L.-Å., 1972, Isolation of proteoglycans from bovine sclera, *Scand. J. Clin. Lab. Invest.* **29**(Suppl. 123):9.

Dansihefsky, I., and Bella, A., 1966, The sulfated mucopolysaccharides from human umbilical cord, *J. Biol. Chem.* **241**:143.

Davidson, E. A., 1970, Glycoprotein and mucopolysaccharide hydrolysis (glycoprotein and mucopolysaccharide hydrolysis in the cell), in: *Metabolic Conjugation and Metabolic Hydrolysis,* Vol. 1 (W. H. Fishman, ed.), p. 327, Academic Press, New York.

Dietrich, C. P., 1970, A heparin sulfamidase from mammalian lymphoid tissues, *Can. J. Biochem.* **48**:725.

Dietrich, C. P., and De Oca, H. M., 1970, Production of heparin related mucopolysaccharides by mammalian cells in culture, *Proc. Soc. Exp. Biol. Med.* **134**:995.

Dietrich, C. P., and De Oca, H. M., 1978, Surface sulfated mucopolysaccharides of primary and permanent mammalian cell lines, *Biochem. Biophys. Res. Commun.* **80:**805.

Dietrich, C. P., Silva, M. E., and Michelacci, Y. M., 1973, Sequential degradation of heparin in *Flavobacterium heparinum:* Purification and properties of five enzymes involved in heparin degradation, *J. Biol. Chem.* **248:**6408.

Di Ferrante, N., Ginsberg, L. C., Donnelly, P. V., Di Ferrante, D. T., and Caskey, C. T., 1978, Deficiencies of glucosamine-6-sulfate or galactosamine-6-sulfate sulfatases are responsible for different mucopolysaccharidoses, *Science* **199:**79.

Di Natale, P., Leder, I. G., and Neufeld, E. F., 1977, A radioactive substrate and assay for α-L-iduronidase, *Clin. Chim. Acta* **77:**211.

Distler, J. J., and Jourdian, G. W., 1973, The purification and properties of β-galactosidase from bovine testes, *J. Biol. Chem.* **248:**6772.

Dodgson, K. S., and Rose, F. A., 1970, Sulfoconjugation and sulfohydrolysis, in: *Metabolic Conjugation and Metabolic Hydrolysis,* Vol. 1 (W. H. Fishman, ed.), p. 239, Academic Press, New York.

Dodgson, K. S., and Spencer, B., 1953, Studies on sulfphatases. I. The choice of substrate for the assay of rat-liver arylsulfatase, *Biochem. J.* **53:**444.

Dodgson, K. S., Spencer, B., and Thomas, J., 1955, Studies on sulphatases. 9. The arysulfatases of mammalian livers, *Biochem. J.* **59:**29.

Dorfman, A., 1969, Biosynthesis of acid mucopolysaccharides (glycosaminoglycans) of connective tissues, Thule International Symposia, *Aging of Connective and Skeletal Tissue,* p. 81, Nordiska Bokhandelns Förlag, Stockholm.

Dorfman, A., 1974, Adventures in viscous solutions, *Mol. Cell. Biochem.* **4:**45.

Dorfman, A., and Matalon, R., 1972, The mucopolysaccharidoses, in: *The Metabolic Basis of Inherited Disease,* 3rd ed. (J. B. Stanbury, J. B. Wyngaarden, and D. S. Fredrickson, eds.), p. 1211, McGraw-Hill, New York.

Dorfman, A., and Matalon, R., 1976, The mucopolysaccharidoses (a review), *Proc. Natl. Acad. Sci. U.S.A.* **73:**630.

Dorfman, A., Matalon, R., Cifonelli, J. A., Thompson, J., and Dawson, G., 1972, The degradation of acid mucopolysaccharides and the mucopolysaccharidoses, in: *Sphingolipids, Sphingolipidoses and Allied Disorders* (B. W. Volk and S. M. Aronson, eds.), p. 195, Plenum Press, New York.

Dorfman, A., Arbogast, B., and Matalon, R., 1976, The enzymic defects in Morquio and Maroteaux–Lamy syndrome, in: *Current Trends in Sphingolipidoses and Allied Disorders* (B. W. Volk and L. Schneck, eds.), p. 261, Plenum Press, New York.

Dulaney, J. T., and Moser, H. W., 1978, Sulfatide lipidosis: Metachromatic leukodystrophy, in: *The Metabolic Basis of Inherited Disease* (J. B. Stanbury, J. B. Wyngaarden, and D. S. Fredrickson, eds.), p. 770, McGraw-Hill, New York.

Fabian, I., Bleiberg, I., and Aronson, M., 1976, Desulphation of heparin by mice and guinea pig leukocytes, *Biochim. Biophys. Acta* **437:**122.

Faltynek, C. R., and Silbert, J. E., 1978, Copolymers of chondroitin 4-sulfate and chondroitin 6-sulfate in chick embryo epiphyses and other cartilage, *J. Biol. Chem.* **253:**7646.

Fisher, D., and Kent, P. W., 1969, Rat liver β-xylosidase, a lysosomal membrane enzyme, *Biochem. J.* **115:**50.

Fisher, D., Higham, M., Kent, P. W., and Pritchard, P., 1966, β-Xylosidases of animal and other sources in relation to the degradation of chondroitin sulphate–peptide complexes, *Biochem. J.* **98:**46P.

Fluharty, A. L., Stevens, R. L., Sanders, D. L., and Kihara, H., 1974, Arylsulfatase B deficiency in Maroteaux–Lamy syndrome cultured fibroblasts, *Biochem. Biophys. Res. Commun.* **59:**455.

Fluharty, A. L., Stevens, R. L., Fung, D., Peak, S., and Kihara, H., 1975, Uridine diphospho-N-acetylgalactosamine-4-sulfate sulfohydrolase activity of human arylsulfatase B and its deficiency in the Maroteaux–Lamy syndrome, *Biochem. Biophys. Res. Commun.* **64**:955.

Foley, T., and Baker, J. R., 1973, Heparan sulphate sulphotransferase: Properties of an enzyme from ox lung, *Biochem. J.* **135**:187.

Forsee, W. T., Belcher, J., and Rodén, L., 1978, Biosynthesis of heparin. The N-acetylglucosaminyl transfer reaction, *Fed. Am. Soc. Exp. Biol. Fed. Proc.* **37**:1777.

Fransson, L.-Å., 1968a, Structure of dermatan sulfate. III. The hybrid structure of dermatan sulfate from umbilical cord, *J. Biol. Chem.* **243**:1504.

Fransson, L.-Å., 1968b, Structure of dermatan sulfate. V. The hybrid structure of dermatan sulfate from hog intestinal mucosa, *Ark. Kemi* **29**:95.

Fransson, L.-Å., and Havsmark, B., 1970, Structure of dermatan sulfate. VII. The copolymeric structure of dermatan sulfate from horse aorta, *J. Biol. Chem.* **245**:4770.

Fransson, L.-Å., and Malmström, A., 1971, Structure of pig skin dermatan sulfate. 1. Distribution of D-glucuronic acid residues, *Eur. J. Biochem.* **18**:422.

Fransson, L.-Å., and Rodén, L., 1967a, Structure of dermatan sulfate. I. Degradation by testicular hyaluronidase, *J. Biol. Chem.* **242**:4161.

Fransson, L.-Å., and Rodén, L., 1967b, Structure of dermatan sulfate. II. Characterization of products obtained by hyaluronidase digestion of dermatan sulfate, *J. Biol. Chem.* **242**:4170.

Fransson, L.-Å., Anseth, A., Antonopoulos, C. A., and Gardell, S., 1970, Structure of dermatan sulfate. VI. The use of cetylpyridinium chloride–cellulose microcolumns for determination of the hybrid structure of dermatan sulfates, *Carbohydr. Res.* **15**:73.

Fransson, L.-Å., Cöster, L., Havsmark, B., Malmström, A., and Sjöberg, I., 1974, The copolymeric structure of pig skin dermatan sulphate. Isolation and characterization of L-idurono-sulphate-containing oligosaccharides from copolymeric chains, *Biochem. J.* **143**:379.

Friedman, R. B., and Weissmann, B., 1972, The phenyl α- and β-L-idopyranosid uronic acids and some other aryl glycopyranosiduronic acids, *Carbohydr. Res.* **24**:123.

Friedman, Y., and Arsenis, C., 1972, The resolution of aryl sulfatase and heparin sulfamidase activities from various rat tissues, *Biochem. Biophys. Res. Commun.* **48**:1133.

Friedman, Y., and Arsenis, C., 1974, Studies on the heparin sulphamidase activity from rat spleen: Intracellular distribution and characterization of the enzyme, *Biochem. J.* **139**:699.

Fukuda, M., Muramatsu, T., Egami, F., Takahashi, N., and Yasuda, Y., 1968, Purification of β-xylosidase and its action on O-β-xylosyl L-serine and stem bromelain glycopeptide, *Biochim. Biophys. Acta* **159**:215.

Galligani, L., Hopwood, J., Schwartz, N. B., and Dorfman, A., 1975, Stimulation of synthesis of free chondroitin sulfate chains by β-D-xylosides in cultured cells, *J. Biol. Chem.* **250**:5400.

Ganschow, R. E., 1973, The genetic control of acid hydrolases, in: *Metabolic Conjugation and Metabolic Hydrolysis*, Vol. 3 (W. H. Fishman, ed.), p. 189, Academic Press, New York.

Glaser, J. H., and Conrad, H. E., 1979, Chondroitin SO_4 catabolism in chick embryo chondrocytes, *J. Biol. Chem.* **254**:2316.

Glaser, L., and Brown, D. H., 1955, The enzymatic synthesis *in vitro* of hyaluronic acid chains, *Proc. Natl. Acad. Sci. U.S.A.* **41**:253.

Glimelius, B., Norling, B., Westermark, B., and Wasteson, Å., 1978, Composition and distribution of glycosaminoglycans in cultures of human normal and malignant glial cells, *Biochem. J.* **172**:443.

Glössl, J., Truppe, W., and Kresse, H., 1979, Purification and properties of N-acetylga-lactosamine 6-sulphate sulphatase from human placenta, *Biochem. J.* **181**:37.

Gniot-Szulzycka, J., and Donnelly, P. V., 1976, Arylsulphatase B (Maroteaux–Lamy factor): A part of the enzyme system responsible for sulphate release from mucopolysaccharide fragment, *FEBS Lett.* **65**:63.

Grebner, E. E., Hall, C. W., and Neufeld, E. F., 1966a, Incorporation of D-xylose-^{14}C into glycoprotein by particles from hen oviduct, *Biochem. Biophys. Res. Commun.* **22**:672.

Grebner, E. E., Hall, C. W., and Neufeld, E. F., 1966b, Glycosylation of serine residues by a uridine disphosphate-xylose:protein xylosyltransferase from mouse mastocytoma, *Arch. Biochem. Biophys.* **116**:391.

Gregory, J. D., and Rodén, L., 1961, Isolation of keratosulfate from chondromucoprotein of bovine nasal septa, *Biochem. Biophys. Res. Commun.* **5**:430.

Gregory, J. D., Laurent, T. C., and Rodén, L., 1964, Enzymatic degradation of chondro-mucoprotein, *J. Biol. Chem.* **239**:3312.

Greiling, H., Momburg, M., and Stuhlsatz, H. W., 1972, Isolation and substrate specificity of a PAPS:chondroitin 6-sulfotransferase from mouse liver, *Scand. J. Clin. Lab. Invest.* **29**(Suppl. 123):37.

Habuchi, H., Yamagata, T., Iwata, H., and Suzuki, S., 1973, The occurrence of a wide variety of dermatan sulfate–chondroitin sulfate copolymers in fibrous cartilage, *J. Biol. Chem.* **248**:6019.

Hagopian, A., Bosmann, H. B., and Eylar, E. H., 1968, Glycoprotein biosynthesis: The localization of polypeptidyl: N-acetylgalactosaminyl, collagen:glucosyl, and glycopro-tein:galactosyl transferases in HeLa cell membrane fractions, *Arch. Biochem. Biophys.* **128**:387.

Hall, C. W., and Neufeld, E. F., 1973, α-L-Iduronidase activity in cultured skin fibroblasts and amniotic fluid cells, *Arch. Biochem. Biophys.* **158**:817.

Hall, C. W., Cantz, M., and Neufeld, E. F., 1973, A β-glucuronidase deficiency mucopo-lysaccharidosis: Studies in cultured fibroblasts, *Arch. Biochem. Biophys.* **155**:32.

Hall, C. W., Liebaers, I., Di Natale, P., and Neufeld, E. F., 1978, Enzymic diagnosis of the genetic mucopolysaccharide storage disorders, *Methods Enzymol.* **50**:439.

Hamerman, D., Rojkind, M., and Sandson, J., 1966, Protein bound to hyaluronate: Chem-ical and immunological studies, *Fed. Am. Soc. Exp. Biol.* **25**:1040.

Handley, C. J., and Lowther, D. A., 1977, Extracellular matrix metabolism by chondro-cytes. III. Modulation of proteoglycan synthesis by extracellular levels of proteoglycan in cartilage cells in culture, *Biochim. Biophys. Acta* **500**:132.

Hardingham, T. E., 1979, The role of link-protein in the structure of cartilage proteoglycan aggregates, *Biochem. J.* **177**:237.

Hardingham, T. E., and Muir, H., 1972, The specific interaction of hyaluronic acid with cartilage proteoglycan, *Biochim. Biophys. Acta* **279**:401.

Hardingham, T. E., and Muir, H., 1974, Hyaluronic acid in cartilage and proteoglycan aggregation, *Biochem. J.* **139**:565.

Hardingham, T. E., and Phelps, C. F., 1968, The tissue content and turnover rates of intermediates in the biosynthesis of glycosaminoglycans in young rat skin, *Biochem. J.* **108**:9.

Hart, G. W., and Lennarz, W. J., 1978, Effects of tunicamyin on the biosynthesis of glycosaminoglycans by embryonic chick cornea, *J. Biol. Chem.* **253**:5795.

Hascall, V. C., 1977, Interaction of cartilage proteoglycans with hyaluronic acid, *J. Supra-mol. Struct.* **7**:101.

Hascall, V. C., and Heinegård, D., 1974, Aggregation of cartilage proteoglycans. I. The role of hyaluronic acid, *J. Biol. Chem.* **249**:4232.

Hascall, V. C., and Sajdera, S. W., 1969, Proteinpolysaccharide complex from bovine nasal cartilage: The function of glycoprotein in the formation of aggregates, *J. Biol. Chem.* **244:**2384.

Hascall, V. C., and Sajdera, S. W., 1970, Physical properties and polydispersity of proteoglycan from bovine nasal cartilage, *J. Biol. Chem.* **245:**4920.

Haug, A., and Larsen, B., 1971, Biosynthesis of alginate. Part II. Polymannuronic acid C-5-epimerase from *Azotobacter vinelandii* (Lipman), *Carbohydr. Res.* **17:**297.

Heidelberger, M., and Kendall, F. E., 1929, A crystalline aldobionic acid derived from gum arabic, *J. Biol. Chem.* **84:**639.

Heinegård, D., 1972, Extraction, fractionation and characterization of proteoglycans from bovine tracheal cartilage, *Biochem. Biophys. Acta* **285:**181.

Helting, T., 1971, Biosynthesis of heparin: Solubilization, partial separation, and purification of uridine diphosphate-galactose: Acceptor galactosyltransferases from mouse mastocytoma, *J. Biol. Chem.* **246:**815.

Helting, T., 1972, Biosynthesis of heparin: Solubilization and partial purification of uridine disphosphate glucuronic acid: Acceptor glucuronosyltransferase from mouse mastocytoma, *J. Biol. Chem.* **247:**4327.

Helting, T., and Lindahl, U., 1971, Occurrence and biosynthesis of β-glucuronidic linkages in heparin, *J. Biol. Chem.* **246:**5442.

Helting, T., and Lindahl, U., 1972, Biosynthesis of heparin. I. Transfer of *N*-acetylglucosamine and glucuronic acid to low-molecular weight heparin fragments, *Acta Chem. Scand.* **26:**3515.

Helting, T., and Rodén, L., 1969a, Biosynthesis of chondroitin sulfate. I. Galactosyl transfer in the formation of the carbohydrate–protein linkage region, *J. Biol. Chem.* **244:**2790.

Helting, T., and Rodén, L., 1969b, Biosynthesis of chondroitin sulfate. II. Glucuronosyl transfer in the formation of the carbohydrate–protein linkage region, *J. Biol. Chem.* **244:**2799.

Helwig, J.-J., Farooqui, A. A., Bollack, C., and Mandel, P., 1977, Purification and some properties of arylsulphatases A and B from rabbit kidney cortex, *Biochem. J.* **165:**127.

Hemming, F. W., 1974, Lipids in glycan biosynthesis, in: *MTP International Review of Science,* Ser. 1, Vol. 4, *Biochemistry of Lipids* (T. W. Goodwin, ed.), p. 39, University Park Press, Baltimore.

Hestrin, S., Avineri-Shapiro, S., and Aschner, M., 1943, The enzymic production of levan, *Biochem. J.* **37:**450.

Himeno, M., Hashiguchi, Y., and Kato, K., 1974, β-Glucuronidase of bovine liver. Purification, properties, carbohydrate composition, *J. Biochem.* **76:**1243.

Himeno, M., Ohhara, H., Arakawa, Y., and Kato, K., 1975, β-Glucuronidase of rat preputial gland, *J. Biochem.* **77:**427.

Himeno, M., Nishimura, Y., Tsuji, H., and Kato, K., 1976, Purification and characterization of microsomal and lysosomal β-glucuronidase from rat liver by use of immunoaffinity chromatography, *Eur. J. Biochem.* **70:**349.

Hirano, S., and Hoffman, P., 1962, The hexosaminidic linkage of hyaluronic acid, *J. Org. Chem.* **27:**395.

Hoffman, P., Linker A., and Meyer, K., 1956a, Uronic acid of chondroitin sulfate B, *Science* **124:**1252.

Hoffman, P., Meyer, K., and Linker, A., 1956b, Transglycosylation during the mixed digestion of hyaluronic acid and chondroitin sulfate by testicular hyaluronidase, *J. Biol. Chem.* **219:**653.

Hoffman, P., Linker A., and Meyer, K., 1957a, The acid mucopolysaccharides of connective tissues. II. Further experiments on chondroitin sulfate B, *Arch. Biochem. Biophys.* **69:**435.

Hoffman, P., Linker, A., Sampson, P., Meyer, K., and Korn, E. D., 1957*b*, The degradation of hyaluronate, the chondroitin sulfates and heparin by bacterial enzymes (flavobacterium), *Biochim. Biophys. Acta* **25**:658.

Hoffman, P., Linker, A., Lippman, V., and Meyer, K., 1960, The structure of chondroitin sulfate B from studies with flavobacterium enzymes, *J. Biol. Chem.* **235**:3066.

Höök, M., Lindahl, U., Bäckström, G., Malmstroṁ, A., and Fransson, L.-Å., 1974, Biosynthesis of heparin. III. Formation of iduronic acid residues, *J. Biol. Chem.* **249**:3908.

Höök, M., Lindahl, U., Hallén, and Bäckström, G., 1975*a*, Biosynthesis of heparin: Studies on the microsomal sulfation process, *J. Biol. Chem.* **250**:6065.

Höök, M., Wasteson, Å., and Oldberg, Å., 1975*b*, A heparan sulfate–degrading endoglycosidase from rat liver tissue, *Biochem. Biophys. Res. Commun.* **67**:1422.

Höök, M., Pettersson, I., and Ögren, S., 1977, A heparin-degrading endoglycosidase from rat spleen, *Thromb. Res.* **10**:857.

Hopwood, J., 1972, Cartilage proteoglycans: Biosynthesis and chemical structure, Ph.D. thesis, Monash University, Clayton, Australia.

Hopwood, J. J., 1979, α-L-Iduronidase, β-D-glucuronidase, and 2-sulfo-L-iduronate 2-sulfatase: Preparation and characterization of radioactive substrates from heparin, *Carbohydr. Res.* **69**:203.

Hopwood, J. J., and Robinson, H. C., 1974, The alkali-labile linkage between keratan sulphate and protein, *Biochem. J.* **141**:57.

Horner, A. A., 1971, Macromolecular heparin from rat skin: Isolation, characterization, and depolymerization with ascorbate, *J. Biol. Chem.* **246**:231.

Horner, A. A., 1972, Enzymic depolymerization of macromolecular heparin as a factor in control of lipoprotein lipase activity, *Proc. Nat. Acad. Sci. U.S.A.* **69**:3469.

Horowitz, M. I., 1977, Gastrointestinal glycoproteins, in: *The Glycoconjugates*, Vol. II (M. I. Horowitz and W. Pigman, eds.), p. 189, Academic Press, New York.

Horwitz, A. L., 1972, Cellular sites and enzymic mechanisms for synthesis of chondromucoprotein of cartilage, Ph.D. thesis, University of Chicago.

Horwitz, A. L., and Dorfman, A., 1978, The enzymic defect in Morquio's disease: The specificity of N-acetylhexosamine sulfatases, *Biochem. Biophys. Res. Commun.* **80**:819.

Ishimoto, N., and Strominger, J. L., 1967, Uridine diphosphate as the sole uridine nucleotide product of hyaluronic acid synthetase in group A streptococci, *Biochim. Biophys. Acta* **148**:296.

Jacobson, B., and Davidson, E. A., 1962, Biosynthesis of uronic acids by skin enzymes. II. Uridine diphosphate-D-glucuronic acid-5-epimerase, *J. Biol. Chem.* **237**:638.

Jacobson, B., and Davidson, E. A., 1963, UDP-D-glucuronic acid-5-epimerase and UDP-N-acetylglucosamine-4-epimerase of rabbit skin, *Biochim. Biophys. Acta* **73**:145.

Jacobsson, I., Bäckström, G., Höök, M., Lindahl, U., Feingold, D. S., Malmström, A., and Rodén, L., 1979, Biosynthesis of heparin: Assay and properties of the microsomal uronosyl C-5 epimerase, *J. Biol. Chem.* **254**:2975.

Jansson, L., Höök, M., Wasteson, Å., and Lindahl, U., 1975, Biosynthesis of heparin. V. Solubilization and partial characterization of N-and O-sulphotransferases, *Biochem. J.* **149**:49.

Jaques, L. B., 1940, Heparinase, *J. Biol. Chem.* **133**:445.

Johnson, A. H., and Baker, J. R., 1973, The enzymatic sulphation of heparin sulphate by hen's uterus, *Biochim. Biophys. Acta* **320**:341.

Kaplan, D., 1969, Classification of the mucopolysaccharidoses based on the pattern of mucopolysacchariduria, *Am. J. Med.* **47**:721.

Keiser, H., Shulman, H. J., and Sandson, J. I., 1972, Immunochemistry of cartilage proteoglycan: Immunodiffusion and gel-electrophoretic studies, *Biochem. J.* **126**:163.

Kiely, M. L., McKnight, G. S., and Schimke, R. T., 1976, Studies on the attachment of carbohydrate to ovalbumin nascent chains in hen oviduct, *J. Biol. Chem.* **251**:5490.

Kindler, A., Klein, U., and von Figura, K., 1977, Characterization of glycosaminoglycans stored in mucopolysaccharidosis. IIIA: Evidence for a generally occurring degradation of heparan sulfate by endoglycosidases, *Hoppe-Seyler's Z. Physiol. Chem.* **358**:1431.

Klein, U., and von Figura, K., 1976a, Demonstration of D-glucuronic acid as reducing terminal of intracellular heparan sulfates, *FEBS Lett.* **71**:266.

Klein, U., and von Figura, K., 1976b, Partial purification and characterization of a heparan sulfate specific endoglucuronidase, *Biochem. Biophys. Res. Commun.* **73**:569.

Klein, U., Kresse, H., and von Figura, K., 1976, Evidence for degradation of heparan sulfate by endoglycosidases: Glucosamine and hexuronic acid are reducing terminals of intracellular heparan sulfate from human skin fibroblasts, *Biochem. Biophys. Res. Commun.* **69**:158.

Klein, U., Kresse, H., and von Figura, K., 1978, Sanfilippo syndrome type C: Deficiency of acetyl-CoA:α-glucosaminide *N*-acetyltransferase in skin fibroblasts, *Proc. Natl. Acad. Sci. U.S.A.* **75**:5185.

Knecht, J., Cifonelli, J. A., and Dorfman, A., 1967, Structural studies on heparitin sulfate of normal and Hurler tissues, *J. Biol. Chem.* **242**:4652.

Kornfeld, R., 1967, Studies on L-glutamine D-fructose 6-phosphate amido-transferase. I. Feedback inhibition by uridine diphosphate-*N*-acetylglucosamine, *J. Biol. Chem.* **242**:3135.

Kornfeld, S., Kornfeld, R., Neufeld, E. F., and O'Brien, P. J., 1964, The feedback control of sugar nucleotide biosynthesis in liver, *Proc. Natl. Acad. Sci. U.S.A.* **52**:371.

Kraemer, P. M., 1971, Heparan sulfates of cultured cells. I. Membrane-associated and cell-sap species in Chinese hamster cells, *Biochemistry* **10**:1437.

Kraemer, P. M., 1977, Heparin releases heparan sulfate from the cell surface, *Biochem. Biophys. Res. Commun.* **78**:1334.

Kresse, H., 1973, Mucopolysaccharidosis III A (Sanfilippo A disease): Deficiency of a heparin sulfamidase in skin fibroblasts and leucocytes, *Biochem. Biophys. Res. Commun.* **54**:1111.

Kresse, H., and Neufeld, E. F., 1972, The Sanfilippo A corrective factor: Purification and mode of action, *J. Biol. Chem.* **247**:2164.

Kresse, H., Wiesmann, U., Cantz, M., Hall, C. W., and Neufeld, E. F., 1971, Biochemical heterogeneity of the Sanfilippo syndrome: Preliminary characterization of two deficient factors, *Biochem. Biophys. Res. Commun.* **42**:892.

Larsen, B., and Haug, A., 1971, Biosynthesis of alginate. Part III. Tritium incorporation wit polymannuronic acid 5-epimerase from *Azotobacter vinelandii, Carbohydr. Res.* **20**:225.

Leaback, D. H., 1970, The metabolic hydrolysis of hexosaminide linkages, in: *Metabolic Conjugation and Metabolic Hydrolysis,* Vol. 2 (W. H. Fishman, ed.), p. 443, Academic Press, New York.

Leloir, L. F., 1972, Biosynthesis of polysaccharides seen from Buenos Aires, in: *Biochemistry of the Glycosidic Linkage* (R. Piras and H. G. Pontis, eds.), p. 1, Academic Press, New York.

Lemaire, A., Picard, J., and Gardais, A., 1967, Le catabolisme de l'héparine: Étude *in vivo* de la désulfatation de l'héparine chez le rat, *C. R. Acad. Sci. Paris* **264**:949.

Lennarz, W. J., 1975, Lipid linked sugars in glycoprotein synthesis, *Science* **188**:986.

Levvy, G. A., and Conchie, J., 1966, β-Glucuronidase and the hydrolysis of gluronides, in: *Glucuronic Acid, Free and Combined* (G. J. Dutton, ed.), p. 301, Academic Press, New York.

Liebaers, I., and Neufeld, E. F., 1976, Iduronate sulfatase activity in serum, lymphocytes, and fibroblasts—Simplified diagnosis of the Hunter syndrome, *Pediatr. Res.* **10**:733.

Liebaers, I., Di Natale, P., and Neufeld, E. F., 1977, Iduronate sulfatase in amniotic fluid: An aid in the prenatal diagnosis of the Hunter syndrome, *J. Pediatr.* **90**:423.

Lim, T. W., Leder, I. G., Bach, G., and Neufeld, E. F., 1974, An assay for iduronate sulfatase (Hunter corrective factor), *Carbohydr. Res.* **37**:103.

Lindahl, U., 1966a, The structures of xylosylserine and galactosylxylosylserine from heparin, *Biochim. Biophys. Acta* **130**:361.

Lindahl, U., 1966b, Further characterization of the heparin–protein linkage region, *Biochim. Biophys. Acta* **130**:368.

Lindahl, U., 1976, Structure and biosynthesis of L-iduronic acid-containing glycosaminoglycans, in: *MTP International Review of Science: Organic Chemistry Series Two— Carbohydrate Chemistry* (G. O. Aspinall, ed.), Vol. 7, p. 283, Butterworths, London.

Lindahl, U., and Axelsson, O., 1971, Identification of iduronic acid as the major sulfated uronic acid of heparin, *J. Biol. Chem.* **246**:74.

Lindahl, U., and Höök, M., 1978, Glycosaminoglycans and their binding to biological macromolecules, *Annu. Rev. Biochem.* **47**:385.

Lindahl, U., and Rodén, L., 1964, The linkage of heparin to protein, *Biochem. Biophys. Res. Commun.* **17**:254.

Lindahl, U., and Rodén, L., 1965, The role of galactose and xylose in the linkage of heparin to protein, *J. Biol. Chem.* **240**:2821.

Lindahl, U., and Rodén, L., 1972, Carbohydrate–peptide linkages in proteoglycans of animal, plant and bacterial origin, in: *Glycoproteins* (A. Gottschalk, ed.), p. 491, Elsevier, Amsterdam.

Lindahl, U., Cifonelli, J. A., Lindahl, B., and Rodén, L., 1965, The role of serine in the linkage of heparin to protein, *J. Biol. Chem.* **240**:2817.

Lindahl, U., Bäckström, G., Malmström, A., and Fransson, L.-Å., 1972, Biosynthesis of L-iduronic acid in heparin: Epimerization of D-glucuronic acid on the polymer level, *Biochem. Biophys. Res. Commun.* **46**:985.

Lindahl, U., Bäckström, G., Jansson, L., and Hallén, A., 1973, Biosynthesis of heparin. II. Formation of sulfamino groups, *J. Biol. Chem.* **248**:7234.

Lindahl, U., Jacobsson, I., Höök, M., Bäckström, G., and Feingold, D. S., 1976, Biosynthesis of heparin. VI. Loss of C-5 hydrogen during conversion of D-glucuronic to L-iduronic acid residues, *Biochem. Biophys. Res. Commun.* **70**:492.

Lindahl, U., Höök, M., Bäckström, G., Jacobsson, I., Riesenfeld, J., Malmström, A., Rodén, L., and Feingold, D. S., 1977, Structure and biosynthesis of heparin-like polysaccharides, *Fed. Proc. Fed. Am. Soc. Exp. Biol.* **36**:19.

Linker, A., Meyer, K., and Weissmann, B., 1955, Enzymatic formation of monosaccharides from hyaluronate, *J. Biol. Chem.* **213**:237.

Linker, A., Meyer, K., and Hoffman, P., 1956, The production of unsaturated uronides by bacterial hyaluronidases, *J. Biol. Chem.* **219**:13.

Lloyd, A. G., Embery, G., Wusterman, F. S., and Dodgson, K. S., 1966, The metabolic fate of ^{35}S-labeled heparin and related compounds, *Biochem. J.* **98**:33.

Lloyd, A. G., Fowler, L. J., Embery, G., and Lau, B. A., 1968, Degradation of [^{35}S]heparin by mammalian and bacterial sulphamidases, *Biochem. J.* **110**:54P.

Malawista, I., and Schubert, M., 1958, Chondromucoprotein: New extraction method and alkaline degradation, *J. Biol. Chem.* **230**:535.

Malmström, A., and Fransson, L.-Å., 1971a, Structure of pig skin dermatan sulfate. 2. Demonstration of sulfated iduronic acid residues, *Eur. J. Biochem.* **18**:431.

Malmström, A., and Fransson, L.-Å., 1971b, Studies on the biosynthesis of dermatan sulfate, *FEBS Lett.* **16**:105.

Malmström, A., Carlstedt, I., Åberg, L., and Fransson, L.-Å., 1975a, The copolymeric structure of dermatan sulphate from cultured human fibroblasts, *Biochem. J.* **151**:477.

Malmström, A., Fransson, L.-Å., Höök, M., and Lindahl, U., 1975b, Biosynthesis of dermatan sulfate. I. Formation of L-iduronic acid residues, *J. Biol. Chem.* **250**:3419.

Malmström, A., Bäckström, G., Feingold, D. S., Höök, M., Jacobsson, I., Lindahl, U., Riesenfeld, J., and Rodën, L., 1976, Biosynthesis of heparin: Purification of the C-5-uronosyl epimerase, *Fed. Proc. Fed. Am. Soc. Exp. Biol.* **35**:1374 (abstract).

Malmström, A., Rodén, L., Feingold, D., Jacobsson, I., Bäckström, G., Höök, M., and Lindahl, U., 1980, Biosynthesis of heparin. Partial purification of the uronosyl C-5 epimerase, *J. Biol. Chem.,* in press.

Marbet, R., and Winterstein, A., 1951, β-Heparin, ein neuer, blutgerinnungshemmender Mucoitinschwefelsäureester, *Helv. Chim. Acta* **34**:2311.

Markovitz, A., and Dorfman, A., 1962, Synthesis of capsular polysaccharide (hyaluronic acid) by protoplast membrane preparations of group A streptococcus, *J. Biol. Chem.* **237**:273.

Markovitz, A., Cifonelli, J. A., and Dorfman, A., 1959, The biosynthesis of hyaluronic acid by group A streptococcus. VI. Biosynthesis from uridine nucleotides in cell-free extracts, *J. Biol. Chem.* **234**:2343.

Matalon, R., and Dorfman, A., 1968, The structure of acid mucopolysaccharides produced by Hurler fibroblasts in tissue culture, *Proc. Natl. Acad. Sci. U.S.A.* **60**:179.

Matalon, R., and Dorfman, A., 1972, Hurler's syndrome, an α-L-iduronidase deficiency, *Biochem. Biophys. Res. Commun.* **47**:959.

Matalon, R., and Dorfman, A., 1973, Sanfilippo A syndrome: A sulfamidase deficiency, *Pediatr. Res.* **7**:384.

Matalon, R., and Dorfman, A., 1974, Sanfilippo A syndrome: Sulfaminidase deficiency in cultured skin fibroblasts and liver, *J. Clin. Invest.* **54**:907.

Matalon, R., Cifonelli, J. A., and Dorfman, A., 1971, L-Iduronidase in cultured human fibroblasts and liver, *Biochem. Biophys. Res. Commun.* **42**:340.

Matalon, R., Arbogast, B., and Dorfman, A., 1974a, Deficiency of chondroitin sulfate N-acetylgalactosamine 4-sulfate sulfatase in Maroteaux–Lamy syndrome, *Biochem. Biophys. Res. Commun.* **61**:1450.

Matalon, R., Arbogast, B., Justice, P., Brandt, I. K., and Dorfman, A., 1974b, Morquio's syndrome: Deficiency of a chondroitin sulfate N-acetylhexosamine sulfate sulfatase, *Biochem. Biophys. Res. Commun.* **61**:759.

Mathews, M. B., 1958, Isomeric chondroitin sulphates, *Nature (London)* **181**:421.

Mathews, M. B., 1975, *Connective Tissue,* Springer-Verlag, New York.

Mathews, M. B., and Dorfman, A., 1955, Inhibition of hyaluronidase, *Physiol. Rev.* **35**:381.

Mathews, M. B., and Lozaityte, I., 1958, Sodium chondroitin sulfate–protein complexes of cartilage. I. Molecular weight and shape, *Arch. Biochem. Biophys.* **74**:158.

McKusick, V. A., Neufeld, E. F., and Kelly, T. E., 1978, The mucopolysaccharide storage diseases, in: *The Metabolic Basis of Inherited Disease,* 4th ed. (J. B. Stanbury, J. B. Wyngaarden, and D. S. Fredrickson, eds.), p. 1282, McGraw-Hill, New York.

Meyer, K., 1970, Reflections on "mucopolysaccharides" and their protein complexes, in: *Chemistry and Molecular Biology of the Intercellular Matrix* (E. A. Balazs, ed.), p. 5, Academic Press, New York.

Meyer, K., and Chaffee, E., 1941, The mucopolysaccharides of skin, *J. Biol. Chem.* **138**:491.

Meyer, K., Davidson, E., Linker, A., and Hoffman, P., 1956, The acid mucopolysaccharides of connective tissue, *Biochim. Biophys. Acta* **21**:506.

Muir, H., and Hardingham, T. E., 1975, Structure of proteoglycans, in: *Biochemistry of Carbohydrates, MTP International Review of Science,* Biochemistry Series One, Vol. 5 (W. J. Whelan, ed.), p. 153, Butterworths, London.

Neufeld, E. F., and Fratantoni, J. C., 1970, Inborn errors of mucopolysaccharide metabolism, *Science* **169**:141.

Neufield, E. F., and Hall, C. W., 1965, Inhibition of UDP-D-glucose dehydrogenase by UDP-D-xylose: A possible regulatory mechanism, *Biochem. Biophys. Res. Commun.* **19:**456.

Neufeld, E. F., Lim, T. W., and Shapiro, L. J., 1975, Inherited disorders of lysosomal metabolism, *Annu. Rev. Biochem.* **44:**357.

Norden, A. G. W., Tennant, L. L., and O'Brien, J. S., 1974, G_{M1} ganglioside β-galactosidase A: Purification and studies of the enzyme from human liver, *J. Biol. Chem.* **249:**7969.

O'Brien, J. S., 1972, Sanfilippo syndrome: Profound deficiency of alpha-acetylglucosaminidase activity in organs and skin fibroblasts from type-B patients, *Proc. Natl. Acad. Sci. U.S.A.* **69:**1720.

O'Brien, J. S., 1978, The gangliosidoses, in: *The Metabolic Basis of Inherited Disease* (J. B. Stanbury, J. B. Wyngaarden, and D. S. Fredrickson, eds.), p. 841, McGraw-Hill, New York.

O'Brien, J. S., Cantz, M., and Spranger, J., 1974, Maroteaux–Lamy disease (mucopolysaccharidosis VI), subtype A: Deficiency of a *N*-acetylgalactosamine-4-sulfatase, *Biochem. Biophys. Res. Commun.* **60:**1170.

O'Brien, J. S., Gugler, E., Giedion, A., Wiessmann, U., Herschkowitz, N., Meier, C., and Leroy, J., 1976, Spondyloepiphyseal dysplasia, corneal clouding, normal intelligence and acid β-galactosidase deficiency, *Clin. Genet.* **9:**495.

Öbrink, B., 1972, Isolation and partial characterization of a dermatan sulphate-proteoglycan from pig skin, *Biochim. Biophys. Acta* **264:**354.

Öckerman, P. A., 1968, Identity of β-glucosidase, β-xylosidase and one of the β-galactosidase activities in human liver when assayed with 4-methylumbelliferyl-β-D-glycosides: Studies in cases of Gaucher's disease, *Biochim. Biophys. Acta* **165:**59.

Oegema, T. R., Hascall, V. C., and Dziewiatkowski, D. D., 1975, Isolation and characterization of proteoglycans from the Swarm rat chondrosarcoma, *J. Biol. Chem.* **250:**6151.

Oegema, T. R., Brown, M., and Dziewiatkowski, D. D., 1977, The link protein in proteoglycan aggregates from the Swarm rat chondrosarcoma, *J. Biol. Chem.* **252:**6470.

Ögren, S., and Lindahl, U., 1975, Cleavage of macromolecular heparin by enzyme and mouse mastocytoma, *J. Biol. Chem.* **250:**2690.

Okayama, M., and Lowther, D. A., 1973, Effects of β-xylosides on the synthesis of chondroitin sulfate–protein complexes in cartilage slices, *Proc. Aust. Biochem. Soc.* **6:**75.

Okayama, M., Kimata, K., and Suzuki, S., 1973, The influence of *p*-nitrophenyl β-D-xyloside on the synthesis of proteochondroitin sulfate by slices of embryonic chick cartilage, *J. Biochem.* **74:**1069.

Oldberg, Å., Höök, M., Öbrink, B., Pertoft, H., and Rubin, K., 1977, Structure and metabolism of rat liver heparan sulphate, *Biochem. J.* **164:**75.

Oldberg, Å., Kjellén, L., and Höök, M., 1979, Cell-surface heparan sulfate: Isolation and characterization of a proteoglycan from rat liver membranes, *J. Biol. Chem.* **254:**8505.

Orkin, R. W., Jackson, G., and Toole, B. P., 1977, Hyaluronidase activity in cultured chick embryo skin fibroblasts, *Biochem. Biophys. Res. Commun.* **77:**132.

Patel, V., 1978, Degradation of glycoproteins, in: *The Glycoconjugates,* Vol. II (M. I. Horowitz and W. Pigman, eds.), p. 185, Academic Press, New York.

Patel, V., and Tappel, A. L., 1969*a*, β-Glucosidase and β-xylosidase of rat kidney, *Biochim. Biophys. Acta* **191:**653.

Patel, V., and Tappel, A. L., 1969*b*, Identity of β-glucosidase and β-xylosidase activities in rat liver lysosomes, *Biochim. Biophys. Acta* **191:**86.

Perlman, R. L., Telser, A., and Dorfman, A., 1964, The biosynthesis of chondroitin sulfate by a cell-free preparation, *J. Biol. Chem.* **239:**3623.

Platt, D., and Dorn, M., 1968, Nachweis, Reinigung und Eigenschaften der Glycosaminoglycano-hydrolasen im menschlichen Hyalinen Knorpel, *Clin. Chim. Acta* **21:**333.

Preston, B. N., 1968, Physical characterization of dermatan sulphate-protein, *Arch. Biochem. Biophys.* **126**:974.

Rapport, M. M., Weissmann, B., Linker, A., and Meyer, K., 1951, Isolation of a crystalline disaccharide, hyalobiuronic acid, from hyaluronic acid, *Nature (London)* **168**:996.

Revell, P. A., and Muir, H., 1972, The excretion and degradation of chondroitin 4-sulphate administered to guinea pigs as free chondroitin sulphate and as proteoglycan, *Biochem. J.* **130**:597.

Ringertz, N. R., 1963, Polysaccharides of neoplastic mast cells, *Ann. N.Y. Acad. Sci.* **103**:209.

Robbins, P. W., Bray, D., Danhert, M., and Wright, A., 1967, Direction of chain growth in polysaccharide synthesis, *Science* **158**:1536.

Robinson, D., and Abrahams, H. E., 1967, β-D-Xylosidase in pig kidney, *Biochim. Biophys. Acta* **132**:212.

Robinson, H. C., Telser, A., and Dorfman, A., 1966, Studies on the biosynthesis of the linkage region of chondroitin–sulfate protein complex, *Proc. Natl. Acad. Sci. U.S.A.* **56**:1859.

Robinson, H. C., Brett, M. J., Tralaggan, P. J., Lowther, D. A., and Okayama, M., 1975, The effect of D-xylose, β-D-xylosides and β-D-galactosides on chondroitin sulphate biosynthesis in embryonic chicken cartilage, *Biochem. J.* **148**:25.

Robinson, H. C., Horner, A. A., Höök, M., Ögren, S., and Lindahl, U., 1978, A proteoglycan form of heparin and its degradation to single-chain molecules, *J. Biol. Chem.* **253**:6687.

Roblin, R., Albert, S. O., Gelb, N. A., and Black, P. H., 1975, Cell surface changes correlated with density-dependent growth inhibition, glycosaminoglycan metabolism in 3T3, SV3T3, and Con A selected revertant cells, *Biochemistry* **14**:347.

Rodén, L., 1970, Biosynthesis of acidic glycosaminoglycans (mucopolysaccharides), in: *Metabolic Conjugation and Metabolic Hydrolysis*, Vol. 2 (W. H. Fishman, ed.), p. 345, Academic Press, New York.

Rodén, L., and Armand, G., 1966, Structure of the chondroitin-4-sulfate–protein linkage region: Isolation and characterization of the disaccharide 3-*O*-β-D-glucuronosyl-D-galactose, *J. Biol. Chem.* **241**:65.

Rodén, L., and Horowitz, M. I., 1978, Proteoglycans and structural glycoproteins, in: *The Glycoconjugates*, Vol. II (M. I. Horowitz and W. Pigman, eds.), p. 3, Academic Press, New York.

Rodén, L., and Schwartz, N. B., 1975. Biosynthesis of connective tissue proteoglycans, in: *Biochemistry of Carbohydrates, MTP International Review of Science*, Biochemistry Series One, Vol. 5 (W. J. Whelan, ed.), p. 95, Butterworths, London.

Rome, L. H., Garvin, A. J., and Neufeld, E. F., 1978, Human kidney α-L-iduronidase: Purification and characterization, *Arch. Biochem. Biophys.* **189**:344.

Roseman, S., and Dorfman, A., 1951, α-Glucosaminidase, *J. Biol. Chem.* **191**:607.

Rosenberg, L., Hellmann, W., and Kleinschmidt, A. K., 1975, Electron microscope studies of proteoglycan aggregates from bovine articular cartilage, *J. Biol. Chem.* **250**:1877.

Roy, A. B., 1953, The sulphatase of ox liver. I. The complex nature of the enzyme, *Biochem. J.* **53**:12.

Roy, A. B., 1958, Comparative studies on the liver sulphatases, *Biochem. J.* **68**:519.

Roy, A. B., 1976, Sulphatases, lysosomes and disease, *Aust. J. Exp. Biol. Med.* **54**:111.

Roy, A. B., and Trudinger, P. A., 1970, *The Biochemistry of Inorganic Compounds of Sulphur*, p. 133, Cambridge, New York.

Sajdera, S. W., and Hascall, V. C., 1969, Proteinpolysaccharide complex from bovine nasal cartilage: A comparison of low and high shear extraction procedures, *J. Biol. Chem.* **244**:77.

Sandson, J., and Hamerman, D., 1962, Isolation of hyaluronate protein from human synovial fluid, *J. Clin. Invest.* **41:**1817.

Schachter, H., 1978, Glycoprotein biosynthesis, in: *The Glycoconjugates,* Vol. II (M. I. Horowitz and W. Pigman, eds.), p. 87, Academic Press, New York.

Schachter, H., and Rodén, L., 1973, The biosynthesis of animal glyco proteins, in: *Metabolic Conjugation and Metabolic Hydrolysis,* Vol. 3 (W. H. Fishman, ed.), p. 1, Academic Press, New York.

Scher, I., and Hamerman, D., 1972, Isolation of human synovial-fluid hyaluronate by density-gradient ultracentrifugation and evaluation of its protein content, *Biochem. J.* **126:**1073.

Schiller, S., and Dorfman, A., 1957, The metabolism of mucopolysaccharides in animals. IV. The influence of insulin, *J. Biol. Chem.* **227:**625.

Schubert, M., and Hamerman, D., 1968, *A Primer on Connective Tissue Biochemistry,* Lea and Febiger, Philadephia.

Schwartz, N. B., 1975, Biosynthesis of chondroitin sulfate: Immunoprecipitation of interaction xylosyltransferase and galactosyltransferase, *FEBS Lett.* **49:**342.

Schwartz, N. B., 1976a, Biosynthesis of chondroitin sulfate: Role of phospholipids in the activity of UDP-D-galactose: D-xylose galactosyltransferase, *J. Biol. Chem.* **251:**285.

Schwartz, N. B., 1976b, Chondroitin sulfate glycosyltransferases in cultured chondrocytes: Turnover, oscillatory change during growth, and suppression by 5-bromodeoxyuridine, *J. Biol. Chem.* **251:**3346.

Schwartz, N. B., 1977, Regulation of chondroitin sulfate synthesis: Effect of β-xylosides on synthesis of chondroitin sulfate proteoglycan, chondroitin sulfate chains, and core protein, *J. Biol. Chem.* **252:**6316.

Schwartz, N. B., and Dorfman, A., 1975, Purification of rat liver chondrosarcoma xylosyltransferase, *Arch. Biochem. Biophys.* **171:**136.

Schwartz, N. B., and Rodén, L., 1974, Biosynthesis of chondroitin sulfate: Purification of UDP-D-xylose:core protein β-D-xylosyltransferase by affinity chromatography, *Carbohyd. Res.* **37:**167.

Schwartz, N. B., and Rodén, L., 1975, Biosynthesis of chondroitin sulfate: Solubilization of chondroitin sulfate glycosyltransferases and partial purification of uridine diphosphate D-galactose: D-xylose galactosyltransferase, *J. Biol. Chem.* **250:**5200.

Schwartz, N. B., Galligani, L., Ho, P.-L., and Dorfman, A., 1974a, Stimulation of synthesis of free chondroitin sulfate chains by β-D-xylosides in cultured cells, *Proc. Natl. Acad. Sci. U.S.A.* **71:**4047.

Schwartz, N. B., Rodén, L., and Dorfman, A., 1974b, Biosynthesis of chondroitin sulfate: Interaction between xylosyltransferase and galactosyltransferase, *Biochem. Biophys. Res. Commun.* **56:**717.

Seno, N., Meyer, K., Anderson, B., Hoffman, P., 1965, Variations in keratosulfates, *J. Biol. Chem.* **240:**1005.

Seno, N., Anno, K., Yaegashi, Y., and Okuyama, T., 1975, Microheterogeneity of chondroitin sulfates from various cartilages, *Connect. Tissue Res.* **3:**87.

Shapiro, L. J., Hall, C. W., Leder, I. G., and Neufeld, E. F., 1976, The relationship of α-L-iduronidase and Hurler corrective factor, *Arch. Biochem. Biophys.* **172:**156.

Shatton, J., and Schubert, M., 1954, Isolation of a mucoprotein from cartilage, *J. Biol. Chem.* **211:**565.

Sheehan, J. K., Nieduszynski, I. A., Phelps, C. F., Muir, H., and Hardingham, T. E., 1978, Self-association of proteoglycan subunits from pig laryngeal cartilage, *Biochem. J.* **171:**109.

Siewert, G., and Strominger, J. L., 1967, Bacitracin: An inhibitor of the dephosphorylation of lipid pyrophosphate, an intermediate in biosynthesis of the peptidoglycan of bacterial cells walls, *Proc. Natl. Acad. Sci. U.S.A.* **57:**767.

Silbert, J. E., 1963, Incorporation of ^{14}C and ^{3}H from nucleotide sugars into a polysaccharide in the presence of a cell-free preparation from mouse mast cell tumors, *J. Biol. Chem.* **238:**3542.

Silbert, J. E., 1964, Incorporation of ^{14}C and ^{3}H from labeled nucleotide sugars into a polysaccharide in the presence of a cell-free preparation from cartilage, *J. Biol. Chem.* **239:**1310.

Silbert, J. E., 1967a, Incorporation of $^{35}SO_4$ into endogenous heparin by a microsomal fraction from mast cell tumors, *J. Biol. Chem.* **242:**2301.

Silbert, J. E., 1967b, Biosynthesis of heparin. III. Formation of a sulfated glycosaminoglycan with a microsomal preparation from mast cell tumors, *J. Biol. Chem.* **242:**5146.

Silbert, J. E., 1967c, Biosynthesis of heparin. IV. N-deacetylation of a precursor glycosaminoglycan, *J. Biol. Chem.* **242:**5153.

Singh, J., Di Ferrante, N., Niebes, P., and Tavella, D., 1976, N-acetylgalactosamine-6-sulfate sulfatase in man: Absence of the enzyme in Morquio disease, *J. Clin. Invest.* **57:**1036.

Sjöberg, I., 1978, Structure and metabolism of glycosaminoglycans in human fibroblasts, M.D. thesis, University of Lund, Sweden.

Sjöberg, I., Fransson, L.-Å., Matalon, R., and Dorfman, A., 1973, Hunter's syndrome: A deficiency of L-idurono-sulfate sulfatase, *Biochem. Biophys. Res. Commun.* **54:**1125.

Sly, W. S., Quinton, B. A., McAlister, W. H., and Rimoin, D. L., 1973, Beta glucuronidase deficiency: Report of clinical, radiologic, and biochemical features of a new mucopolysaccharidosis, *J. Pediatr.* **82:**249.

Smith, H., and Gallop, R. C., 1953, The "acid polysaccharides" of hog gastric mucosa, *Biochem. J.* **53:**666.

Spiro, M. J., 1977, Presence of a glucuronic acid–containing carbohydrate unit in human thyroglobulin, *J. Biol. Chem.* **252:**5424.

Spranger, J. W., 1977, Beta galactosidase and the Morquio syndrome, *Am. J. Med. Gen.* **1:**207.

Srivastava, R. M., Hudson, N., Seymour, F. R., and Weissmann, B., 1978, Preparation of (aryl α-L-idopyranosid)uronic acids, *Carbohydr. Res.* **60:**315.

Stern, E. L., Lindahl, B., and Rodén, L., 1971, The linkage of dermatan sulfate to protein. I. Monosaccharide sequence of the linkage region, *J. Biol. Chem.* **246:**5707.

Stevens, R. L., Fluharty, A. L., Killgrove, A. R., and Kihara, H., 1977, Arylsulfatases of human tissue: Studies on a form of arylsulfatase B found predominantly in brain, *Biochim. Biophys. Acta* **481:**549.

Stoffyn, P. J., and Jeanloz, R. W., 1960, The identification of the uronic acid component of dermatan sulfate (β-heparin, chondroitin sulfate B), *J. Biol. Chem.* **235:**2507.

Stoolmiller, A. C., and Dorfman, A., 1969a, The biosynthesis of hyaluronic acid by *Streptococcus, J. Biol. Chem.* **244:**236.

Stoolmiller, A. C., and Dorfman, A., 1969b, The metabolism of glycosaminoglycans, in: *Comprehensive Biochemistry,* Vol. 17 (M. Florkin and E. H. Stotz, eds.), p. 241, Elsevier, Amsterdam.

Stoolmiller, A. C., Horwitz, A. L., and Dorfman, A., 1972, Biosynthesis of the chondroitin sulfate proteoglycan: Purification and properties of xylosyltransferase, *J. Biol. Chem.* **247:**3525.

Stumpf, D. A., Austin, J. H., Crocker, A. C., and LaFrance, M., 1973, Mucopolysaccharidosis type VI (Maroteaux–Lamy syndrome). I. Sulfatase B deficiency in tissues, *Am. J. Dis. Child.* **126:**747.

Suzuki, S., 1960, Isolation of novel disaccharides from chondroitin sulfates, *J. Biol. Chem.* **235:**3580.

Suzuki, S., and Strominger, J. L., 1960, Enzymatic sulfation of mucopolysaccharides in hen oviduct. II. Mechanism of the reaction studied with oligosaccharides and mono-saccharides as acceptors, *J. Biol. Chem.* **235**:267.

Telser, A., Robinson, H. C., and Dorfman, A., 1966, The biosynthesis of chondroitin sulfate, *Arch. Biochem. Biophys.* **116**:458.

Thomas, L., 1956, Reversible collapse of rabbit ears after intravenous papain, and prevention of recovery by cortisone, *J. Exp. Med.* **104**:245.

Thompson, J. N., 1978, Substrates for the assay of α-L-iduronidase, *Clin. Chim. Acta* **89**:435.

Thompson, J. N., Stoolmiller, A. C., Matalon, R., and Dorfman, A., 1973, N-acetyl-β-hexosaminidase: Role in the degradation of glycosaminoglycans, *Science* **181**:866.

Tominaga, F., Oka, K., and Yoshida, H., 1965, The isolation and identification of O-xylosyl-serine and S-methylcysteine sulfoxide from human urine, *J. Biochem.* **57**:717.

Tomino, S., Paigen, K., Tulsiani, D. R. P., and Touster, O., 1975, Purification and chemical properties of mouse liver lysosomal (L form) β-glucuronidase, *J. Biol. Chem.* **250**:8503.

Toole, B. P., and Lowther, D. A., 1965, The isolation of a dermatan sulphate–protein complex from bovine heart valves, *Biochim. Biophys. Acta* **101**:364.

Toole, B. P., and Lowther, D. A., 1968, Dermatan sulfate–protein: Isolation from and interaction with collagen, *Arch. Biochem. Biophys.* **128**:567.

Truppe, W., Basner, R., von Figura, K., and Kresse, H., 1977, Uptake of hyaluronate by cultured cells, *Biochem. Biophys. Res. Commun.* **78**:713.

Tsay, G. C., and Dawson, G., 1973, Structure of the "keratosulfate-like" material in liver from a patient with G_{M1}-gangliosidosis (β-D-galactosidase deficiency), *Biochem. Biophys. Res. Commun.* **52**:759.

Tulsiani, D. R. P., Keller, R. K., and Touster, O., 1975, The preparation and chemical composition of the multiple forms of β-glucuronidase from the female rate preputial gland, *J. Biol. Chem.* **250**:4770.

Turco, S. J., and Heath, E. C., 1977, Glucuronosyl-N-acetylglucosaminyl pyrophosphoryldolichol: Formation in SV_{40}-transformed human lung fibroblasts and biosynthesis in rat lung microsomal preparations, *J. Biol. Chem.* **252**:2918.

von Figura, K., 1977a, Human α-N-acetylglucosaminidase. 1. Purification and properties, *Eur. J. Biochem.* **80**:525.

von Figura, K., 1977b, Human α-N-acetylglucosaminidase. 2. Activity towards natural substrates and multiple recognition forms, *Eur. J. Biochem.* **80**:535.

von Figura, K., and Kresse, H., 1972, The Sanfilippo B corrective factor: A N-acetyl-α-D-glucosaminidase, *Biochem. Biophys. Res. Commun.* **48**:262.

von Figura, K., and Kresse, H., 1976, Sanfilippo disease type B: Presence of material cross reacting with antibodies against α-N-acetylglucosaminidase, *Eur. J. Biochem.* **61**:581.

Wakabayashi, M., 1970, β-Glucuronidases in metabolic hydrolysis, in: *Metabolic Conjugation and Metabolic Hydrolysis,* Vol. 2 (W. H. Fishman, ed.), p. 520, Academic Press, New York.

Wasserman, S. I., and Austen, K. F., 1977, Identification and characterization of arylsulfatase A and B of the rat basophil leukemia tumor, *J. Biol. Chem.* **252**:7074.

Wasteson, Å., Uthne, K., and Westermark, B., 1973, A novel assay for the biosynthesis of sulphated polysaccharide and its application to studies on the effects of somatomedin on cultured cells, *Biochem. J.* **136**:1069.

Wasteson, Å., Amado, R., Ingmar, B., and Heldin, C.-H., 1975, Degradation of chondroitin sulphate by lysosomal enzymes from embryonic chick cartilage, in: *Protides of the Biological Fluids, 22nd Colloquium* (H. Peeters, ed.), p. 431, Pergamon, Oxford.

Wasteson, Å., Höök, M., and Westermark, B., 1976, Demonstration of a platelet enzyme, degrading heparan sulphate, *FEBS Lett.* **64**:218.

Wasteson, Å., Glimelius, B., Busch, C., Westermark, B., Heldin, C.-H., and Norling, B.,

1977, Effect of a platelet endoglycosidase on cell surface associated heparan sulphate of human cultured endothelial and glial cells, *Thromb. Res.* **11**:309.

Weissmann, B., and Meyer, K., 1952, Structure of hyaluronic acid: The glucuronidic linkage, *J. Am. Chem. Soc.* **74**:4729.

Weissmann, B., and Meyer, K., 1954, The structure of hyalobiuronic acid and of hyaluronic acid from umbilical cord, *J. Am. Chem. Soc.* **76**:1753.

Weissmann, B., and Santiago, R., 1972, α-L-Iduronidase in lysosomal extracts, *Biochem. Biophys. Res. Commun.* **46**:1430.

Weissmann, B., Rapport, M. M., Linker, A., and Meyer, K., 1953, Isolation of the aldobionic acid of umbilical cord hyaluronic acid, *J. Biol. Chem.* **205**:205.

Weissmann, B., Meyer, K., Sampson, P., and Linker, A., 1954, Isolation of oligosaccharides enzymatically produced from hyaluronic acid, *J. Biol. Chem.* **208**:417.

Weissmann, B., Rowin, G., Marshall, J., and Friederici, D., 1967, Mammalian α-acetylglucosaminidase. Enzymic properties, tissue distribution and intracellular localization, *Biochemistry* **6**:207.

Weissmann, B., Cashman, D. C., and Santiago, R., 1975, Concerted action of β-acetylglucosaminidase on hyaluronodextrins, *Connect. Tissue Res.* **3**:7.

Wolfe, L. S., Senior, R. G., and Kin, N. M. K. N. Y., 1974, The structures of oligosaccharides accumulating in the liver of G_{MI}-gangliosidosis, type I, *J. Biol. Chem.* **249**:1828.

Wortman, B., 1961, Enzymic sulfation of corneal mucopolysaccharides by beef cornea epithelial extract, *J. Biol. Chem.* **236**:974.

Zechmeister, L., Toth, G., and Vajda, E., 1939, Chromatographie der in der Chitinreihe wirksamen Enzyme der Weinbergschneck (*Helix pomatia*), *Enzymologia* **7**:170.

Index